Creating and Restoring Wetlands

Creating and Restoring Wetlands

From Theory to Practice

Christopher Craft
**Janet Duey Professor of Rural Land Policy,
School of Public and Environmental Affairs,
Indiana University, Bloomington, Indiana**

ELSEVIER

AMSTERDAM • BOSTON • HEIDELBERG • LONDON • NEW YORK • OXFORD
PARIS • SAN DIEGO • SAN FRANCISCO • SINGAPORE • SYDNEY • TOKYO

Elsevier
Radarweg 29, PO Box 211, 1000 AE Amsterdam, Netherlands
The Boulevard, Langford Lane, Kidlington, Oxford OX5 1GB, UK
225 Wyman Street, Waltham, MA 02451, USA

Notices
Knowledge and best practice in this field are constantly changing. As new research and
experience broaden our understanding, changes in research methods, professional practices, or
medical treatment may become necessary.

Practitioners and researchers must always rely on their own experience and knowledge in
evaluating and using any information, methods, compounds, or experiments described herein.
In using such information or methods they should be mindful of their own safety and the
safety of others, including parties for whom they have a professional responsibility.

To the fullest extent of the law, neither the Publisher nor the authors, contributors, or editors,
assume any liability for any injury and/or damage to persons or property as a matter of
products liability, negligence or otherwise, or from any use or operation of any methods,
products, instructions, or ideas contained in the material herein.

ISBN: 978-0-12-407232-9

British Library Cataloguing in Publication Data
A catalogue record for this book is available from the British Library

Library of Congress Cataloging-in-Publication Data
A catalog record for this book is available from the Library of Congress

For Information on all Elsevier publications
visit our website at http://store.elsevier.com/

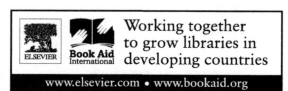
Working together
to grow libraries in
developing countries

www.elsevier.com • www.bookaid.org

Contents

Acknowledgments

This book would not be realized without the efforts of many people. Patricia (Pat) Combs worked tirelessly to acquire references, edit, format, and proofread text and she served as my liaison with Elsevier. Kelsey Thetonia, Kate Drake, Jenna Nawrocki, Michelle Ruan, Kristin Ricigliano, Nate Barnett and Elizabeth Oliver of the School of Public and Environmental Affairs (SPEA) made and remade figures and graphs. SPEA PhD student, Ellen Herbert, kept my lab afloat during the 3 years the book took to complete and, for that, I am grateful. SPEA Dean David Reingold made the book a reality by providing me time, through an extended sabbatical, and resources. I thank my wife, Teresa, and daughter, Rachel, who have put up with me for 33 and 24 years, respectively. Last but not least, I thank my father, William Hugh (Bill) Craft, who, when he was not working to raise nine children, was, in his heart, a crackerjack botanist and teacher. Thanks everyone!

Part One

Foundations

Introduction

1

Chapter Outline

Wetlands, where water and land meet, have a unique place in the development of civilization. Rice, a wetland plant, feeds 3.5 billion people worldwide (Seck et al., 2012). Fish, associated with aquatic littoral zones and wetlands, is the primary source of protein for 2.9 billion people (Smith et al., 2010). Rice (*Oryza sativa*) was first cultivated in India, Southeast Asia, and China (Chang, 1976), and fish were raised among the rice paddies, providing needed protein (Kangmin, 1988). Along the Nile River, early societies were sustained by fish caught from the floodplains and coastal lagoons of the delta (Sahrhage, 2008). Civilization prospered along rivers and deltas of the Yangtze and Yellow Rivers, China; the Irrawaddy, Ganges, and Indus of India; the Nile of Egypt, and the Mesopotamian marshes of Iraq. Later, cities were established where land and water meet, on rivers, lakes, and at the sea's edge, where they were hubs of transport and commerce. As cities grew, it was convenient to drain or fill the low, wet, swampy, and marshy areas, the wetlands, to expand.

With the Industrial Revolution in the eighteenth century and its mechanization of farming and abiotic synthesis of nitrogen fertilizer, large-scale agriculture became feasible. The inevitable result of population growth and the Industrial Revolution was the widespread drainage of freshwater wetlands to grow food crops. Extensive wetlands in regions such as the Midwest US Corn Belt and the interior valleys of California were drained and farmed. Later, large-scale aquaculture, especially shrimp farms, was carved from the extensive mangrove forests of the tropics. During the twentieth century, loss of coastal and freshwater wetlands in temperate regions such as the US, Europe, and China, was extensive. Developing regions of the tropics were not far behind with widespread conversion of mangroves and other wetlands to forest plantations and aquaculture ponds later in the century.

Today, the cumulative loss of wetlands in the US, including Alaska, since European settlement is greater than 30% with much greater losses in the Midwest and California where more than 80% of the original acreage has been lost (Dahl, 1990). Worldwide, loss of mangroves, tropical coastal wetlands, is on the order of 20–50% (Valiela et al., 2001; FAO, 2007). In the past 35 years, more than 30% of coastal wetlands and 25% of freshwater swamps in China, where development has been rapid, have been lost (An et al., 2007; He et al., 2014). Delta regions are particularly susceptible to wetland loss as large areas are converted to agriculture (Coleman et al., 2008). Even peatlands are not immune as extractive industries such as peat harvesting and fossil fuel extraction, including oil sands of Canada and fossil fuel extraction in Siberia, *eat* away at the natural resource.

By the 1970s, increasing recognition of the alarming rate of wetland loss led to laws such as the Clean Water Act of 1972 in the US, created to protect the nation's aquatic resources, including wetlands. A key component of the law was the restoration of degraded wetlands or creation of entirely new ones to compensate for their loss. Today, government programs such as the Wetlands Reserve and Conservation Reserve Programs of the U.S. Department of Agriculture offer financial incentives to restore wetlands. In the Glaciated Interior Plains of the American Midwest, more than 110,000 ha of wetland and riparian buffers were restored between 2000 and 2007 (Fennessy and Craft, 2011). Restoration of freshwater wetlands on former agricultural land has been implemented in Europe and elsewhere to improve water quality and increase landscape diversity (Comin et al., 2001). Wetlands also are created and restored to compensate for their loss from developmental activities such as road building and urban/suburban construction. Globally, while not legally binding, the Ramsar convention encourages protection and restoration of wetlands of international importance (see Chapter 2, Definitions).

Whereas the science of wetland restoration is relatively new, people have been restoring for years. The earliest restoration projects were reforestation schemes, planting mangroves for fuel and timber. In Indochina, large-scale mangrove afforestation dates to the late 1800s or earlier (Chowdhury and Ahmed, 1994). Nearly 100 years ago, salt marsh vegetation was planted in Western Europe, the US, Australia, and New Zealand to reclaim land from the sea and to slow coastal erosion (Ranwell, 1967; Knutson et al., 1981; Chung, 2006). At the same time, freshwater wetlands were being reflooded to provide waterfowl habitat (Weller, 1994). This was done by government agencies such as the U.S. Fish and Wildlife Service and by nongovernmental organizations like Ducks Unlimited. These early restoration activities—reforestation, shoreline protection, waterfowl habitat—focused on restoring a particular function such as productivity. Restoration today consists of reestablishing a variety of ecological attributes including community structure (species diversity and habitat) and ecosystem processes (energy flow and nutrient cycling), and the broad spectrum of goods and services delivered by healthy, functioning wetlands.

Webster's Dictionary (http://www.merriam-webster.com) defines restoration as *the act or process of returning something to its original condition.* In the book, *Restoration of Aquatic Ecosystems* (1992), the U.S. National Research Council (NRC) defines restoration as *the act of bringing an ecosystem back into, as nearly as possible, its original condition.* In this book, I expand on the NRC definition to define restoration as *the act of bringing an ecosystem back into, as nearly as possible, its* *original condition **faster than nature does it on its own***. This definition contains two key points. Restoration aims to accelerate succession and ecosystem development by deliberate means, spreading propagules, seeds, seedlings, and transplants, and amending the soil with essential nutrients (N) and, sometimes, organic matter. The second point, from the NRC definition, recognizes that often it is not possible to restore a wetland to its original, pre-disturbance condition because stressors that degrade the system cannot be completely eliminated. Many stressors that affect aquatic ecosystems and wetlands, such as flow mistiming, nutrient enrichment, salinity, and other soluble materials (Palmer et al., 2010), originate off-site and propagate downhill and downstream where they cause damage. Other stressors, many related to hydrology, occur on-site and are easier to ameliorate. These include levees, ditches, or placement of spoil atop the site that can be breached, filled, and removed, respectively.

This book introduces the science and practice of restoring wetlands: freshwater marshes, floodplain forests, peatlands, tidal marshes, and mangroves. Globally, wetland restoration is driven by policies such as the Ramsar convention on wetlands of international importance, the Clean Water Act of the US, the Water Framework Directive of the European Union, and others. Arguably, the science of wetland restoration, using ecological theory to guide the process, lags behind practice. Wetland restoration, historically, was more of a *cut and fit* process, applying well-developed techniques used by agronomy and forestry. These techniques were initially employed on surface-mined terrestrial lands where the goal was to reclaim the land for forestry, rangeland, or wildlife habitat. In these mostly terrestrial ecosystems, lack of freshwater often slowed the restoration process and so the idea of flooded or saturated soil hydrology was seldom considered. From a scientific perspective, ecological concepts such as disturbance, succession, and ecosystem development provide a framework to understand what is needed (or not needed) to successfully restore wetlands and other ecosystems. An understanding of ecosystem dynamics, energy flow and nutrient cycling, and the natural history of wetland plants and animals also is critical. Last but not least, one cannot understate the role that humans, through activities that disturb and degrade natural systems and their efforts to repair the damage, play in restoring wetlands.

Why Restore Wetlands?

Why the interest in restoring wetlands? There are two reasons. (1) There has been dramatic and widespread decline in wetland area as noted above. Nearly all of the losses are caused by human activities, drainage, placement of fill, nutrient overenrichment, and other waterborne pollutants. Extractive activities such as peat harvesting and mining of sand and other construction materials also contribute to the loss. There is an old saying that you do not appreciate something until it's gone, and with wetlands there is truth to that. (2) The benefits that wetlands provide to society (Table 1.1). Mostly unappreciated in the past, it is widely recognized that wetlands provide valuable services such as high levels of biological productivity, both fisheries and waterfowl, disturbance regulation including shoreline protection and floodwater storage,

Table 1.1 Ecological Functions and Services of Various Types of Wetlands

Floodplain/riparian	Water quality improvement (sediment trapping, denitrification) Biological productivity (including C export to aquatic ecosystems) Floodwater storage Biological dispersal corridors Biodiversity
Freshwater marshes	Biological productivity (waterfowl) Biodiversity
Peatlands	Carbon sequestration Biodiversity
Tidal marshes	Shoreline protection Biological productivity (finfish and shellfish, outwelling of nutrients) Water quality improvement
Mangroves	Shoreline protection Biological productivity (finfish, shellfish, outwelling) Water quality improvement

water quality improvement through sediment trapping and denitrification, and habitat and biodiversity.

It is recognized that different types of wetlands provide different kinds and levels of ecosystem services. Wetlands with strong connections to aquatic ecosystems such as floodplains, tidal marshes, and mangroves maintain and enhance water quality by filtering pollutants. They also regulate natural disturbances and perturbations by storing floodwaters, dissipating wave energy, and protecting shorelines. Some wetlands possess high levels of biological productivity that support commercial and recreation finfish populations, shellfish harvesting, and breeding waterfowl populations. Freshwater marshes of the prairie pothole region in the north central US and Canada are critical breeding habitat for North American ducks (Batt et al., 1989). Wetlands of the far north in Canada and Siberia are essential to breeding populations of cranes (Kanai et al., 2002; Chavez-Ramirez and Wehtje, 2012). Coastal wetlands, saline tidal marshes, and mangroves, contribute to aquatic food webs by serving as habitat for fish and crustaceans and by outwelling or exporting organic matter that supports heterotrophic food webs. Forested wetlands, riparian areas, and floodplain forests, support food webs of aquatic ecosystems, including streams and rivers. Wetlands that lack strong surface water connections such as peatlands sequester large amounts of carbon and support high levels of plant biodiversity.

Wetland restoration projects vary in their goals, scope, and costs. It is difficult to evaluate costs versus benefits of wetland restoration projects because it is hard to assess the economic value of various ecosystem services (see Chapter 2, Definitions). Bernhardt et al. (2005) reviewed the number and cost of various aquatic ecosystems and wetland restoration projects in the US. Most projects were associated with water quality management, followed by riparian management, bank stabilization, flow modification, and floodplain reconnection. Water quality management using riparian buffers and bank stabilization were among the cheaper techniques ($19,000–$41,000 per project) whereas flow modification and floodplain reconnection were much larger and more expensive projects ($198,000–$207,000). In Louisiana, where the scale and pace of wetland loss is staggering, the costs to benefits of restoration measures range from $900 for small-scale plantings to $2000–$4000 for large-scale freshwater and sediment diversions (Merino et al., 2011) (see Chapter 12, Restoration on a Grand Scale).

Fundamental Characteristics of Wetlands

Wetlands are defined by three distinct characteristics, hydrology, vegetation, and soils, which differ from terrestrial and aquatic ecosystems (Figure 1.1). Wetland hydrology is described by the depth, duration, frequency, and timing or seasonality of flooding or soil saturation. Different types of wetlands possess different hydrological regimes, from tidal marshes and mangroves that are flooded twice daily by the astronomical tides to peat bogs that may never flood but whose soils are nearly permanently saturated. Wetlands that receive most of their water from precipitation such as depressional wetlands and vernal pools dry out for extended periods and may be dry longer than they are wet. Depending on the type of wetland, the presence of water may be permanent or it may be fleeting. The common thread is that they are flooded or saturated long enough during the growing season, when the vegetation is active and growing, to produce soils and plant communities unique to wetland ecosystems.

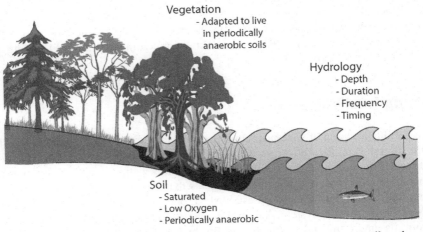

Figure 1.1 The three defining characteristics of wetlands: wetland hydrology, soil, and vegetation.

Wetlands also differ in the source(s) of water that flood or saturate them. Inundation may be the result of surface flow, overbank flooding from rivers and streams, or tidal inundation in estuaries. Groundwater may be a significant water source as it occurs in the case of seepage wetlands and fens. A third source of water is precipitation, rain, and snow, which contribute to the hydrology of nearly all wetlands. In many cases, all three sources of water contribute to wetland hydrology in varying proportions. The source(s) of water have a powerful effect on wetland water and soil chemistry and on propagation of off-site stressors into the system.

When soils become flooded or saturated with water, they shift from aerobic to anaerobic conditions. Once flooded, microorganisms in the soil quickly consume the limited oxygen in the pore space to support respiration for cell growth, maintenance, and reproduction. Plants and animals, that also require oxygen to live, are affected by anaerobic soil conditions as well. Since many animals are mobile, they move elsewhere to avoid the oxygen-poor conditions. Plants, however, are sedentary and must adapt or perish. Plants adapted to the wetland environment possess adaptations, both morphological and metabolic, not found in terrestrial vegetation that enable them to maintain the flow of air-rich oxygen to the roots and to survive and thrive in anaerobic soils.

Wetland soils also possess characteristics that are distinct from terrestrial soils. Lack of oxygen also inhibits aerobic microorganisms so that decomposition of organic matter produced by vegetation is much slower in wetlands than in terrestrial soils. The result is accumulation of partially or undecomposed organic matter that produces distinctive dark-colored layers or horizons in wetland soils. The extreme case of organic matter accumulation is the formation of peat, a soil of biogenic origin consisting of mostly dead plant remains. A defining characteristic of many mineral soil wetlands is the reduction of oxidized iron (Fe^{3+}) by microorganisms that use it for respiration in the absence of oxygen. Soils containing oxidized Fe exhibit rustlike colors, red, orange, and yellow, that often are observed in terrestrial soils. When flooded, microorganisms reduce oxidized ferric Fe^{3+} to ferrous Fe^{2+}, producing soils that are gray in color. Under conditions of permanent flooding, mineral soils may take on a greenish or bluish color indicating continuous flooding and complete absence of oxygen.

Setting Realistic Goals

Successful restoration of wetlands requires setting explicit goals at the outset (Zedler, 1995) (Figure 1.2). Ideally, the goal is to reestablish the suite of ecological functions observed in nature for a given wetland type. However, this is not always possible so, in some situations, one must identify goals that are achievable and aim for them (Ehrenfeld, 2000). Once goals are established, one must identify and ameliorate the stressors impacting the system. A thorough understanding of the dynamics of the ecosystem, its environmental template, and life history traits of the species to be reintroduced, is needed to know which species will prosper and which ones will not. Techniques such as seeding, planting, and amendments may be implemented to

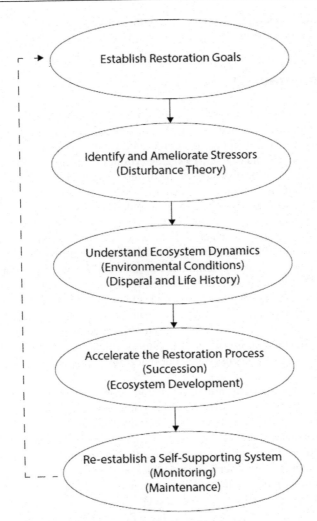

Figure 1.2 Five key steps for successful restoration of wetlands.

accelerate succession and ecosystem development. Establishing small-scale experiments to test various restoration techniques is useful as it can identify better methods for improving future restoration efforts (Zedler, 2005). Reestablishing a self-supporting wetland also requires monitoring and sometimes maintenance to direct the wetlands toward the desired endpoint community.

Sometimes, goals may need to be reevaluated when off-site stressors cannot be ameliorated or invasive species colonize the site. Two goals that are not mutually compatible are biodiversity support and water quality improvement. In nutrient-enriched environments, restoration of wetlands for nutrient removal will inevitably lead to loss of biodiversity (Zedler, 2003). Restoration of biodiversity should target areas where nutrient loading is not a problem and where the restoration provides continuity with existing healthy, intact wetland, upland, and aquatic habitats.

Theory and Practice

Restoration ecology and wetland restoration are buttressed by an understanding of the key ecological processes—disturbance, succession, and ecosystem development—that structure plant and animal communities. Disturbance, its size, frequency, and intensity (Connell and Slatyer, 1977), determines the pace of ecosystem development following restoration. The size of the disturbance determines how quickly a site will be colonized by propagules with faster reestablishment of vegetation on smaller than larger sites. The intensity of a disturbance determines the degree to which colonization occurs from propagules on-site, either from the seed bank or from vegetative fragments. The frequency at which a disturbance occurs determines the amount of time in which succession can occur on a site before it is disturbed again. In wetlands, altered hydrology is the most common disturbance and it tends to be chronic (Turner and Lewis, 1997). That is, the disturbance presses continuously on the system so there is no frequency or recurrence interval. Other chronic stressors affecting wetlands include nutrient enrichment, grazing, and encroachment by invasive species. Disturbances may originate on-site or off-site. For aquatic ecosystems and the wetlands connected to them, it is critical to ameliorate disturbances that originate off-site in upstream and terrestrial ecosystems but that propagate downstream and downhill (Loucks, 1992).

Succession, how plant (and animal) communities change over time, proceeds slowly on large sites with intense disturbance. Succession theory consists of two camps: the organismic view of Clements (1916) and the individualistic view of Gleason (1917). According to Clements, succession is the orderly, predictable change in plant communities over time. Early colonizers improve the environment, paving the way for succeeding organisms. The Clementsian model certainly applies to peatlands where the plant community modifies the soil environment by building peat that alters hydrology and determines soil chemical properties. Gleasonian models, including inhibition and tolerance (Connell and Slatyer, 1977), relay floristics (Egler, 1954), and assembly rules (Weiher and Keddy, 1995), posit that environmental conditions and stochastic or random events determine who the initial colonizers are and which species, if any, will colonize later. Support for Gleasonian models includes tidal marshes and mangroves where environmental stress, flooding, and salinity, are high.

Ecosystem development describes how energy, often expressed as carbon, flows and nutrient cycles change over time. In *The Strategy of Ecosystem Development*, Odum (1969) made predictions of how community energetics, community structure, life history traits, nutrient cycling, and other attributes change as an ecosystem ages from a young system to a mature one. Odum's ideas clearly tend toward the organismic view of Clements. In his paper, Odum introduced the idea of pulse stability, that ecosystems with regular predictable disturbance and whose organisms are adapted to it are maintained at an intermediate stage of succession with the optimal benefits of young ecosystems (high productivity) and mature ones (high diversity). Odum described these young and mature systems, respectively, as production and protection systems. From a restoration perspective, pulse stability is applicable to a number of wetland types, including tidal marshes, mangroves, and floodplain wetlands.

The development of ecosystems depends on both biological and physicochemical processes. Biological processes, especially those related to nutrient accumulation, are essential for the development of a properly functioning ecosystem (Dobson et al., 1997). This is especially true for organic carbon and nitrogen that accumulate in the soil. Nearly all N is stored as organic N in soil organic matter which is slowly mineralized to ammonium (NH_4^+) and nitrate (NO_3^-) by microorganisms and then used by plants. Organic C is essential to support the largely heterotrophic food webs of wetland and terrestrial forest ecosystems. Biological processes develop faster than physical processes (Table 1.2). Immigration and establishment of plant species occur relatively quickly and processes that accompany their arrival, sedimentation and accumulation of soil organic matter and N, do as well. Soil flora and fauna arrive once soil properties, especially organic matter, begin to develop. Physicochemical properties that drive long-term soil development take longer. In wetlands, the pervasiveness of water leads to rapid leaching of soluble materials, especially reduced forms of iron (see Chapter 2, Definitions). Other processes such as rock weathering (characteristic of all soils), release of inorganic nutrients, and formation of soil horizons take decades to centuries.

Understanding ecosystem dynamics, including the natural history and environmental requirements of organisms (especially plants), is critical to identify which species

Table 1.2 Important Biological and Physical Processes Involved in the Development of Wetland Ecosystems and Their Timescales of Development

	Timescale (Years)
Biological Processes	
Immigration of appropriate plant species	1–20
Establishment of appropriate plant species	1–20
Accumulation of sediment and inorganic nutrients	1–20
Accumulation of nitrogen by biological fixation	10–50
Accumulation of soil organic matter	10–100
Immigration of soil flora and fauna	1–20
Physical Processes	
Accumulation of soil particles by rock weathering	10–1000
Release of inorganic nutrients from soil minerals	10–1000
Leaching of soluble materials	1–100
Formation of soil horizons	10–1000

Modified from Dobson et al. (1997).

will disperse to the site and, once there, will thrive. Grime (1977) describes three strategies of plants for surviving and thriving under different environmental conditions. Ruderal species are among the first colonizers of disturbed sites and are similar to the r-strategists described by Odum (1969). Stress tolerators are slow-growing species that exist in high-stress environments. Some common wetland species fit this definition, notably some members of the genus *Schoenoplectus* and *Juncus* (Boutin and Keddy, 1993). Competitor species exist in low disturbance, low-stress environments, and often tend to dominate the site. Competitors, also known as clonal dominants, include many aggressive and invasive wetland plants such as *Phragmites*, *Phalaris*, *Typha*, and *Lythrum*.

Disturbance: Identifying and Ameliorating Stressors

The first step to restore a wetland is to identify and ameliorate the stressors that impair it. A degraded site is a disturbed site and disturbance theory, the size, intensity, and frequency of the disturbance, informs the steps and efforts needed to restore it (see Chapter 3, Ecological Theory and Restoration). Stressors may be physical, chemical, or biological (Table 1.3). Physical stressors involve the delivery of water that determines hydrology. Chemical stressors affect the chemical composition and quality of water. Biological stressors involve the introduction or colonization of alien, aggressive, or weedy species (plants and animals) that alter community structure, function, and ecosystem services.

Stressors may originate on-site or off-site. On-site stressors usually involve alterations to wetland hydrology. Hydrology—the frequency, depth, duration, and timing or seasonality of flooding—is fundamental to a healthy functioning wetland. Without reintroducing the proper hydrology first, all restoration projects will fail. Reintroducing hydrology consists of blocking or filling ditches or removing fill. Sometimes hydrology is altered by building levees, including sea defenses that isolate wetlands from their water source. Other on-site stressors include grazing and silvicultural activities that affect plant communities. Paradoxically, periodic disturbance in the form of grazing or mowing may be needed to maintain species richness of some wetlands as in

Table 1.3 Stressors to Ameliorate When Restoring Wetlands

	Stressor	Examples
Physical	Hydrology (altered depth, duration, frequency of inundation or soil saturation; timing and seasonality of flooding)	Ditches that promote drainage Levees that restrict flow Placement of fill
Chemical	Water quality	Nutrients (N, P), sediment, salinity Other contaminants
Biological	Invasive species	*Phragmites, Phalaris, Typha,* and others
	Grazing	Domesticated livestock

the case of wet grasslands (Joyce, 2014). The key is to try to re-create the conditions on-site that are needed to meet the goals of the restoration.

Off-site hydrological alterations involve changes to the magnitude and timing of flooding. Flow mistiming occurs when dams constructed upstream alter or mute the seasonal flood pulse that occurs following snowmelt or during the "wet" season. A pervasive stressor that originates off-site is chemical pollution that leads to degradation of water quality. Nutrient overenrichment from agricultural and urban–suburban fertilizer use is a widespread problem affecting wetlands and other aquatic ecosystems (Craft et al., 2007; NRC, 2000). Excess nutrients supply *too much of a good thing* as enrichment stimulates plant productivity. Often it is fast growing, aggressive, and invasive species that are the beneficiaries. Other stressors, salinization of freshwater wetlands, heavy metals, thermal pollution and others, may affect some sites. But, hydrologic alteration, nutrient enrichment, and invasive species are the most chronic and widespread problems and, without intervention, the goals of the restoration will not be achieved (Parker, 1997).

Understanding Ecosystem Dynamics

Wetland restoration requires a thorough understanding of the ecosystem dynamics of the system one is working to restore (Hobbs, 2007). To repair a degraded ecosystem requires not only ameliorating the stressors that impact the system, but recognizing how the intact, functioning wetland works. This includes a comprehensive understanding of the environmental requirements of the plants (and animals), their preferred depth, duration, and frequency of flooding, nutrient condition (oligotrophy vs mesotrophy), light and temperature requirements, and other factors. This is done by observing the conditions in intact, undisturbed wetlands of the same type. Identifying the different ways that propagules disperse, be it by wind, water, fowl, or other animals is important. Hydrochory, the dispersal of wetland propagules by water, may be especially important for colonization of riverine and tidal wetlands (Nilsson et al., 2010). Understanding their dormancy and germination requirements is needed to know which species are likely to reach the site, germinate, become established, and prosper and which ones will not. Just as a watchmaker knows how a timepiece works and how to repair a broken one, restoring wetlands requires a thorough understanding of how the natural ecosystem functions.

Accelerating Restoration: Succession and Ecosystem Development

Once proper environmental conditions, especially hydrology, are reestablished, the ecosystem is repaired by restoring the appropriate soil conditions and reintroducing the characteristic vegetation of the site (Table 1.4). Some sites colonize naturally. Species that produce large numbers of wind-dispersed seeds are among the first to arrive. Most restoration projects require deliberate reintroduction of at least some species by introducing seeds, seedlings, or other types of propagules. Sites that are exposed to wave and wind action such as tidal marshes and mangroves often require planting (Figure 1.3(a)). Other sites that periodically dry down can be seeded, enabling seeds to

**Table 1.4 Common Methods to Accelerate Succession
and Ecosystem Development of Restored Wetland**

	Amendment/Addition	Examples
Soils	Nutrients	Nitrogen, phosphorus, lime
	Organic matter	Topsoil, compost, peat, manures
	Sediment	Thin layer placement of dredge material
	Topsoil removal[a]	Sod cutting
Vegetation	Propagules	Seeds, fragments (rhizomes), seedlings, saplings

[a]To remove excess nutrients.

(a)

(b)

Figure 1.3 (a) Seedlings of *Spartina alterniflora* planted on dredge material, North Carolina, USA. (b) Freshwater marsh mitigation wetland established by seeding, Indiana, USA. Photo credit: (a) Steve Broome.

germinate before flooding resumes (Figure 1.3(b)). Some species may be introduced initially to colonize the site before undesirable species recruit from outside. Species that are keystone components of the ecosystem may not readily colonize and often must be introduced.

A common approach to restoring many terrestrial and wetland ecosystems is to introduce species important for restoring ecosystem function and those that are major components of the desired endpoint community (Dobson et al., 1997). Other species that make up the overall biodiversity of the endpoint community are left to colonize on their own. The question of planting depends on who you ask. Some argue that natural colonization or self-design is preferred because, as Mitsch et al. (2000) said, *we do not know enough to play the role of nature*. For mangroves, natural recolonization is a viable technique if there is a nearby source of propagules (Lewis, 2005). Others suggest that planting is necessary to produce a diverse plant community and keep invasive species from colonizing (Streever and Zedler, 2000). Planting also provides additional benefits such as erosion control and can provide a nurse crop to facilitate colonization by desired species (Lewis, 1982; Clewell and Rieger, 1997).

Various amendments are used to accelerate ecosystem development. Nitrogen (N) and sometimes phosphorus (P) are added to jumpstart growth of vegetation so that it quickly colonizes the site (Figure 1.3(b)). Sometimes soils are amended with organic matter (topsoil, peat, compost, *green* manure such as alfalfa, or biochar) to improve physical properties (porosity), enhance fertility, and support heterotrophic activity.

Once vegetation establishes, community structure and ecosystem functions begin to develop. Different attributes of the ecosystem develop along different trajectories and at different rates (see Chapter 10, Performance Standards and Trajectories of Ecosystem Development) as the site matures (Figure 1.4(a)). Sometimes, structure and functions of the restored site follow an entirely different trajectory, leading to an alternative stable state (Figure 1.4(b)). This may be because the proper environmental conditions, usually hydrology, were not reestablished or the stressors were not ameliorated. A contributing factor is the history of the site, especially disturbance (Hobbs, 2007; Higgs et al., 2014) including land-use legacies such as nutrient enrichment and subsidence. The availability of propagules and the stochastic nature of dispersal also may lead to a different stable state.

Alternative stable states have been observed in terrestrial ecosystems, for example, grasslands where the removal of livestock does not lead to reestablishment of the original plant community (Hobbs and Norton, 1996). Although wetland restoration relies much on the Clementsian view of succession, disturbances and stochastic events, especially dispersal and recruitment, may lead to alternative trajectories and alternative stable states (Palmer et al., 1997). This may be true for some forested wetlands where recruitment of key species does not occur because there is no nearby seed source (Allen, 1997; Haynes, 2004) or where subsidence leads to permanent flooding so that seedlings cannot establish (Doyle et al., 2007).

Reestablishing a Self-Supporting System

Once wetland vegetation is reestablished, the site inevitably will require some effort to maintain it in its desired state. Biodiversity, a common goal of many wetland restoration projects, requires constant vigilance to combat encroachment by invasive species. New colonizers may come to dominate a restoration site, altering energy flows and nutrient cycling, leading to an alternative stable state (Figure 1.4(b)). This has

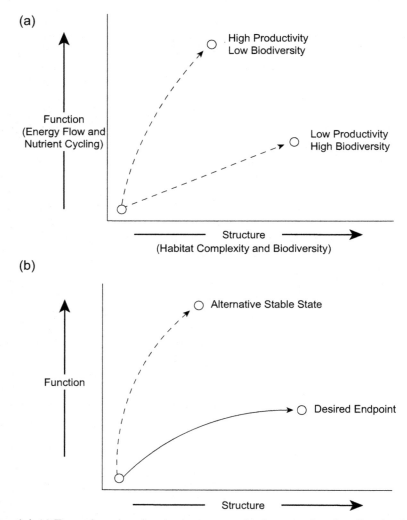

Figure 1.4 (a) Two trajectories of wetland ecosystem development. One describes development of a highly productive wetland. The second describes a wetland with high biodiversity. (b) Trajectories of desired versus alternative stable state wetlands. The alternative stable state often is characterized by a highly productive, low diversity wetland as occurs when an invasive species dominate.

been shown in terrestrial ecosystems where N-fixing invaders (*Myrica*) dramatically alter N cycling (Vitousek and Walker, 1989) but also in wetlands where *Phragmites* and *Typha* alter nutrient (N) and C cycles through their high levels of aboveground biomass and litter production(Windham and Ehrenfeld, 2003; Larkin et al., 2012). Arguably, maintaining species diversity is one of the biggest challenges facing wetland restoration and restoration ecology today.

To determine whether a restoration project is successful requires monitoring before and following restoration. Some attributes of community structure and ecosystem function take years or more to develop. Ideally, monitoring before and after restoration should be performed to gauge how quickly the benefits of restoration develop. It is also useful to employ reference wetlands, intact, functioning, natural wetlands of the same type as the one that is degraded, to gauge if and how quickly the wetland develops toward a well-functioning system (Brinson and Rheinhardt, 1996) (see Chapter 10, Performance Standards and Trajectories of Ecosystem Development). Ideally, multiple reference wetlands are monitored to account for inherent spatial and temporal variability of natural systems (Pickett and Parker, 1994; Parker, 1997; Clewell and Rieger, 1997) driven by stochastic events such as disturbance and colonization. This *flux of nature* should be recognized, embraced, and incorporated into wetland restoration and monitoring protocols.

There also is a need to periodically evaluate older restored wetlands to inform future restoration projects. This can help us understand which restoration practices work, which did not, and what can be done to improve success of future projects (Clewell and Rieger, 1997). A meta-analysis of 621 restored wetlands worldwide shows that, on average, restored wetlands have 26% less biological (plant community) structure and 23% less carbon storage in soils relative to comparable natural reference wetlands (Moreno-Mateos et al., 2012). In this analysis, larger wetlands (>100 ha) and wetlands restored in temperate and tropical regions developed more quickly than smaller wetlands and wetlands in cold climates. Not surprisingly, wetlands with strong connections to surface waters, riverine, and tidal wetlands, developed faster than precipitation-driven depressional wetlands.

In spite of these shortcomings, there has been much progress in understanding how to restore wetlands. Zedler (2000) identified a number of ecological principles to guide wetland restoration. They include landscape context and position (Chapter 4 of this book), reference wetlands to evaluate success (Chapter 10), establishing proper hydrology (Chapter 2), the role of seed banks and propagule dispersal, environmental conditions, and life history traits (Chapters 5–9), succession and ecosystem development (Chapter 3), and trajectories as restored wetlands mature (Chapter 10). A key attribute of ecological restoration that is often unappreciated is the importance of humans in the process (Cairns and Heckman, 1996; Hobbs and Norton, 1996). Constraints imposed by society such as availability of water resources or antecedent conditions such as land-use legacies may hinder restoration efforts (Simenstad et al., 2006). On the other hand, involvement and "buy-in" of the local community is essential for long-term success of most if not all restoration projects (Field, 1998; Geist and Galatowitsch, 1999; Pfadenhauer, 2001; Comin et al., 2005; Higgs et al., 2014).

A number of books about wetland restoration have been published to date (Lewis, 1982; Zelazny and Feierabend, 1988; Kusler and Kentula, 1989; Galatowitsch and van der Valk, 1994; Wheeler et al., 1995; Joyce and Wade, 1998; Middleton, 1999; Quinty and Rochefort, 2003) but they tend to focus on a specific wetland habitat or wetlands in a particular geographic region. Restoration of coastal wetlands, saline tidal marshes, and mangroves, has received widespread attention (Lewis, 1982;

Thayer, 1992; Field, 1996; Turner and Streever, 2002; Perillo et al., 2009; Roman and Burdick, 2012; Lewis and Brown, 2014), perhaps because of their value as habitat for commercial and recreational fisheries and shoreline protection. In the US, restoration as a means to compensate for wetland loss under the Clean Water Act produced two books by the National Research Council (1992, 2001). Finally, several books provide detailed guidance including case studies for restoring wetlands, especially tidal marshes (Zedler, 1996, 2001), mangroves (Field, 1996; Lewis and Brown, 2014), and peatlands (Quinty and Rochefort, 2003).

Creating and Restoring Wetlands brings together the ecological theory and restorationist's practice to create and restore wetlands. Restoration of five common wetland habitats—freshwater marsh, peatland, floodplain forest, tidal marsh, and mangrove—are presented in detail. The wetland habitats differ in their landscape position, hydrology, environmental conditions, species assemblage, and rates of succession and ecosystem development. The book describes key characteristics that constitute wetlands, ecological theories behind restoration, trajectories of succession and ecosystem development, performance standards to gauge success, and, for the five wetland habitats, it offers keys to ensure success. The book also covers watershed and landscape considerations, restoration at a larger (grand) scale, and the future of wetland restoration.

I am indebted to those who came before me and on whose shoulders I stand. They include the team from North Carolina State University, led by W.W. Woodhouse Jr., Ernest D. Seneca, and Stephen W. Broome, for their efforts to restore tidal marshes. Roy R. (Robin) Lewis of Florida and Colin Field of Australia directed the development of mangrove restoration strategies. Line Rochefort and coworkers in Canada laid much of the groundwork for understanding how to restore peatlands. Joy B. Zedler of the University of Wisconsin bridged restoration of tidal wetlands with inland freshwater wetlands and encouraged us to think about wetland restoration at watershed and landscape scales. Last but not least, Curtis Richardson taught me to think bigger about wetlands, wetland ecosystem services, and wetland restoration. Because of them, I am able see farther. I hope this book reflects that.

References

Allen, J.A., 1997. Reforestation of bottomland hardwoods and the issue of woody species diversity. Restoration Ecology 5, 125–134.

An, S., Li, H., Guan, B., Zhou, C., Wang, Z., Deng, Z., Zhi, Y., Liu, Y., Xu, C., Fang, S., Jiang, J., Li, H., 2007. China's natural wetlands: past problems, current status and future challenges. AMBIO 36, 335–342.

Batt, B.D.J., Anderson, M.G., Anderson, C.D., Caswell, F.D., 1989. The use of prairie potholes by North American ducks. In: van der Valk, A. (Ed.), Northern Prairie Wetlands. Iowa State University Press, Ames, IA, pp. 204–227.

Bernhardt, E.S., Palmer, M.A., Allan, J.D., Alexander, G., Barnas, K., Brooks, S., Carr, J., Clayton, S., Dahm, C., Follstad-Shah, J., Galat, D., Gloss, S., Goodwin, P., Hart, D., Hassett, B., Jenkinson, R., Katz, S., Kondolf, G.M., Lake, P.S., Lave, R., Meyer, J.L., O'Donnell, T.K., Pagano, L., Powell, B., Sudduth, E., 2005. Synthesizing U.S. river restoration efforts. Science 308, 636–637.

Boutin, C., Keddy, P.A., 1993. A functional classification of wetland plants. Journal of Vegetation Science 4, 591–600.

Brinson, M.M., Rheinhardt, R., 1996. The role of reference wetlands in functional assessment and mitigation. Ecological Applications 6, 69–76.

Cairns Jr., J., Heckman, J.R., 1996. Restoration ecology: the state of an emerging field. Annual Review of Energy and the Environment 21, 167–189.

Chang, T.T., 1976. The rice cultures. Philosophical Transactions of the Royal Society of London Series B: Biological Sciences 275, 143–157.

Chavez-Ramirez, F., Wehtje, W., 2012. Potential impact of climate change scenarios on whooping crane life history. Wetlands 32, 11–20.

Chowdhury, R.A., Ahmed, I., 1994. History of forest management. In: Hussain, Z., Acharya, G. (Eds.), Mangroves of the Sundarbans, vol. II: Bangladesh. IUCN, Bangkok, Thailand, pp. 155–180.

Chung, C.H., 2006. Forty years of ecological engineering with *Spartina* plantations in China. Ecological Engineering 27, 49–57.

Clements, F.E., 1916. Plant Succession. Publication 242. Carnegie Institute, Washington, DC.

Clewell, A., Rieger, J.P., 1997. What practitioners need from restoration ecologists. Restoration Ecology 5, 350–354.

Coleman, J.M., Huh, O.K., Braud Jr., D., 2008. Wetland loss in world deltas. Journal of Coastal Research 24, 1–14.

Comin, F.A., Romero, J.A., Hernadez, O., Menendez, M., 2001. Restoration of wetlands from abandoned rice fields for nutrient removal, and biological community and landscape diversity. Restoration Ecology 9, 201–208.

Comin, F.A., Menendez, M., Pedrocchi, C., Moreno, S., Sorando, R., Cabezas, A., Garcia, M., Rosas, V., Moreno, D., Gonzalez, E., Gallardo, B., Herrera, J.A., Ciancarelli, C., 2005. Wetland restoration: integrating scientific, technical and social perspectives. Ecological Restoration 23, 182–186.

Connell, J.H., Slatyer, R.O., 1977. Mechanisms of succession in natural communities and their role in community stability and organization. American Naturalist 111, 1119–1144.

Craft, C., Krull, K., Graham, S., 2007. Ecological indicators of nutrient condition, freshwater wetlands, Midwestern United States (U.S.). Ecological Indicators 7, 733–750.

Dahl, T.E., 1990. Wetlands Losses in the United States 1780's to 1980's. U.S. Department of the Interior, Fish and Wildlife Service, Washington, DC.

Dobson, A.P., Bradshaw, A.D., Baker, A.J.M., 1997. Hopes for the future: restoration ecology and conservation biology. Science 277, 515–522.

Doyle, T.W., O'Neil, C.P., Melder, M.P.V., From, A.S., Palta, M.M., 2007. Tidal freshwater swamps of the southeastern United States: effects of land use, hurricanes, sea-level rise, and climate change. In: Conner, W.H., Doyle, T.W., Krauss, K.W. (Eds.), Ecology of Tidal Freshwater Forested Wetlands of the Southeastern United States. Springer, Dordrecht, The Netherlands, pp. 1–28.

Egler, F.E., 1954. Vegetation science concepts I. Initial floristic composition, a factor in old-field vegetation development. Vegetatio 4, 412–417.

Ehrenfeld, J.G., 2000. Defining the limits of restoration: the need for realistic goals. Restoration Ecology 8, 2–9.

Fennessy, S., Craft, C., 2011. Effects of agricultural conservation practices on wetland ecosystem services in the Interior Glaciated Plains. Ecological Applications (Supplement) 21, S49–S64.

Field, C.D., 1996. Restoration of Mangrove Ecosystems. International Society for Mangrove Ecosystems, Okinawa, Japan.

Field, C.D., 1998. Rehabilitation of mangrove ecosystems: an overview. Marine Pollution Bulletin 37, 383–392.

Food and Agricultural Organization, 2007. The World's Mangroves: 1980–2005. Food and Agricultural Organization of the United Nations, Rome, Italy.

Galatowitsch, S.M., van der Valk, A.G., 1994. Restoring Prairie Wetlands: An Ecological Approach. Iowa State University Press, Ames, IA.

Geist, C., Galatowitsch, S.M., 1999. Reciprocal model for meeting ecological and human needs in restoration projects. Conservation Biology 13, 970–979.

Gleason, H.A., 1917. The structure and development of the plant association. Bulletin of the Torrey Botanical Club 43, 463–481.

Grime, J.P., 1977. Evidence for the existence of three primary strategies in plants and its relevance to ecological and evolutionary theory. American Naturalist 111, 1169–1194.

Haynes, R.J., 2004. The development of bottomland forest restoration in the lower Mississippi alluvial valley. Ecological Restoration 22, 170–182.

He, Q., Bertness, M.D., Bruno, J.F., Li, B., Chen, G., Coverdale, T.C., Altieri, A.H., Bai, J., Sun, T., Pennings, S.C., Liu, J., Ehrlich, P.R., Cui, B., 2014. Economic development and coastal ecosystem change in China. Scientific Reports 4. http://dx.doi.org/10.1038/srep05995.

Higgs, E., Falk, D.A., Guerrini, A., Hall, M., Harris, J., Hobbs, R.J., Jackson, S.T., Rhemtulla, J.M., Throop, W., 2014. The changing role of history in restoration ecology. Frontiers in Ecology and the Environment 12, 499–506.

Hobbs, R.J., 2007. Setting effective and realistic restoration goals: key directions for research. Restoration Ecology 15, 354–357.

Hobbs, R.J., Norton, D.A., 1996. Towards a conceptual framework for restoration ecology. Restoration Ecology 4, 93–110.

Joyce, C.B., 2014. Ecological consequences and restoration potential of abandoned wet grasslands. Ecological Engineering 66, 91–102.

Joyce, C.B., Wade, P.M., 1998. European Wet Grasslands: Biodiversity, Management, and Restoration. John Wiley and Sons, Chichester, UK.

Kanai, Y., Ueta, M., Germogenov, N., Nagendran, M., Mita, N., Higuchi, H., 2002. Migration routes and important resting areas of Siberian cranes (*Grus leucogeranus*) between northeastern Siberia and China as revealed by satellite tracking. Biological Conservation 106, 339–346.

Kangmin, L., 1988. Rice-fish culture in China: a review. Aquaculture 71, 173–186.

Knutson, P.L., Ford, J.C., Inskeep, M.R., 1981. National survey of planted salt marshes (vegetative stabilization and wave stress). Wetlands 1, 129–157.

Kusler, J.A., Kentula, M.E. (Eds.), 1989. Wetland Creation and Restoration: The Status of the Science. U.S. Environmental Protection Agency, Corvallis, Oregon. EPA/600/3089/038.

Larkin, D.J., Freyman, M.J., Lishawa, S.C., Geddes, P., Tuchman, N.C., 2012. Mechanisms of dominance by the invasive hybrid cattail *Typha×glauca*. Biological Invasions 14, 65–77.

Lewis III, R.R., 1982. Creation and Restoration of Coastal Plant Communities. CRC Press, Boca Raton, FL.

Lewis III, R.R., 2005. Ecological engineering for successful management and restoration of mangrove forests. Ecological Engineering 24, 403–418.

Lewis III, R.R., Brown, B., 2014. Ecological Mangrove Rehabilitation. A Field Manual for Practitioners. http://www.mangroverestoration.com.

Loucks, O.L., 1992. Predictive tools for rehabilitating linkages between land and wetland ecosystems. In: Wali, M.K. (Ed.), Ecosystem Rehabilitation. Ecosystem Analysis and Synthesis, vol. 2. SPB Academic Publishing, The Hague, The Netherlands, pp. 297–308.

Merino, J., Aust, C., Caffrey, R., 2011. Cost-efficacy of wetland restoration projects in coastal Louisiana. Wetlands 31, 367–375.

Middleton, B., 1999. Wetland Restoration: Flood Pulsing and Disturbance Dynamics. John Wiley and Sons, New York.

Mitsch, W.J., Wu, X.B., Nairn, R.W., Wang, N., 2000. To plant or not to plant: a response by Mitsch et al. BioScience 50, 189–190.

Moreno-Mateos, D., Power, M.E., Comin, F.A., Yockteng, R., 2012. Structural and functional loss in restored wetland ecosystems. PLoS Biology 10, e1001247.

National Research Council, 1992. Restoration of Aquatic Ecosystems. National Research Council, Washington, DC.

National Research Council, 2000. Clean Coastal Waters: Understanding and Reducing the Effects of Nutrient Pollution. National Academy Press, Washington, DC.

National Research Council, 2001. Compensating for Wetland Losses under the Clean Water Act. National Academy of Sciences, Washington, DC.

Nilsson, C., Brown, R.L., Jansson, R., Merritt, D.M., 2010. The role of hydrochory in structuring riparian and wetland vegetation. Biological Reviews 85, 837–858.

Odum, E.P., 1969. The strategy of ecosystem development. Science 164, 262–270.

Palmer, M.A., Ambrose, R.F., Poff, N.L., 1997. Ecological theory and community restoration ecology. Restoration Ecology 5, 291–300.

Palmer, M.A., Menninger, H.L., Bernhardt, E.S., 2010. River restoration, habitat heterogeneity and biodiversity: a failure of theory or practice? Freshwater Biology 55 (S1), 205–222.

Parker, V.T., 1997. The scale of successional models and restoration objectives. Restoration Ecology 5, 301–306.

Perillo, G.M.E., Wolanski, E., Cahoon, D.R., Brinson, M.M. (Eds.), 2009. Coastal Wetlands: An Integrated Ecosystem Approach. Elsevier, Amsterdam, The Netherlands.

Pfadenhauer, J., 2001. Some remarks on the socio-cultural background of restoration ecology. Restoration Ecology 9, 220–229.

Pickett, S.T.A., Parker, V.T., 1994. Avoiding the old pitfalls: opportunities in a new discipline. Restoration Ecology 2, 75–79.

Quinty, F., Rochefort, L., 2003. Peatland Restoration Guide, Second ed. Canadian Sphagnum Peat Moss Association and New Brunswick Department of Natural Resources and Energy, Québec.

Ranwell, D.S., 1967. World resources of *Spartina townsendii* (*sensu lato*) and economic use of *Spartina* marshland. Journal of Applied Ecology 4, 239–256.

Roman, C.T., Burdick, D.M. (Eds.), 2012. Tidal Marsh Restoration: A Synthesis of Science and Management. Island Press, Washington, DC.

Sahrhage, D., 2008. Fishing in ancient Egypt. In: Encyclopedia of the History of Science, Technology and Medicine in Non-Western Cultures. Springer, The Netherlands, pp. 922–927.

Seck, P.A., Diagne, A., Mohanty, S., Wopereis, M.C.S., 2012. Crops that feed the world 7: rice. Food Security 4, 7–24.

Simenstad, C., Reed, D., Ford, M., 2006. When is restoration not? Incorporating landscape-scale processes to restore self-sustaining ecosystems in coastal wetland restoration. Ecological Engineering 26, 27–39.

Smith, M.D., Roheim, C.A., Crowder, L.B., Halpern, B.S., Turnipseed, M., Anderson, J.L., Asche, F., Bourillon, L., Guttormsen, A.G., Khan, A., Liguori, L.A., McNevin, A., O'Conner, M.I., Squires, D., Tyedmers, P., Browstein, C., Carden, K., Klinger, D.H., Sagarin, R., Selkoe, K.A., 2010. Sustainability and global seafood. Science 327, 784–786.

Streever, B., Zedler, J., 2000. To plant or not to plant. BioScience 50, 188–189.

Thayer, G.W., 1992. Restoring the Nation's Marine Environment. Maryland Sea Grant, College Park, MD.

Turner, R.E., Lewis III, R.R., 1997. Hydrologic restoration of coastal wetlands. Wetlands Ecology and Management 4, 65–72.

Turner, R.E., Streever, B., 2002. Approaches to Coastal Wetland Restoration: Northern Gulf of Mexico. SPB Academic Publishing bv, The Hague, The Netherlands.

Valiela, I., Bowen, J.L., York, J.K., 2001. Mangrove forests: one of the world's threatened ecosystems. BioScience 51, 807–815.

Vitousek, P.M., Walker, L.R., 1989. Biological invasion by *Myrica faya* in Hawaii: plant demography, nitrogen fixation, ecosystem effects. Ecological Monographs 59, 247–265.

Weller, M.W., 1994. Freshwater Marshes: Ecology and Wildlife Management. University of Minnesota Press, Minneapolis, MN.

Wheeler, B.D., Shaw, S.C., Fojt, W.J., Robertson, R.A. (Eds.), 1995. Restoration of Temperate Wetlands. John Wiley and Sons, Chichester, UK.

Weiher, E., Keddy, P.A., 1995. The assembly of experimental wetland plant communities. Oikos 73, 323–335.

Windham, L., Ehrenfeld, J.G., 2003. Net impact of plant invasion on nitrogen-cycling processes within a brackish tidal marsh. Ecological Applications 13, 883–897.

Zedler, J.B., 1995. Salt marsh restoration: lessons from California. In: Cairns Jr., J. (Ed.), Rehabilitating Damaged Ecosystems. Lewis Publishers, Boca Raton, FL, pp. 75–95.

Zedler, J.B., 1996. Tidal Wetland Restoration: A Scientific Perspective and Southern California Focus. Publication No. T-38. California Sea Grant College System, University of California, La Jolla, CA.

Zedler, J.B., 2000. Progress in wetland restoration ecology. Trends in Ecology & Evolution 15, 402–407.

Zedler, J.B., 2001. Handbook of Restoring Tidal Wetlands. CRC Press, Boca Raton, FL.

Zedler, J.B., 2003. Wetlands at your service: reducing impacts of agriculture at the watershed scale. Frontiers in Ecology and the Environment 1, 65–72.

Zedler, J.B., 2005. Restoring wetland plant diversity: a comparison of existing and adaptive approaches. Wetlands Ecology and Management 13, 5–14.

Zelazny, J., Feierabend, J.S. (Eds.), 1988. Increasing Our Wetlands Resources. Wetlands Conference Papers. National Wildlife Federation, Washington, DC.

Definitions

2

Chapter Outline

Introduction

Wetlands are "edge" ecosystems, lands transitional between dry, terrestrial lands and permanently flooded waters. They are defined and delineated in different ways in different countries. The common thread is that they are inundated with shallow water or saturated for an extended period of time, sometimes permanently, but long enough to influence the vegetation that grows there. In the early days, hydrology, especially the presence of surface water, was their defining characteristic as it was critical to support natural resources associated with food, waterfowl, and fish. In the twentieth century, wetlands were mostly recognized for their biological productivity. The breeding grounds of waterfowl or duck "factories" of the upper Mississippi River and the prairie pothole region of the US and Canada spurred the purchase and protection of freshwater wetlands by the U.S. Fish and Wildlife Service (USFWS) (http://www.fws.gov) and Ducks Unlimited (http://www.ducks.org). Wetlands also were important to the fur industry with the harvest of beaver, muskrat, and nutria. It was much later that wetlands became recognized for other reasons: their high levels of nongame biological production, ability to cleanse water by trapping pollutants, sequester carbon, maintain high levels of biodiversity, and more.

Creating and Restoring Wetlands. http://dx.doi.org/10.1016/B978-0-12-407232-9.00002-6

In addition to hydrology, wetlands possess other unique characteristics especially vegetation and soils that differ from terrestrial and aquatic ecosystems and that contribute to the provisioning of these benefits. Most terrestrial plants, food crops such as corn and soybean and commercially important forest species, cannot survive in permanently to semipermanently flooded or saturated soil. Yet, plants such as cattail, sedges, and some woody trees and shrubs thrive there if the flooding is not too deep. Flooding leads to anaerobic soil conditions as the water-filled pores of the soil inhibit diffusion of oxygen into them. These anaerobic conditions act as an environmental sieve or filter that restricts colonizers to those species that can adapt (van der Valk, 1981; Keddy, 2010). Flood-intolerant plants lack such adaptations, morphological and physiological, to acquire oxygen from the air to support cell growth and maintenance. In contrast, wetland plants possess adaptations to keep oxygen flowing to the roots where much respiration occurs.

While the value of wetlands is recognized around the world, the degree of protection afforded them varies tremendously. The United States arguably has the most rigorous methodology to define wetlands, assess their functional benefits, and restore them. Most of this is codified by the US law through the Clean Water Act of 1972 and its amendments, especially Section 404 that regulates placement of fill material in wetlands. In other countries, laws and means for wetland protection is less well defined, but international instruments such as the Ramsar Convention on Wetlands of International Importance afford protection to them (http://www.ramsar.org).

In the 1980s, increased public awareness of the benefits that wetlands provide to people led to the assessment and valuation of their ecosystem services including biological productivity, water quality maintenance, disturbance regulation, and others. Compared to other ecosystems, wetlands and other "edge" ecosystems, such as seagrass beds and coral reefs, contribute disproportionately to the global delivery of ecosystem services (Costanza et al., 1997, 2014), reenforcing the need to protect, manage, and restore them.

Wetland Characteristics

Hydrology

Hydrology describes the spatial and temporal patterns of flooding in a wetland. Wetlands may be inundated, as evidenced by surface flooding, or they may be saturated, the pore spaces in the soil are filled with water. Pattern of flooding can be described with a hydrograph, a two-dimensional figure that illustrates the depth, duration, frequency, and seasonality of inundation or saturation. Different types of wetlands, such as tidal marshes, floodplain forests, bogs, and fens, exhibit varying patterns of flooding (Figure 2.1). Tidal marshes often are flooded twice daily by the astronomical tides and the depth of flooding is relatively shallow, less than 1 m and often much less than that. Floodplain wetlands are inundated several times a year often to a depth of several meters or more. Bogs, a type of peatland, usually are not inundated. Rather, the peat is saturated. The water table is below the surface of the peat but the capillary action of

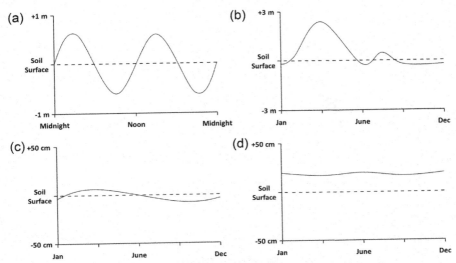

Figure 2.1 Hydrographs describing the depth, frequency, and duration of inundation of (a) saline tidal marsh, (b) floodplain forested wetland, (c) bog peatland, and (d) seepage wetland.

the pores brings water near to the surface. In fens, a type of peatland fed by ground-water, hydrology is relatively constant so that the water table is relatively stable over time. With a hydrograph, annual patterns of inundation can be illustrated to compare hydrology among different wetland types. For example, in Figure 2.1, the salt marsh is inundated 45% of the time, the floodplain forest 29%, the bog 20%, while the fen is inundated year-round, 100%.

Timing or seasonality of inundation is important. Inundation and soil saturation must occur long enough during the growing season to function as an environmental sieve, effectively excluding those species that lack adaptations to survive and thrive in the periodically to continuously waterlogged soil. While not directly evident from a hydrograph, the source of water that a wetland receives determines its hydroperiod and chemical composition of its water. Wetlands receive water from three potential sources: precipitation, surface flow, and groundwater (Figure 2.2; Brinson, 1993). Wetlands that receive most of their water from surface flow include wetlands in floodplains and estuaries, including riverine wetlands, tidal marshes, and mangroves. Those that receive mostly precipitation include bogs and wetlands in closed or iso-lated depressions. Wetlands where groundwater is a major water source include seeps and fens. The source of water, together with local soils and geology, determines the chemical characteristics of the floodwaters. Precipitation, essentially water distilled by atmospheric processes, contains little in the way of nutrients and dissolved materi-als such that precipitation-driven wetlands such as bogs tend to be nutrient (and mate-rial) poor (see Chapter 7, Peatlands). Wetlands that receive surface flow from tidal and nontidal sources often contain large amounts of dissolved materials, especially if the floodwaters are rich in eroded sediment. Tidal wetlands, especially saline marshes and mangroves, contain many dissolved salts, courtesy of the salinity in seawater (see

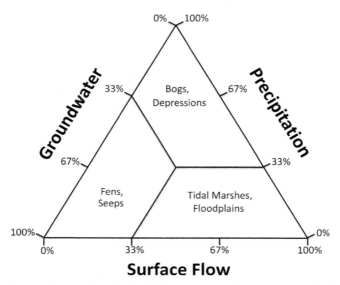

Figure 2.2 Diagram showing the relationship between water source—precipitation, surface water, groundwater—and wetland vegetation/plant communities.
Redrawn from Brinson (1993).

Chapter 8, Tidal Marshes). Groundwater-fed wetlands, seeps and fens, also may contain ample dissolved materials, especially if the underlying geology is composed of limestone or other calcareous materials.

Hydrodynamics, the direction of flow, also varies depending on the water source (Brinson, 1993). Flow may be horizontal or lateral or it may be vertical (Figure 2.3). Vertical flow often is associated with precipitation or sometimes groundwater. Horizontal flow may be unidirectional as in the downstream direction in riverines and floodplain wetlands. Or it may be bidirectional as in tidal marsh and mangroves.

Vegetation

To survive and thrive in wetlands, vegetation must be adapted to grow and reproduce in soils that are, at least, periodically inundated or saturated. Adaptations are morphological (a change to the physical structure of the plant) or physiological (a change in metabolism) (Table 2.1). Morphological adaptations consist of alterations to the structure of roots, stems, or leaves to promote transport of oxygen to the roots. A common feature of many herbaceous wetland plants is the presence of aerenchyma (Crawford, 1983), spongy tissue in the shoots and roots that is filled mostly with air. An aerenchymous stem, when sliced in cross section, is mostly hollow and serves as a conduit for transport of air and oxygen to the roots. Stems of some species, such as water lilies (*Nymphaea, Nuphar*), are completely hollow. In the case of *Nuphar*, gradients in temperature and pressure cause oxygen to diffuse into the younger shoots while forcing waste gases such as ethylene (discussed below) out through the older stems, a process known as pressurized ventilation (Dacey, 1980). Many emergent species exhibit a

(a)

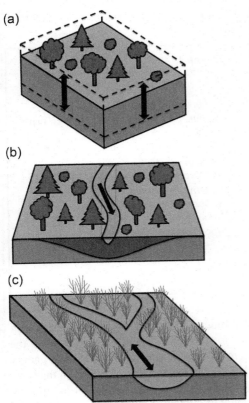

Figure 2.3 Hydrodynamics, showing (a) vertical flow as in the case of bogs and closed depressions, (b) horizontal, unidirectional flow (floodplain wetlands), and (c) horizontal and bidirectional flow (tidal wetlands, mangroves).

(b)

(c)

similar mechanism of gas transport known as convective throughflow (Armstrong and Armstrong, 1991). This is especially true for plants with cylindrical stems or linear leaves (Brix et al., 1992).

Another morphological adaptation is the production of adventitious roots (Crawford, 1983). They are roots that originate aboveground and are common in plants exposed to long periods of inundation. Rapid stem elongation, especially after overwintering, occurs in herbaceous species following flooding (Summers et al., 2000). Other adaptations of herbaceous plants include abundant (hypertrophied) lenticels and multiple trunks that are induced by flooding (Crawford, 1983). Many of these morphological changes are mediated by ethylene (Jackson, 1985) and improve the flow of oxygen to the roots.

Woody plants, while lacking aerenchyma, often have buttressed or swollen trunks to increase surface area and hypertrophied lenticels or pores to enable diffusion of oxygen from aboveground to belowground tissues. Swollen trunks are a common feature in many species that are exposed to prolonged flooding. Pneumatophores, kneelike structures that extend from the root aboveground, also increase surface area that promotes oxygen transport to belowground tissues. Bald cypress, *Taxodium distichum*, and the many species of mangroves possess such attributes.

Table 2.1 Morphological and Physiological Adaptations of Wetland Plants to Anaerobic Soil Conditions

Adaptation	Example
Morphological	
Aerenchyma	*Typha*, many rushes
Pressurized ventilation/convective throughflow	*Nuphar, Phragmites, Typha, Juncus*, many others
Buttressed (swollen) trunks	Bald cypress (*Taxodium distichum*), many deciduous trees
Pneumatophores	Bald cypress, mangroves, including *Rhizophora, Avicennia*
Adventitious roots	Herbaceous and woody species
Hypertrophied lenticels	Woody species
Multiple trunks	Woody species
Physiological	
Anaerobic (ethanol) metabolism	*Spartina* spp.
Accumulation of malic acid	*Spartina alterniflora*

Plants, like animals and many but not all microorganisms, require oxygen to support aerobic respiration to produce energy for cell growth and maintenance. In the absence of oxygen, they must turn to anaerobic respiration that produces less energy along with waste products such as acetaldehyde and ethanol that are toxic. In the presence of flooding, many wetland plants produce alcohol dehydrogenase, an enzyme that catalyzes the conversion of acetaldehyde to ethanol (Mendelssohn et al., 1981). Ethanol, which disperses faster than acetaldehyde, diffuses from the root, lessening its toxic effect. While these metabolic adaptations alleviate the stress of anoxia temporarily, eventually the plants will succumb unless they are able to maintain the flow of oxygen to the roots. Cronk and Fennessy (2001) offer a thorough summary of plant adaptations to anaerobic conditions.

Clearly, some plants are more tolerant to anaerobic soils than others. In the US, a classification system was developed to identify plant species tolerance to flooding and inundation. The classification system consists of five categories from flood-tolerant to flood-intolerant species (Lichvar, 2013; Table 2.2). A plant's tolerance on the scale is known as its hydrophytic indicator status. Obligate and facultative wet species are typically found in wetlands whereas facultative upland and upland species occupy mostly terrestrial lands. The predominance of hydrophytic vegetation is one of the three criteria, the others being hydrology and (anaerobic) soils, that serve as the basis for the jurisdictional or legal definition of wetlands in the US.

Soils

Wetland soils differ from terrestrial soils in that they are anaerobic. The absence of oxygen produces characteristics, especially differences in soil color and texture

Table 2.2 Classification of Vegetation According to Their Hydrophytic Indicator Status (Lichvar, 2013)

Category	Definition	Example
Obligate	Naturally found (not planted) >99% of the time in a wetland	*Typha* spp., *Juncus* spp., *Spartina alterniflora, Taxodium distichum*, many *Carex, Cyperus* spp.
Facultative wet	Occurs 67–99% in a wetland	*Calamagrostis* spp., *Impatiens capensis, Fraxinus pennsylvanica*, many *Carex, Cyperus* spp.
Facultative	Equally likely (33–67% of the time) to be found in a wetland and upland	*Acer rubrum, Nyssa sylvatica, Solidago* spp.
Facultative upland	Occurs 1–33% of the time in a wetland	*Juniperus virginiana, Sassafras albidum*
Upland	Seldom if ever (<1% of the time) in a wetland	*Zea mays* (corn), *Glycine max* (soybean)

that are uniquely different from aerobic, terrestrial soils. In anaerobic soils, a shift in microbial metabolism occurs, from one of aerobic, oxygen-driven metabolism to one driven by other energy-producing compounds. Unlike plants and animals that require oxygen (i.e., they are obligate aerobes) to support metabolism, many microorganisms are facultative aerobes. In the absence of oxygen, they use a different element or compound as a terminal electron acceptor to support respiration. These terminal electron acceptors include nitrate (NO_3), oxidized forms of iron (Fe^{3+}) and manganese (Mn^{4+}), sulfate (SO_4^{2-}), and some organic compounds. Terrestrial soils, especially those that are fine-textured (i.e., they contain much silt and clay), contain large amounts of oxidized Fe that in aerobic environments give soils a yellow, orange, or reddish color depending on the form of oxidized Fe present. When soils are flooded, oxidized Fe^{3+} (ferric) is reduced to Fe^{2+} (ferrous) by the microbes to support respiration, and soil changes from yellow, orange, or red to a gray color (Vepraskas, 1994; USDA, 2010.). The gray color, or low chroma, is indicative of the presence of anaerobic or hydric soil. Low chroma also is evident in dark colors, blacks and browns, indicative of accumulating organic matter, another characteristic of hydric soil (discussed below). Mineral soils that are continuously inundated or saturated may exhibit uniform gray color, also known as gley. Sometimes, soil takes on hues of green or blue that indicates complete reduction of Fe^{3+} in the soil matrix. In wetlands that dry down periodically, reduced Fe can reoxidize and the soil may take on a mottled color, with areas of red (oxidized Fe) and gray (reduced Fe). Thus, soil color reveals the presence of anaerobic conditions and is a useful indicator of the occurrence of flooding and saturation and, qualitatively, the duration of time in which it occurs. The presence of reduced iron can be detected using dipyridyl dye that reacts with Fe^{2+} (Vepraskas, 1994). IRIS—indicator of reduction in soil—tubes that are coated with oxidized Fe also are useful (Jenkinson and Franzmeier, 2006;

Castenson and Rabenhorst, 2006). Placed in the soil for an extended period, IRIS tubes can be used to infer the position of the water table since the oxidized Fe coating is reduced and dissolved below this depth.

Coarse-textured or sandy soils do not contain much Fe. So, the shift in color from red to gray is not necessarily a good indicator of hydric soil conditions in these situations. Here, enrichment of surface and subsurface layers with organic matter is used to infer hydric soil conditions (USDA, 2010). A consequence of anaerobic soil conditions is slowed decomposition of organic matter with the result being enrichment of wetland soil with organic matter, especially compared to terrestrial soils. In the presence of oxygen as occurs in terrestrial soils, microorganisms completely decompose organic matter to produce energy with the end products being carbon dioxide (CO_2), water, and energy. Under anaerobic conditions, decomposition of organic matter slows dramatically, in part because the terminal electron acceptors such as Fe yield less energy for metabolism than oxygen. Another reason is that anaerobic soils lack large numbers of the strictly aerobic bacteria as well as fungi (Thormann, 2006) that also require oxygen and that mediate decomposition in terrestrial soils. In some cases, thick deposits of organic matter accumulate over time leading to the development of soils formed exclusively from dead and decaying vegetation. These organic soils or histosols (Buol et al., 1980) are more commonly known as peat (see Chapter 7, Peatlands). Characteristics indicative of hydric soils develop relatively quickly once flooding is introduced. Within 5 years following hydrologic restoration, both low chroma and organic matter enrichment are visually evident (Vepraskas et al., 1999).

Definitions

Wetlands often are described by their vegetation. A marsh consists of emergent vegetation dominated by graminoid (grasslike) species (Figure 2.4(a)). A swamp or carr refers to a wetland dominated by trees (Figure 2.4(b)). A bog is a peat-forming wetland dominated by mosses, *Sphagnum*, and often ericaceous shrubs and coniferous trees (Figure 2.4(c)). Another common peatland is a fen whose vegetation is dominated by graminoids (Figure 2.4(d)).

One of the first comprehensive definitions was that of Shaw and Fredine (1956) published by the USFWS.

> *Wetland refers to lowlands covered by shallow and sometimes temporary or intermittent waters. They are referred to by such names as marshes, swamps, bogs, wet meadows, potholes, sloughs and river over-flowed lands. Shallow lakes and ponds, usually with emergent vegetation as a conspicuous feature, are included in the definition, but the permanent waters of streams, reservoirs and deep lakes are not included.*

In 1979, the USFWS published a more comprehensive definition in *Classification of Wetlands and Deepwater Habitats of the United States* (Cowardin et al., 1992). This definition took into account the presence of not only hydrology and vegetation but also soils.

Figure 2.4 Photographs of (a) saline tidal marsh, (b) floodplain swamp forest, (c) bog, and (d) fen.

> *Wetlands are transitional between terrestrial and aquatic systems where the water table is usually at or near the surface or the land is covered by shallow water...Wetlands must have one or more of the following three attributes: (1) at least periodically, the land supports predominantly hydrophytes, (2) the substrate is predominantly undrained hydric soil, and (3) the substrate is nonsoil and is saturated with water or covered by shallow water at some time during the growing season each year.*

Because of the USFWS focus on protection and management of waterfowl and fisheries, the 1979 definition emphasized areas, including rocky, shallow water habitats, used by these organisms. It is important to note that the presence of water is an essential part of the USFWS definition, but not necessarily vegetation or soils. The Canadian definition of wetlands is similar to that of the USFWS, emphasizing hydrology, soils, and hydrophytic vegetation but also emphasizes processes unique to wetlands (Tarnocai et al., 1988; Zoltai and Vitt, 1995).

> *Land that is saturated with water long enough to promote wetland or aquatic processes as indicated by poorly drained soils, hydrophytic vegetation and various kinds of biological activity which are adapted to a wetland environment.*

There is no single unifying definition of wetlands accepted throughout the world. However, the Ramsar Convention on Wetlands of International Importance held in Ramsar, Iran in 1971, offers perhaps the broadest definition (http://www.ramsar.org).

Areas of marsh, fen, peatland or water, whether natural or artificial, permanent
or temporary, with water that is static or flowing, fresh, brackish or salt including
areas of marine water, the depth of which at low tide does not exceed 6 meters.

The limits defined by the Ramsar Convention are broader than that of the US and Canadian definitions in that the lower limit of wetland habitat is 6 m whereas with the US and Canadian definitions, the lower limit is 2 m (Cowardin et al., 1992). Ramsar also does not explicitly mention hydrophytic vegetation and hydric soils in its definition.

Classification Systems

Classification systems developed for wetlands are based on features that are relatively easy to observe. Vegetation type, soils, landscape position, and water chemistry all have been used in classification schemes. At the highest level, wetlands typically are classified based on water source, inundation with freshwater versus seawater. Salinity, the concentration and ionic composition of seawater, consists mostly of sodium (Na) and chloride (Cl) though in inland saline waters, sulfate (SO_4), carbonate (CO_3), and base cations such as Ca and Mg may dominate (see Chapter 8, Tidal Marshes). Compared to freshwater, salinity exerts additional stress on wetland plants. Only those species that can tolerate flooding with saline water will survive and thrive. Below this level, landscape position and vegetation type are used to classify different types of wetlands.

Probably the most comprehensive system is the USFWS, *Classification of Wetland and Deepwater Habitats of the United States* (Table 2.3, Cowardin et al., 1992). This effort was an outgrowth of USFWS Circular 39 that originally identified 20 wetland classes (Shaw and Fredine, 1956). In the 1979 USFWS classification, wetlands were separated at the system level based on salinity while landscape position was used to separate wetlands at the subsystem level. At the next subordinate (class) level, wetlands were separated based on substrate composition and vegetation type. Again, the USFWS preoccupation with wetlands as habitat for waterfowl and fisheries produced a classification system focusing on shallow water habitats that might or might not contain wetland vegetation and soil. Using the definition that includes wetland vegetation and hydric soils, only moss-lichen, emergent, scrub-shrub, and forested wetland (Table 2.3) would be considered a true wetland in the 1979 USFWS classification.

Classification systems have been developed for other countries, including Canada (Zoltai and Pollett, 1983; Tarnocai et al., 1988; Zoltai and Vitt, 1995), Russia (Zhulidov et al., 1997), and China (Lu, 1995). The Canadian classification system is based on five classes reflecting differences in hydrology (water source and location of the water table) and vegetation type (Environment Canada, 1996; National Wetlands Working Group, 1997). The five classes are (1) bog, (2) fen, (3) marshes dominated by herbaceous vegetation, (4) swamps dominated by forested vegetation, and (5) shallow open water. Below this level, wetlands are classified based on surface morphology and pattern, water type, soil properties, and vegetation. The Canadian system has a strong focus on classifying

Table 2.3 Classification of Wetland and Deepwater Habitats of the US

System	Subsystem	Class
Marine	Subtidal	Rock bottom, unconsolidated bottom aquatic bed, reef
	Intertidal	Aquatic bed, reef, rocky shore, unconsolidated shore
Estuarine	Subtidal	Rock bottom, unconsolidated bottom aquatic bed, reef
	Intertidal	Aquatic bed, reef, rocky shore, unconsolidated shore, streambed, *emergent wetland, scrub-shrub wetland, forested wetland*
Riverine	Tidal	Rock bottom, unconsolidated bottom, aquatic bed, streambed, rocky shore, unconsolidated shore, *emergent wetland*
	Lower perennial	Rock bottom, unconsolidated bottom, aquatic bed, rocky shore, unconsolidated shore, *emergent wetland*
	Upper perennial	Rock bottom, unconsolidated bottom, aquatic bed, rocky shore, unconsolidated shore
	Intermittent	Streambed
Lacustrine	Limnetic	Rock bottom, unconsolidated bottom, aquatic bed
	Littoral	Rock bottom, unconsolidated bottom, aquatic bed, rocky shore, unconsolidated shore, *emergent wetland*
Palustrine	None	Rock bottom, unconsolidated bottom, aquatic bed, unconsolidated shore, *moss-lichen, emergent wetland, scrub-shrub wetland, forested wetland*

True wetlands are shown in *italics*.
From Cowardin et al. (1992).

peatlands, mostly because they are the most abundant freshwater wetlands in cold climates. Likewise, the Russian system is heavily weighted toward peatlands (Botch and Masing, 1983). The Russian system focuses on water source and chemistry (eutrophic, mesotrophic, oligotrophic) and its effect on vegetation and peat type.

In China, wetlands are classified into four systems: (1) coastal and estuarine wetlands, (2) riverine and lacustrine wetlands, (3) peat bogs, and (4) artificial wetlands (Lu, 1995). The Chinese also developed a classification system for coastal wetlands that is compatible with the Ramsar system (Zuo et al., 2013). The coastal classification is similar to the USFWS classification system in that both rely on vegetation type and substrate to classify wetlands. The Chinese classifications are different in that the lower limit of wetland habitat is deeper (5–6 m) and that the emphasis is on artificial wetlands, which are common in that country.

The Ramsar Convention developed a classification system that has many similarities to the USFWS system, relying on water source (fresh versus saline), substrate, and

Table 2.4 Classification System Developed by Ramsar, the Convention of Wetlands of International Importance

Marine and Coastal Wetlands	Shallow marine water, marine beds, coral reefs, rocky shores, sand/shingle shores, estuarine waters, tidal mud flats, salt marshes, mangrove, lagoons, deltas
Inland wetlands	Rivers, streams and creeks (permanent and intermittent), freshwater lakes (permanent and intermittent), saline lakes (permanent and intermittent), freshwater marshes, shrub wetlands, forested wetlands, peatlands, tundra and alpine wetlands, freshwater springs, geothermal wetlands
Artificial wetlands	Aquaculture ponds, farm ponds, irrigated land (rice), seasonally flooded agricultural land, salt pans, reservoirs, borrow pits, sewage farms, canals

From Matthews (1993).

vegetation to classify wetland types (Matthews, 1993). The Ramsar system contains 35 classes, grouped into marine and coastal wetlands, inland wetlands, and artificial wetlands (Table 2.4). Because of its greater lower limit (6 m below mean low water) and inclusion of artificial habitats, the Ramsar system includes habitats such as coral reefs and deltas as well as artificial habitats such as reservoirs, aquaculture ponds, and canals.

In the 1990s, the U.S. Army Corps of Engineers (The Corps) created a classification system based on describing wetland functions. The hydrogeomorphic (HGM) classification for wetlands (Brinson, 1993) was based on geomorphic setting, water source (Figure 2.2), and hydrodynamics (Figure 2.3). The idea of the classification scheme was to evaluate wetland-dependent functions pertaining to hydrology, biogeochemistry, plant community, and food web/habitat. It is widely recognized that different kinds of wetlands provide different types and levels of functions that depend on landscape position and connectivity to adjacent terrestrial and aquatic ecosystems.

The HGM system is composed of seven classes that differ in their dominant sources of water and hydrodynamics (Smith et al., 1995; Table 2.5). The combination of geomorphic setting and water source determines, in large part, a wetland's ability to provide certain functions. For example, riverine systems, because of their connectivity to the river channel, are able to intercept pollutants such as sediment from overbank flow. In contrast, depressions and flats that receive mostly precipitation are ineffective in removing sediment and other pollutants because they lack strong connections to surface water bodies that are source of these materials. Fringe wetlands, on estuaries and lakes, also may intercept pollutants, but because flow is bidirectional they may be less effective for pollutant removal than riverine systems. Precipitation-dominated systems such as organic soil flats may be ineffective for pollutant removal but very effective for sequestering carbon.

Semeniuk and Semeniuk (1995) developed a classification system for Australian wetlands that is similar to HGM, incorporating both landform and hydroperiod. Five

Table 2.5 Hydrogeomorphic Classes of Wetlands and Their Relationship with Water Sources (Figure 2.2) and Hydrodynamics (Figure 2.3; Smith et al., 1995)

Hydrogeomorphic Class	Dominant Water Source	Dominant Hydrodynamics
Riverine	Overbank flow from channel	Unidirectional, horizontal
Depressional	Return flow from groundwater, interflow, precipitation	Vertical
Slope	Return flow from groundwater	Unidirectional, horizontal
Mineral soil flats	Precipitation	Vertical
Organic soil flats	Precipitation	Vertical
Estuarine fringe	Overbank flow from estuary	Bidirectional, horizontal
Lacustrine fringe	Overbank flow from lake	Bidirectional, horizontal

landform (basins, channels, flats, slopes, hills) and four hydroperiod (permanent inundation, seasonal inundation, intermittent inundation, seasonal waterlogging) classes are used to produce 13 primary types of common inland wetlands. The classification system has been tested in arid and humid regions of Australia, South Africa, and northern Europe.

The advantage of classification systems such as the HGM, Australian, and Canadian systems over traditional vegetation and soil-based systems is that they provide a meaningful framework for semiquantitative characterization of wetland functions and ecosystem services. The disadvantage, of course, is that they lack the much finer-scale characterization of wetland habitats that vegetation-based classification systems offer.

Legal Frameworks

The US

In the US, one of the first laws to protect wetlands and other aquatic habitats was the Fish and Wildlife Coordination Act of 1934 and its amendments (USFWS, 2013b). The Act authorized the secretaries of agriculture and commerce to provide assistance to federal and state agencies to *"protect, rear, stock, and increase the supply of game and fur-bearing animals."* It also directed the Bureau of Fisheries to *"use impounded waters for fish culture-stations and migratory-bird resting and nesting areas."*

It was with the Federal Water Pollution Control Act of 1948, amended in 1972 that wetlands began to be recognized as aquatic ecosystems and subject to federal regulation. The 1972 Act set forth broad national objectives *"to restore and maintain the chemical, physical and biological integrity of the Nation's waters"* (USFWS, 2013a). Section 404, in particular, was important for wetlands protection since it authorized the U.S. Army Corps of Engineers (The Corps) to issue permits for the discharge of dredge or fill into *navigable waters*. In 1977, the Federal Water Pollution Control Act was renamed the Clean Water Act (CWA) (Environmental Law Institute, 2007) and, at this time, the term *navigable waters* was replaced by *Waters of the United States*. With the new definition and the permitting of dredge and fill activities under Section 404, wetlands in the US were increasingly subject to regulation and protection.

Prior to the 1970s, wetlands were drained for agriculture or filled for urban development. Now, these activities fall under the jurisdiction of Section 404 of the CWA. While the new law did not prohibit wetlands from being drained or filled, it required the approval and issuance of a permit by the Corps. The Corps could deny the permit or issue one after requiring that the environmental damage be offset by avoiding, minimizing, or mitigating the damage by creating or restoring wetlands (National Research Council, 2001). This idea was strengthened when, in 1989, President H.W. Bush stated that *no net loss* of wetlands was a goal of his administration (National Research Council, 2001).

Mitigation may consist of wetland creation, restoration, preservation, and/or other activities such as improving public access to the natural resource such as a river or estuary or some combination thereof. The Corps solicits input from other federal agencies such as the U.S. Environmental Protection Agency (USEPA) and the USFWS before issuing the permit. Thus, regulating and protecting wetlands became a complicated process and, for this reason, a precise legal or jurisdictional definition of wetlands was needed. The process of developing such a definition, some of which is described below, took years and, in fact, continues to this day.

In the US, meeting the definition of a jurisdictional wetland is a two-step process. First, a wetland must be considered a *Water of the United States* as interpreted by the CWA. If it meets this test, then an ecological definition, based on the presence of wetland hydrology, vegetation, and soils determined in the field, must be met. From the legal (CWA) perspective, wetlands protection requires their ability to *maintain the physical, chemical, and biological integrity of Waters of the US*. But, what constitutes a *Water of the US*? Involvement of the Corps in water issues originated several hundred years ago with the military's protection of the nation's waterways to support interstate commerce. Later, with the Rivers and Harbors Act of 1899, the Corps was empowered to regulate dredging and filling of navigable waters (USFWS, 2013c). This typically meant maintaining navigability of rivers, estuaries, and large streams whereby boats could ply their trade, carrying goods from place to place. Because rivers and estuaries clearly were waters of the US, wetlands such as tidal marshes and floodplains that are directly connected to these waters during high tide or river flooding were protected by the law.

From the mid-1970s to 2000, the definition of *Waters of the US* was broadened to include protection of nearly all wetlands listed in Table 2.4. However, since 2001, with several legal challenges decided by the U.S. Supreme Court, the definition narrowed to include only those wetlands that were adjacent to or abutting traditional navigable waters (TNWs) or tributaries of TNWs (USACOE, 2007). Today, the legal definition of *navigable waters of the United States*, as defined by the Corps and USEPA, is that *"the water body is (a) subject to the ebb and flow of the tide and/or (b) the water body is presently used or has been used in the past, or may be susceptible to use (with or without reasonable improvements) to transport interstate or foreign commerce"* (USACOE, 2007). This definition includes tidal wetlands and wetlands on rivers and perennial streams but not wetlands isolated from flowing waters. Wetlands not connected to navigable waters are not protected by the CWA unless they have a *significant nexus* with TNWs. By significant nexus, the wetland must affect the *physical, chemical, and/or biological integrity* of the TNW (USACOE, 2007). Under current interpretation of the law, many depressions and organic and mineral soil flats listed in Table 2.4 are no longer protected.

The ecological definition of a wetland is more tractable than the legal definition, but a precise definition was needed, one that could be quantitatively documented in the field. The Corps, being responsible for protection of wetlands under the CWA, created the Wetlands Delineation Manual in 1987 (USACOE, 1987). The manual laid out explicit field-based indicators of hydrology (USACOE, 1987), hydrophytic vegetation (Lichvar, 2013), and hydric soils (USDA, 2010) that needed to be met to be considered for protection under the CWA. Commonly known as the *three test rule*, all three criteria must be met for a site to qualify as a jurisdictional wetland. As the legal definition of a wetland was sharpened through the years, the field indicators were refined to the point where indicators have been developed for specific geographic regions (USACOE, 2010).

In the US, regulation of wetlands has created many opportunities for wetland restoration and creation. This is because Section 404 of the CWA requires mitigation when 404 permits are issued for dredge and fill activities (USACOE and USEPA, 1990). The result is the growth of a vibrant private environmental consulting sector to aid landowners and developers when applying for permits and developing mitigation plans. Unfortunately, the proportion of mitigation wetlands that are successful is low, leading to continued loss of wetlands and wetland-based functions (Race and Fonseca, 1996).

Canada

In the 1990s, the Canadian government established a policy on wetland conservation (Government of Canada, 1991; Environment Canada, 1996) consisting of one objective and a set of goals. The objective was to *"promote the conservation of Canada's wetlands to sustain their ecological and socio-economic functions, now and in the future."* The seven goals were:

- Maintenance of wetland functions and values.
- No net loss of wetland functions on federal land.
- Enhancement and rehabilitation of wetlands in areas of loss and degradation.
- Recognition of wetland functions with regard to federal programs.
- Securement of wetlands of significance in Canada.
- Recognition of sound, sustainable management practices in forestry and agriculture that make a positive contribution to wetland conservation while also achieving wise use of wetland resources.
- Utilization of wetlands that enhances their sustained and productive use in the future.

The policy does not require the protection and regulation of wetlands but encourages it, especially on federal lands. The goals also promote the concepts of *no net loss*, as is encouraged in the US, and the *wise use of wetlands*, that is an important component of the Ramsar Convention.

European Union

The European Union (EU) set forth several directives to protect the environment, including water policy and biodiversity. The Water Framework Directive (WFD) (2000) proposes policy to protect aquatic ecosystems and with regard to their water needs, terrestrial ecosystems, and wetlands directly depending on aquatic ecosystems. The policy encompasses surface waters, groundwater, and protected areas but focuses on pollution and water quality, not water quantity (Meyerhoff and Dehnhardt, 2007). The WFD also does not clearly state how wetland protection and restoration will be employed to achieve these goals nor does it quantify the economic value of wetlands (Meyerhoff and Dehnhardt, 2007). This second point is important because the WFD requires an economic analysis to prioritize implementation of water management schemes (Meyerhoff and Dehnhardt, 2007). Another EU directive, 2007/60/EC, further seeks to assess and manage flood risks, including opportunities to restore wetlands (McInnes et al., 2013). The Birds and Habitat Directives seek to reduce or rectify damage to natural habitats and their species (McInnes et al., 2013). The Habitats Directive was established to protect "Europe's finest wildlife areas" (European Commission, 2000). Known as Natura 2000, these protected areas include habitat for birds and, therefore, protect some wetlands. The directive also addresses "compensatory measures" including improving and recreating habitat when Natura 2000 sites are adversely affected by human activities.

Many European countries have directives, strategies, and policies to protect nature, particularly biodiversity. For example, in Germany, the Budesnaturschultgesetz (Federal Nature Conservation) Act of 1976 serves as the framework for nature conservation which is enacted through Lander legislation, which is similar to the states in the US (Stoll-Kleemann, 2010). It differs from US federal and state legislation in that there is no national consensus on how to carry it out. A number of wetland restoration and nature conservation projects have been implemented under it. As part of the legislation, water level regulation and planting have been employed to restore sedge marshes, swamp forests, and reed beds (Bruns and Gilcher, 1995).

Other Countries

Russia, like Canada, has extensive wetlands, especially peatlands, and laws that specifically mention wetland protection. The Water Code of the Russian Federation regulates use of water bodies, including bogs (Russian Federation Council, 2006). Water bodies may have multiple uses, including peat extraction. The code (Article 52) requires reclamation following peat harvesting by "water impoundment and artificial waterlogging". Article 65 establishes water protection zones along streams, rivers, and lakes and includes adjacent wetlands.

Australia developed recommendations for protecting rivers, wetlands, and estuaries of high importance (Kingsford et al., 2005). In 1994, the Council of Australian Governments agreed that the environment was a legitimate user of water and this paved the way for guidelines to protect aquatic water bodies including wetlands. Recommendations include identifying candidate river basins as Australian Heritage Rivers, developing environment flow regimes for rivers, establishing protected areas, creating statutory resource and land-use plans, and creating tax incentives for landowners to protect such areas. Australian law also provides guidance on restoration and environmental offsets for biodiversity, including endemic and rare species and important fauna habitat (Australian Environmental Protection Authority, 2008). But, there is no explicit mention of wetlands in the guidance document.

International

In much of the world, the Convention on Wetlands of International Importance, signed at Ramsar, Iran in 1971 was, and continues to be, the principal vehicle for wetlands protection. The Convention focuses on the importance of wetlands for waterfowl habitat (http://www.ramsar.org). The town of Ramsar located on the Caspian Sea, and an important flyway for migratory waterfowl in Asia, was an appropriate setting for the signing.

The original signatories in 1971 included 18 countries. The first site was established in Australia in 1974 (Frazier, 1999). The lead administrator of the Convention is the International Union for the Conservation of Nature and Natural Resources (IUCN) along with its sister organization, the International Wildfowl Research Bureau (IWRB).

The Convention contains 12 articles. Article 1 provides a definition of wetlands and of waterfowl while Articles 2–7 assign responsibilities to the Contracting Parties. A brief summary of the first seven articles is provided below.

Article 1: A definition of wetlands (*see Definitions*) and waterfowl (birds that are ecologically dependent on wetlands).
Article 2: The Contracting Parties will designate suitable wetlands within their territory for inclusion on the List of Wetlands of International Importance (i.e., the List).
Article 3: The Parties shall formulate and implement planning to promote conservation of wetlands on the List and the wise use of wetlands in their territories.
Article 4: The Parties shall promote the conservation of wetlands and waterfowl by establishing nature reserves on wetlands.
Article 5: The Parties shall consult with each other about implementing obligations, especially where wetlands or water are shared between Parties.
Article 6: The Parties shall convene conferences on the conservation of wetlands and waterfowl.
Article 7: Representatives of the Contracting Parties at such conferences shall include persons who are experts on wetlands and waterfowl.

The remaining five articles, Articles 8–12, address issues pertaining to oversight of the convention and procedures for joining and withdrawing from the convention.

An important aspect of the convention is its promotion of the *wise use* of wetlands. This refers to the full benefits and values that wetlands provide. They include sediment and erosion control, flood control, water quality improvement, support for fisheries, grazing and agriculture, outdoor recreation and education, contribution to climate stability, and, the most important, provision of habitat for wildlife, especially water birds (Ramsar Convention Secretariat, 2010).

Ramsar created a handbook to aid Contracting Parties in their efforts to develop a national wetland policy (Ramsar Convention Secretariat, 2010). Many nations have developed Ramsar-based national strategies or policies for wetland protection. They include Australia, Canada, Costa Rica, France, Jamaica, Malaysia, Peru, Trinidad and Tobago, and Uganda (Ramsar Convention Secretariat, 2010). Common themes of the final (or draft) documents include increasing public education/awareness, developing partnerships among different levels of government, developing legislation for land and water use policies, implementing wetland site management responsibilities based on sound science, and developing logistical and financial capacity to implement policy. Ramsar also developed (draft) resolutions to mitigate and compensate for wetland loss of wetlands of international importance by the Contracting Parties. Resolution XI.9 calls for a commitment to avoiding negative impacts on the ecological character of Ramsar sites. If such avoidance is not feasible, appropriate mitigation/compensation actions should be implemented (Gardner et al., 2012). Ramsar also issued guidance on the benefits of wetland restoration, recognizing that while restoration is important it is not a substitute for preservation (Alexander and McInnes, 2012).

As of September 2014, the number of Contracting Parties is 168. The number of sites designated for the Ramsar list is 2186, with an area of more than 208 million hectares (http://www.ramsar.org). The effectiveness of the Ramsar Convention, of course, depends on the commitment to legal protection and management by the member parties (Pittock et al., 2006).

Wetland Ecosystem Services

Natural ecosystems (forests, grasslands, rivers, and streams) and wetlands provide direct and indirect benefits to society. For example, harvesting fish, foodstuffs, and wood has direct tangible benefits that can easily be assigned a value in the marketplace. Other benefits such as water purification or climate regulation also are essential to the human life support system but it is more difficult to assign a market value to them. Ecosystem services stem from functions that natural and managed ecosystems provide. Costanza et al. (1997) define ecosystem services as *"benefits human populations derive, directly or indirectly, from ecosystem functions"* and list 17 services provided by natural and managed ecosystems (Table 2.6). In the Millennium Ecosystem Assessment of Wetlands and Waters (2005), 20 ecosystem services corresponding to the same provisioning, regulating, cultural, and supporting functions described by Costanza et al. are listed (Table 2.6).

Compared to other ecosystems, wetlands along with other *edge* ecosystems including seagrasses, estuaries, and coral reefs, on per unit area basis provide the highest levels of ecosystem services (Costanza et al., 1997). Tidal marshes and mangroves contribute much to disturbance regulation and waste treatment (Costanza et al., 1997). Floodplain forests contribute to disturbance regulation through flood control and provide waste treatment and water supply as well. Costanza et al. (1997) estimated that tidal marshes and mangroves contribute an equivalent of nearly US$10,000/ha/year of these services. The value of ecosystem services provided by floodplain forests is more,

Table 2.6 Types of Ecosystem Services Provided by Inland and Coastal Wetlands

	Costanza et al.	Millennium Ecosystem Assessment
Provisioning	Food production	Food
	Water supply	Freshwater
	Raw materials	Fiber and fuel
	—[a]	Biochemical products
	Genetic resources	Genetic materials
Regulating	Climate regulation, gas regulation	Climate regulation
	Water regulation	Hydrologic regimes
	Waste treatment	Pollution control and detoxification
	Erosion control	Erosion protection
	Disturbance regulation	Natural hazards
	Biological control	—
Cultural	—[b]	Spiritual and inspirational
	Recreational	Recreation
	—[b]	Aesthetics
	—[b]	Educational
Supporting	—	Biodiversity
	Soil formation	Soil formation
	Nutrient cycling	Nutrient cycling
	Pollination	Pollination
	Refugia	—

[a]Could be subsumed under *genetic resources*.
[b]Covered under *cultural* services.
From Miillennium Ecosystem Assessment (2005) and Costanza et al. (1997).

nearly US$20,000/ha/year. De Groot et al. (2012) compiled more than 320 publications in a reevaluation of goods and services produced by ecosystems. In their review, coastal wetlands ($193,845/ha/year) ranked second only to coral reefs ($352,249) in delivery of ecosystem services. Coastal systems and inland wetlands ranked third ($28,917/ha/year) and fourth ($25,682), respectively, but well above valuations of terrestrial biomes, forests, and grasslands. In a recent paper, Costanza et al. (2014) estimated the value of tidal marsh and mangroves from $13,786/ha/year (in 2007) to $193,843/ha/year based on new studies of their value in storm protection, erosion control, and waste treatment. While the economic value of coastal wetlands is large, the cost of restoring these ecosystems is higher than that for restoration of inland wetlands (De Groot et al., 2013). And, when valuing restored wetlands, much importance is ascribed to waste treatment (Jenkins et al., 2010).

While these market values are subject to much uncertainty, they highlight the importance of wetlands and wetland restoration to the support of society and human well-being. They also illustrate that different types of wetlands provide different kinds and levels of ecosystem services. For example, if the primary goal of restoration is water quality improvement, then it makes sense to restore wetlands adjacent to and strongly connected with other aquatic and terrestrial ecosystems on the landscape. If carbon sequestration is the primary goal, then connectivity is not as important and may even be detrimental to achieving the goal (see Chapter 4, Consideration of the Landscape).

When restoring wetlands, the goal should be to reestablish as many functions and services and at the highest possible level. It also is important to recognize that, depending on landscape context, it may not be possible to restore some services. This may be because of its geomorphic position in the landscape or its water source. The intensity and extent of site degradation also determines kinds and levels of services that can be restored since it may not be possible to remove or ameliorate some stressors. This is especially true for restoration in urban environments where hydrology is dramatically altered, or in agricultural landscapes where ameliorating nonpoint nutrient runoff is not possible. More on these constraints and limitations can be found in Chapter 4, Considerations of the Landscape.

References

Australian Environmental Protection Authority, 2008. Guidance for the Assessment of Environmental Factors (In Accordance with the Environmental Protection Act 1986). Environmental Offsets – Biodiversity, No. 19. http://www.epa.wa.gov.au/EPADocLib/2783_GS19OffsetsBiodiv18808.pdf (accessed 23.03.15.).

Alexander, S., McInnes, R., 2012. The Benefits of Wetland Restoration. Ramsar Convention on Wetlands. Scientific and Technical Review Panel, Briefing Note Number 4. http://www.ramsar.org/sites/default/files/documents/bn/bn4.pdf (accessed 23.03.15.).

Armstrong, J., Armstrong, W., 1991. A convective through-flow of gases in *Phragmites australis* (Cav.) Trin. ex Steud. Aquatic Botany 39, 75–88.

Botch, M.S., Masing, V.V., 1983. Mire ecosystems in the U.S.S.R.. In: Gore, A.J.P. (Ed.), Mires, Swamp, Bog, Fen and Moor, Ecosystems of the World, vol. 4B. Elsevier Science, Amsterdam, Netherlands, pp. 95–152.

Brinson, M.M., 1993. A Hydrogeomorphic Classification for Wetlands. Technical Report WRP-DE-4. U.S. Army Corps of Engineers, Washington, DC.

Brix, H., Sorrell, B.K., Orr, P.T., 1992. Internal pressurization and convective flow in some emergent freshwater macrophytes. Limnology and Oceanography 37, 1420–1433.

Bruns, D., Gilcher, S., 1995. Restoration and management of ecosystems for nature conservation in Germany. In: Cairns Jr., J. (Ed.), Rehabilitating Damaged Ecosystems. Lewis Publishers, Boca Raton, Florida, pp. 133–164.

Buol, S.W., Hole, F., McCracken, R.J., 1980. Soil Genesis and Classification. The Iowa State University Press, Ames, Iowa.

Castenson, K.L., Rabenhorst, M.C., 2006. Indicator of reduction in soil (IRIS): Evaluation of a new approach for assessing reduced conditions in soil. Soil Science Society of America Journal 70, 1222–1226.

Costanza, R., de Groot, R., Sutton, P., van der Ploeg, S., Anderson, S.J., Kubiszewski, I., Farber, S., Turner, R.K., 2014. Changes in the global value of ecosystem services. Global Environmental Change 26, 152–158.

Costanza, R., d'Arge, R., de Groot, R., Farber, S., Grasso, M., Hannon, B., Limburg, K., Haeem, S., O'Neill, R.V., Paruelo, J., Raskin, R.G., Sutton, P., van den Belt, M., 1997. The value of the world's ecosystem services and natural capital. Nature 387, 253–260.

Cowardin, L.M., Carter, V., Golet, F.C., LaRoe, E.T., 1992. Classification of Wetlands and Deepwater Habitats of the United States. United States Fish and Wildlife Service. Biological Services Program. FWS/OBS-79/31.

Crawford, R.M.M., 1983. Root survival in flooded soils. In: Gore, A.J.P. (Ed.), Mires, Swamp, Bog, Fen and Moor, Ecosystems of the World, vol. 4A. Elsevier Science, Amsterdam, Netherlands, pp. 257–283.

Cronk, J.K., Fennessy, M.S., 2001. Wetland Plants: Biology and Ecology. Lewis Publishers, Boca Raton, FL.

Dacey, J.W.H., 1980. Internal winds in water lilies: an adaptation to life in anaerobic sediments. Science 210, 1017–1019.

De Groot, R., Blignaut, J., van der Ploeg, S., Aronson, J., Elmqvist, T., Farley, J., 2013. Benefits of investing in ecosystem restoration. Biological Conservation 27, 1286–1293.

De Groot, R., Brander, L., van der Ploeg, S., Costanza, R., Bernard, F., Braat, L., Christie, M., Crossman, N., Ghermandi, A., Hein, L., Hussain, S., Kumar, P., McVittie, A., Portela, R., Rodriguez, L.C., ten Brink, P., van Beukering, P., 2012. Global estimates of the value of ecosystems. Ecosystem Services 1, 50–61.

Environment Canada, 1996. The Federal Policy on Wetland Conservation Implementation Guide for Federal Land Managers. Wildlife Conservation Branch, Canadian Wildlife Service. Environment Canada, Ottawa, Ontario.

Environmental Law Institute, 2007. The Clean Water Act Jurisdictional Handbook. Environmental Law Institute, Washington, DC.

European Commission, 2000. Managing Natura 2000 Sites: The Provisions of Article 6 of the 'Habitats' Directive 92/43/EEC. http://ec.europa.eu/environment/nature/natura2000/management/docs/art6/provision_of_art6_en.pdf (accessed 23.03.15.).

Frazier, S., 1999. Ramsar Sites Overview: A Synopsis of the World's Wetlands of International Importance. Wetlands International, Lelystad, The Netherlands.

Gardner, R.C., Bonells, M., Okuno, E., Zarama, J.M., 2012. Avoiding, Mitigating, and Compensating for Loss and Degradation of Wetlands in National Laws and Policies. Ramsar Convention on Wetlands. Scientific and Technical Review Panel, Briefing Note Number 3 http://www.ramsar.org/sites/default/files/documents/bn/bn3.pdf (accessed 23.03.15.).

Government of Canada, 1991. The Federal Policy on Wetlands Conservation. Minister of Environment. Minister of Supply and Services, Canada. http://www.publications.gc.ca/site/eng/100725/publication.html (accessed 23.03.15.).

Jackson, M.B., 1985. Ethylene and responses of plants to soil waterlogging and submergence. Annual Review of Plant Physiology 36, 145–174.

Jenkins, W.A., Murray, B.C., Kramer, R.A., Faulkner, S.P., 2010. Valuing ecosystem services from wetlands restoration in the Mississippi alluvial valley. Ecological Economics 69, 1051–1061.

Jenkinson, B.J., Franzmeier, D.P., 2006. Development and evaluation of iron-coated tubes that indicate reduction in soils. Soil Science Society of America Journal 70, 183–191.

Keddy, P.A., 2010. Wetland Ecology: Principles and Conservation. Oxford University Press, Cambridge, UK.

Kingsford, R.T., Dunn, H., Love, D., Neville, J., Stein, J., Tait, J., 2005. Protecting Australia's Rivers, Wetlands, and Estuaries of High Conservation Value. Department of Environment and Heritage, Australia, Canberra. Product Number PR050823.

Lichvar, R.W., 2013. The national wetland plant list. 2013 wetland ratings. Phytoneuron 49, 1–241.

Lu, J., 1995. Ecological significance and classification of Chinese wetlands. Vegetatio 118, 49–56.

McInnes, R., Joyce, C., Comin, F.A., Andersson, K., 2013. Perspectives on European wetland restoration. Wetland Science and Practice 30, 4–11.

Matthews, G.V.T., 1993. The Ramsar Convention on Wetlands: Its History and Development. Ramsar Convention Bureau, Gland, Switzerland.

Mendelssohn, I.A., McKee, K.L., Patrick Jr., W.H., 1981. Oxygen deficiency in *Spartina alterniflora* roots; metabolic adaptation to anoxia. Science 214, 439–441.

Meyerhoff, J., Dehnhardt, A., 2007. The European Water Framework Directive and economic valuation of wetlands: the restoration of floodplains along the river Elbe. European Environment 17, 18–36.

Millennium Ecosystem Assessment, 2005. Ecosystems and human well-being: wetlands and water synthesis. World Resources Institute, Washington, DC.

National Research Council, 2001. Compensating for Wetland Losses under the Clean Water Act. National Academy of Sciences, Washington, DC.

National Wetlands Working Group, 1997. In: Warner, B.G., Rubec, C.D.A. (Eds.), The Canadian Wetland Classification System, second ed. Wetlands Research Centre, University of Waterloo, Waterloo, Ontario.

Pittock, J., Lehner, B., Lifeng, L., 2006. River basin management to conserve wetlands and water resources. In: Bobbink, R., Beltman, B., Verhoeven, J.T.A., Whigham, D.F. (Eds.), Wetlands: Functioning, Biodiversity Conservation, and Restoration. Ecological Studies, vol. 191. Springer-Verlag, Berlin, Germany, pp. 169–196.

Race, M.S., Fonseca, M.S., 1996. Fixing compensatory mitigation: what will it take? Ecological Applications 6, 94–101.

Ramsar Convention Secretariat, 2010. Handbook 2: National Wetland Policies: Developing and Implementing National Wetland Policies, fourth ed. Ramsar Handbooks for the Wise Use of Wetlands, vol. 2. Ramsar Convention Secretariat, Gland, Switzerland.

Russian Federation Council, 2006. Water code of the Russian Federation. http://www.wise-rtd.info/en/info/water-code-russian-federation (accessed 23.03.15.).

Semeniuk, C.A., Semeniuk, V., 1995. A geomorphic approach to global classification for inland wetlands. Vegetatio 118, 103–124.

Shaw, S.P., Fredine, C.G., 1956. Wetlands of the United States: Their Extent and Their Value to Other Wildlife. Circular 39. United States Department of the Interior, Fish and Wildlife Service, Washington, DC.

Smith, R.D., Ammann, A., Bartoldus, C., Brinson, M.M., 1995. An Approach for Assessing Wetland Functions Using Hydrogeomorphic Classification, Reference Wetlands, and Functional Indices. Technical Report WRP-DE-9. U.S. Army Corps of Engineers, Vicksburg, MS.

Stoll-Kleeman, S., 2010. Reconciling opposition to protected areas management in Europe: the German experience. Environment: Science and Policy for Sustainable Development 43, 32–44.

Summers, J.E., Ratcliffe, R.G., Jackson, M.B., 2000. Anoxia tolerance in the aquatic monocot *Potamogeton pectinatus*: absence of oxygen stimulates elongation in association with an unusually large Pasteur effect. Journal of Experimental Botany 51, 1413–1422.

Tarnocai, C., Adams, G.D., Glooschenko, V., Glooschenko, W.A., Grondin, P., Rubec, C.D.A., Wells, E.D., Zoltai, S.C., 1988. The Canadian wetlands classification system. In: National Wetlands Working Group (Ed.), Wetlands of Canada. Ecological Land Classification Series 24. Environment Canada, Ottawa, Ontario and Polyscience Publications, Montreal, Quebec.

Thormann, M.N., 2006. Diversity and function of fungi in peatlands: a carbon cycling perspective. Canadian Journal of Soil Science 86, 281–293.

U.S. Army Corps of Engineers, 1987. Wetlands Delineation Manual. Technical Report Y-87-1 Environmental Laboratory, U.S. Army Corps of Engineers. Waterways Experiment Station, Vicksburg, MS. http://el.erdc.usace.army.mil/elpubs/pdf/wlman87.pdf (accessed 23.03.15.).

U.S. Army Corps of Engineers, 2010. Regional Supplement to the Corps of Engineers Wetland Delineation Manual: Midwest Region (Version 2.0). ERDC/EL TR-10–16 U.S. Army Corps of Engineers. U.S. Army Engineer Research and Development Center, Vicksburg, MS. http://el.erdc.usace.army.mil/elpubs/pdf/trel10-16.pdf (accessed 23.03.15.).

U.S. Army Corps of Engineers and U.S. Environmental Protection Agency, 1990. The Determination of Mitigation under the Clean Water Act Section 404(b)(1) Guidelines. Memorandum of Agreement. http://water.epa.gov/lawsregs/guidance/wetlands/mitigate.cfm (accessed 23.03.15.).

U.S. Army Corps of Engineers and U.S. Environmental Protection Agency, 2007. U.S. Army Corps of Engineers Jurisdictional Determination Form Instructional Guidebook. http://www.usace.army.mil/Portals/2/docs/civilworks/regulatory/cwa_guide/jd_guidebook_051207final.pdf (accessed 23.03.15.).

U.S. Fish and Wildlife Service, 2013a. Digest of Federal Resource Laws of Interest to the U.S. Fish and Wildlife Service. Federal Water Pollution Control Act (Clean Water Act) http://www.fws.gov/laws/lawsdigest/fwatrpo.html (accessed 23.03.15.).

U.S. Fish and Wildlife Service, 2013b. Digest of Federal Resource Laws of Interest to the U.S. Fish and Wildlife Service. Fish and Wildlife Coordination Act http://www.fws.gov/laws/lawsdigest/RIV1899.html (accessed 23.03.15.).

U.S. Fish and Wildlife Service, 2013c. Digest of Federal Resource Laws of Interest to the U.S. Fish and Wildlife Service. Rivers and Harbors Appropriation Act of 1899 http://www.fws.gov/laws/lawsdigest/riv1899.html (accessed 23.03.15.).

U.S. Department of Agriculture (USDA), Natural Resources Conservation Service, 2010. In: Vasilas, L.M., Hurt, G.W., Noble, C.V. (Eds.), Field Indicators of Hydric Soils in the United States, Version 7.0. USDA (NRCS in cooperation with the National Technical Committee for Hydric Soils).

van der Valk, A.G., 1981. Succession in wetlands: a Gleasonian approach. Ecology 62, 688–696.

Vepraskas, M.J., 1994. Redoximorphic Features for Identifying Aquic Conditions. Technical Bulletin 301. North Carolina Agricultural Research Service. North Carolina State University, Raleigh, NC.

Vepraskas, M.J., Richardson, J.L., Tandarich, J.P., Teets, S.J., 1999. Dynamics of hydric soil formation across the edge of a created deep marsh. Wetlands 19, 78–89.

Water Framework Directive, 2000. Directive 2000/60/EC, of the European Parliament and of the Council of 23 October 2000 Establishing a Framework for Community Action in the Field of Water Policy, 2000. O.J. (L327). Water Framework Directive. http://eur-lex.europa.eu/legal-content/EN/LSU/?uri=CELEX:32000L0060 (accessed 23.03.15.).

Zhulidov, A.V., Headley, J.V., Roberts, R.D., Nikanorov, A.N., Ischenko, A.A., 1997. In: Branned, M.J. (Ed.), Atlas of Russian Wetlands. Environment Canada, National Hydrology Research Institute. Translated by Y.V. Flingeffman and O.V. Zhulidov. Saskatoon, Saskatchewan.

Zoltai, S.C., Pollett, F.C., 1983. Wetlands in Canada: their classification, distribution, and use. In: Gore, A.J.P. (Ed.), Mires, Swamp, Bog, Fen and Moor, Ecosystems of the World, vol. 4B. Elsevier Science, Amsterdam, Netherlands, pp. 245–268.

Zoltai, S.C., Vitt, D.H., 1995. Canadian wetlands: environmental gradients and classification. Vegetatio 118, 131–137.

Zuo, P., Li, Y., Liu, C., Zhao, S., Guan, D., 2013. Coastal wetlands of China: changes from 1970's to 2007 based on a new wetland classification system. Estuaries and Coasts 36, 390–400.

Ecological Theory and Restoration

Chapter Outline

Introduction

The theoretical underpinnings of restoration ecology are guided by three concepts: disturbance, colonization and persistence, and succession and ecosystem development. While other lines of theory, including island biogeography (i.e., dispersal limitation), niches (safe sites), population ecology (competition, metapopulation dynamics), and food webs and trophic groups (herbivory, keystone species) also inform ecological restoration (Zedler, 2000), it is disturbance and succession theories that are the most important when it comes to restoring ecosystems.

Ecosystem restoration requires ameliorating stressors that disturb the system, then reestablishing community structure and ecosystem function. The size and intensity of the *disturbance* dictates how large and degraded the site is and how much effort and time will be needed to restore it. Life history traits that enable a plant to colonize the site and persist over time determine which species will colonize a site, which will need help, and which may invade and become a problem. Succession, how plant and animal communities change over time, reveals the pathways or trajectories of restoration and how quickly the system will recover. Another relevant concept is the rate of development of ecosystem functions, energy flow, and nutrient cycling, over time. Finally, a key item to consider in nearly all restorations is the problem of invasive species.

Candidate sites for wetland restoration often are degraded by activities such as tillage, soil removal, and placement of fill, and, thus, are especially susceptible to new, often unwanted colonizers. Many plant invaders possess traits that enable them to readily disperse to the site and, once there, to outcompete native species.

Disturbance

Restoration, by its nature, takes degraded sites, restores the environmental template of hydrology and soils, and then accelerates the introduction, colonization, and spread of native species relative to nature (Odum and Barrett, 2005). Restoration sites are disturbed by human activities and, in the case of wetlands, alteration of hydrology and soils are the key problems. In their discussion of succession, Connell and Slatyer (1977) defined disturbance as *a perturbation that opens up a relatively large space*. With respect to restoration, disturbance is the alteration of ecological structure and function caused by human activities that degrade the site from its natural condition.

A disturbance, either natural or artificial, is described by four characteristics: (1) size, (2) intensity, (3) frequency, and (4) duration (Keddy, 2010). Most natural disturbances such as fire, flood, and wind generally are described by the first three characteristics. Drought is one natural disturbance where duration is important. Duration often is described by stressors that occur as short duration *pulses* or long duration *presses* (Smith et al., 2009). Natural stressors combined with anthropogenic stress can lead to an *alternative stable state* or a *dynamic or catastrophic regime shift* that is difficult to reverse without extraordinary effort (Scheffer et al., 2001; Scheffer and Carpenter, 2003). For example, in dry regions, drought combined with agricultural practices such as grazing and tillage lead to desertification (Kassas, 1995). Once the vegetation is removed, the dry, windy conditions make reestablishment nearly impossible (Suding and Hobbs, 2009). Other well-documented regime shifts include the conversion of semiarid grasslands to shrub land by grazing and drought (Mayer and Rietkerk, 2004; Peters et al., 2006) and woody encroachment into grasslands when fire and grazing are excluded (Scheffer et al., 2001). A classic example of a regime shift in aquatic ecosystems is the conversion of a shallow, clear water lake to a turbid lake caused by eutrophication. As nutrient loadings to the lake increase, an abrupt shift from transparency to turbid water occurs leading to the loss of submerged aquatic vegetation (Scheffer et al., 1993) that may or may not be reversible (Carpenter et al., 1999).

Dynamic regime shifts occur in wetlands as well. A classic example is the Florida Everglades where cattail (*Typha domingensis* Pers.) displaced native sawgrass (*Cladium jamaicense* Crantz) and slough communities in areas of phosphorus eutrophication (Hagerthey et al., 2008). Another example is grazing by lesser snow geese (*Chen caerulescens caerulescens*) that denuded salt marsh vegetation in Hudson Bay marshes of Canada, creating unvegetated mudflat that does not recolonize (Abraham et al., 2005; Jefferies et al., 2006). In the southwestern US, desert streams exhibit two alternative stable states: slow moving sluggish streams with extensive wetland vegetation and deeply incised unvegetated channels that were produced by excessive sedimentation caused by overgrazing (Heffernan, 2008).

There is evidence to suggest that alternative stable states or dynamic regime shifts are more likely to occur in environments with strong underlying abiotic regimes such as wetlands and arid regions (Didham and Watts, 2005). Another school of thought posits that environments that are structured by competition are more likely to exhibit alternative stable states (Fukama and Lee, 2006). Regardless, alternative stable states exist in wetlands (Table 3.1), and active and sometimes continuing efforts may be needed to push the system, in its undesirable state, over the threshold into its desired regime. This transition often is nonlinear and may require more resources, effort, and time than expected (Mayer and Rietkerk, 2004). From a restoration perspective, pushing an ecosystem from an undesirable state to a desirable one involves deliberate efforts like ameliorating external stressors and accelerating ecosystem development with amendments or plantings. Restoration projects fail when efforts are insufficient to push the ecosystem across the threshold or efforts are not concentrated at the appropriate scale (Mayer and Rietkerk, 2004).

In wetlands, alternative stable states are strongly linked to changes in flooding and nutrients that contribute to a shift from native to invasive species (see Section Biological Invasions). With reversal of these environmental conditions and much management (burning, mowing, herbicides, biological control), these undesirable states can be reversed with limited success (Blossey et al., 2001). Logging of forested wetlands can lead to conversion to marsh (Doyle et al., 2007) or open water (Faulkner et al., 2007). Again, this shift may be irreversible without water level management since harvesting destroys the root mat, lowering the soil surface, increasing flooding depth, and inhibiting germination of woody seeds. Hobbs and Norton (1996) posited that, once a restoration transitions to an alternative stable state, massive management effort may be needed to restore the site to its original condition.

There is concern that existing and new restorations may be difficult to maintain or that novel ecosystems, those with no past analogue (Hobbs and Cramer, 2008) may develop stemming from the rapid pace of global change such as agricultural and urban land use, climate change and biotic factors, species invasion and extinction (Hobbs et al., 2009, 2011, 2014). In the future, creation of novel, nonanalogue, or emerging

Table 3.1 Examples of Desired and Alternative Stable States in Wetlands

Example	Desired State	Alternative Stable State	Stressor
Florida Everglades	Native sawgrass and slough	Cattail (*Typha domingensis*)	Phosphorus enrichment
U.S. East and Gulf Coasts	Tidal freshwater forest	Marsh, open water	Logging, rice cultivation
Hudson Bay Lowlands	Estuarine marsh	Mudflat	Grazing by snow geese
U.S. desert southwest	Stream wetland	Gravel wash	Grazing and excessive sedimentation

ecosystems will require moving away from traditional approaches that focus on restoring existing or historical species assemblages (Choi, 2004; Hobbs et al., 2009, 2011).

Restoration sites typically are exposed to chronic stress or a press on the system that must be ameliorated before the restoration can proceed. Examples of chronic disturbances to wetlands include long-term alteration of hydrology caused by drainage or placement of fill, eutrophication, and soil contamination with metals or volatile organic compounds.

In addition to chronic stress, the size and intensity of the disturbance must be taken into account. Connell and Slatyer (1977) describe how disturbance size and intensity determines the pace of succession. Four outcomes of succession occur depending on the size and intensity of the disturbance (Figure 3.1). In a large area that is intensely disturbed, succession will be slow and will take a long time (Figure 3.1(a)). The intense disturbance leaves no viable propagules on the site and it is slow and difficult for propagules to disperse to the center of the site. Moderate succession occurs on large sites affected by slight disturbance as some existing propagules survive and re-colonize the site (Figure 3.1(b)). Some

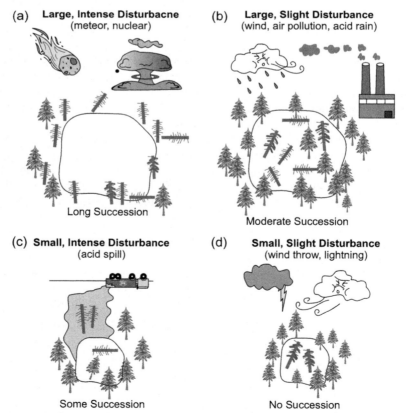

Figure 3.1 Schematic showing the relationships between the size and intensity of a disturbance and succession as described by Connell and Slatyer (1977).

succession also occurs on small sites affected by an intense disturbance (Figure 3.1(c)) as the surrounding vegetation grows into the site and propagules easily disperse to the site. The fourth outcome is a small area that is only slightly disturbed and where no succession occurs (Figure 3.1(d)) because the small gap is quickly filled by growth of surrounding adults. In the case of wetland restoration, the first three outcomes (large area, intense disturbance; large area, slight disturbance; and small area, intense disturbance) are relevant.

Some disturbances, however, can aid in ecosystem restoration. For example, low intensity, periodic disturbance enhances species richness in terrestrial ecosystems (Huston, 1979) and wetlands (Joyce, 2014), reducing the dominance of other species. An added benefit is that it can limit the spread of invasive species. Long-term studies of wetland grasslands suggest that mowing can increase species richness and reduce dominance by the aggressive species reed canary grass, *Phalaris arundinacea* (Straskrabova and Prach, 1998).

Dispersal and Colonization

Reestablishment of vegetation on disturbed sites depends on propagules that disperse to the site and on revegetation by propagules of surviving species. On sites that were intensively disturbed or newly created wetlands, the bare soil presents an ideal opportunity for colonization. Establishment of propagules from beyond the site depends on four sequential steps: dispersal, germination, survival, and adulthood. Species that produce large numbers of wind- or water-dispersed seeds are among the first to arrive. Dispersal into wetlands is different from terrestrial ecosystems such as old fields in that the inundated or saturated soil serves as an environmental sieve (van der Valk, 1981). The anaerobic soil conditions inhibit germination of terrestrial species.

Classification schemes have been developed to describe a species' ability to colonize a site. r-strategists, for example, are species that produce large numbers of offspring (seeds), readily disperse (they are small in size), and complete their life cycle in a short amount of time (Harper, 1965; Pianka, 1970; Odum and Barrett, 2005). With r-strategists, the percent of seeds that germinate often is low but is offset by the large number of seeds produced. Thus, r-strategists tend to be species that colonize the site without assistance from humans. K-strategists, on the other hand, produce fewer, often larger seeds that do not readily disperse (Pianka, 1970; Odum and Barrett, 2005). The germination rate of their seeds is greater than seeds of r-strategists. K-strategists often need assistance, such as seeding or planting, to colonize the site.

Grime (1977) extended this idea to include three strategies of colonization and persistence, (1) ruderal, (2) competitive, and (3) stress-tolerant species, and related their distribution to the intensity of disturbance and stress (Table 3.2). Ruderal species are favored in environments with *high disturbance and low stress*. Ruderal species, like r-strategists, have short life cycles with a large amount of energy devoted to seed production (Table 3.3). They are annual species with rapid growth and flowering that occurs at the end of a temporarily favorable season, as with a wet or monsoon season. In *low-stress, low-disturbance* environments, competitive species dominate. They consist of perennials, herbs, shrubs, and trees that produce much biomass and often

Table 3.2 **The Three Strategies of Plant Colonization and Persistence as Related to Disturbance Stress**

Environmental Condition	Plant Strategy	Example Life Form
Low stress, high disturbance	Ruderal	Annuals
Low stress, low disturbance	Competitive	Perennials, shrubs, trees
High stress, low disturbance	Stress tolerant	Lichens, mosses, some perennials, shrubs, trees
High stress, high disturbance	No viable strategy	

Modified from Grime (1977).

Table 3.3 **Some Characteristics of Competitive, Stress-Tolerant, and Ruderal Species**

	Competitive	Stress Tolerant	Ruderal
Morphology			
Shoots	High, dense canopy of leaves	Wide range of growth forms	Small stature and spread
Leaves	Robust, often mesomorphic	Small or leathery or needlelike	Various, often mesomorphic
Litter	Copious, persistent	Sparse, sometimes persistent	Sparse, not persistent
Growth			
Relative growth rate	Rapid	Slow	Rapid
Leaf longevity	Relatively short	Long	Short
Life History			
Life form	Perennial herbs, shrubs, and trees	Lichens, perennial herbs, shrubs, and trees	Annuals
Production devoted to seeds	Small	Small	Large

Modified from Grime (1977).

high, dense leaves that shade out other species (Table 3.3). Competitive species also produce abundant litter that is persistent and inhibits germination of seeds. Most production goes into biomass with proportionally little devoted to seeds. Clonal wetland species, such as *Phragmites, Typha*, and *Phalaris* that dominate some wetlands, are competitor species. They achieve dominance by shading and producing a thick litter layer that inhibits germination (Farrer and Goldberg, 2009; Vaccaro et al., 2009; Larkin et al., 2012).

Stress-tolerant species are those that exist in high stress environments, including arid, Arctic–alpine, shaded, and low-nutrient habitats (Table 3.3). Life forms of

Table 3.4 **Classification of Wetland Plant According to Functional Groups**

Functional Group		Example Species
Ruderal	Obligate annual	*Bidens* spp., other Asteraceae
	Facultative annual	*Eupatorium* spp., other Asteraceae
Matrix	Clonal dominant	*Typha* spp., *Phalaris arundinacea*, *Scirpus* spp.
	Clonal stress-tolerator	*Scirpus* spp.
Interstitial	Clonal	*Juncus* spp., *Glyceria* spp.
	Reed	*Juncus* spp., *Eleocharis* spp.
	Tussock	*Carex* spp., *Calamagrostis canadensis*

Modified from Boutin and Keddy (1993).

stress-tolerant species are diverse, including lichens, mosses, perennial herbs, shrubs, and trees. Leaves are often small, leathery, or needlelike to reduce exposure to temperature extremes and to reduce water loss. Like competitors, stress-tolerant species devote a small proportion of annual production to seeds.

Boutin and Keddy (1993) expanded on Grime's work to develop a functional classification of wetland plants. Based on *performance* traits such as life span and measures of growth (height, biomass, photosynthetic area, relative growth rate), they grouped plants into matrix, interstitial, and ruderal species (Table 3.4). Matrix species consist of vigorous and deep-rooted plants that include clonal dominants and Grime's competitor and stress-tolerant species. Interstitial species are characterized by compact growth forms and shallow rooting and include clonal, reed, and tussock species such as some species of *Carex*. Ruderal species consist of obligate and facultative annuals.

The importance of life history traits in guiding succession has been echoed by many prominent ecologists, including Chapin et al. (1994), Huston and Smith (1987), McCook (1994), and others. Life history traits that drive succession include growth rates, size, shade tolerance, longevity, as well as mechanisms of dispersal such as propagule availability, vegetative regeneration potential, seed storage, and defenses against herbivory and physical disturbance (McCook, 1994).

To summarize, no single theory of succession perfectly predicts a species place in the plant community and how it will perform. Nevertheless, each informs our knowledge regarding a species ability to reach a restoration site, colonize and persist there, and whether it may become invasive with time.

Succession

Species that colonize a site may or may not persist as the plant and animal community may change over time, a process known as *succession*. Two schools of thought shape concepts regarding succession. The first is the *organismic* or facilitation view of Clements. The second is the *individualist* view of Gleason. Both views are important to understand how to reestablish the desired plant communities following restoration.

The Organismic View—Facilitation

One of the first theories of succession was put forth by Frederick Clements. Clements
(1916) postulated that plant communities change over time in an orderly and predictable
way, eventually leading to an endpoint or climax community (Figure 3.2(a)). Clements
viewed the plant community as an organic entity characterized by four stages: formation,
growth, maturity, and death. The first species to arrive on a site are known as pioneers
and consist mostly of annuals and other ruderal species. Over time, pioneer species are
replaced by different assemblages of species. The replacement of species assemblages
continues until eventually the plant community achieves some sort of equilibrium with

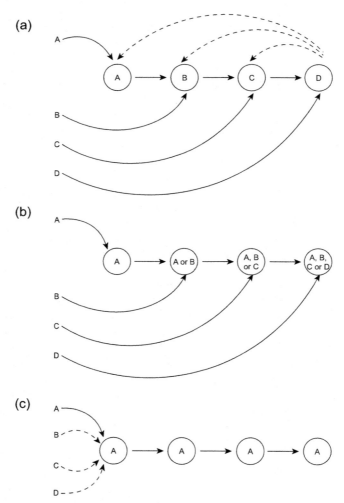

Figure 3.2 Pathways showing the three mechanisms of succession: (a) facilitation, (b) tolerance,
and (c) inhibition of succession. The dashed lines in (a) represent disturbance that moves the
community to an earlier stage of succession. See text for an explanation of the pathways.

the environment. This endpoint community is known as a climax. According to Clements (1916), "the developmental study of vegetation necessarily rests upon the assumption that the unit or climax formation is an organic entity. As an organism, the formation arises, grows, matures and dies."

Succession occurs on primary, previously unvegetated habitats such as sand dunes, mudflats, lava flows, glacial till and outwash. It also occurs on secondary habitats such as old fields and sites previously degraded by human activities. The transition from old field to shrub land and eventually to forest probably is the most enduring example of succession. Other well-studied examples of succession include dunes of the Great Lakes (Cowles, 1899; Olson, 1958) and receding glaciers (Crocker and Major, 1955).

Clements' view of succession is best described as *organismic*, driven by biological processes. According to Clements, certain early succession species colonize the site and then modify the environment, making it more suitable for later successional species to colonize. On sites devoid of vegetation, pioneer species add organic matter to the soil, improving its water holding capacity and nitrogen content (Olson, 1958; Crocker and Major, 1955). Pioneer species also stabilize the blowing sand grains and loose soils. Clements' organismic or autogenic view of succession dominated ecologists' impression of succession for many years (Connell and Slatyer, 1977) and today is referred to as the *facilitation* model since early successional species facilitate colonization of later successional species.

Many wetlands, especially those that build peat, are best described by the facilitation model. Hydrarch succession, the transition from pond or open water to minerotrophic (groundwater-fed) fen and eventually to ombrotrophic (precipitation-fed) bog dominated by peatmosses, *Sphagnum* (Lindeman, 1941; Moore and Bellamy, 1974), is a classic example of facilitation. The process also is referred to as terrestrialization (Glaser, 1987). Another example of facilitation is establishment of pioneer species, usually herbaceous emergent vegetation, on mudflats that trap sediment, increase surface elevation, and pave the way for woody vegetation.

Facilitation and Nurse Plants

In restoration ecology, the use of *nurse plants* is a classic example of facilitation in action. A nurse plant is an adult plant whose close spatial association with seedlings of young plants of another species has a positive effect on them (Padilla and Pugnaire, 2006). Most studies of species interactions tend to focus on negative effects such as competition (see the individualist model of Gleason below). During the past 20 years, however, there is renewed interest in the role of positive plant interactions. Nurse plants are more common in environments where abiotic factors or herbivory limit plant growth and reproduction (Padilla and Pugnaire, 2006). Such environments include arid regions (Niering et al., 1963; Flores and Jurado, 2003), other (fresh) water-limited environments such as salt marshes (Bertness and Hacker, 1994; Bertness and Callaway, 1994) and mangroves (Lewis, 2005; Lewis and Gilmore, 2007; McKee et al., 2007), and alpine habitats (Cavieres et al., 2006). In these environments, nurse plants, many of them shrubs, buffer adverse environmental conditions through shading, protection from frost, concentrating nutrients (and water) in soils. and protection

Table 3.5 **Examples of Nurse Plants in Wetland Restoration**

Nurse Plant	Target Species	Benefit
Juncus and other species	Many species	Soil aeration
Spartina, Batis, Sesuvium	Mangrove	Propagule trapping, structural support, elevation building
Spartina alterniflora	*Baccharis halimifolia*	Improved microclimate (lower temperature)
Polytrichum	*Sphagnum* (peatlands)	Improved microclimate (greater humidity, lower temperature)

See text for detailed explanation.

against grazing (Franco and Nobel, 1988; Padilla and Pugnaire, 2006). The benefits of nurse plants seem to be greatest when target species are young or small (Padilla and Pugnaire, 2006).

Nurse plants are used in a variety of terrestrial restorations. Often seeds of N-fixing legumes are sown to enhance N capital of the soil (Bradshaw and Chadwick, 1980). In wetlands, there is widespread evidence for the use of nurse plants to aid in establishment of target species (Table 3.5). For example, flood-tolerant species such as *Juncus* may serve as a nurse plant to less flood-tolerant species by aerating the soil (Schat, 1984). A number of species, *Spartina alterniflora*, *Batis maritima*, and *Sesuvium portulacastrum*, enhance recruitment of red mangrove (*Rhizophora mangle*) seedlings (Lewis, 2005; Lewis and Gilmore, 2007; McKee et al., 2007). The mechanisms are varied but consist of a combination of trapping propagules, providing structural support, and increasing surface elevation. The use of cordgrass, *S. alterniflora*, has been shown to facilitate growth of the shrub, *Baccharis halimifolia*, in created saline marshes (Egerova et al., 2003). Survival and growth of *Baccharis* was greater when growing among clones of *Spartina* as compared to gaps between clones. In bogs, the moss, *Polytrichum*, has been shown to promote growth of *Sphagnum* on restored peatlands by creating a microclimate of greater humidity and lower temperature (Groeneveld et al., 2007).

Facilitation and Mycorrhizae

Establishment of native species also is enhanced by inoculating soils or seedlings with beneficial mycorrhizae. Mycorrhizae are fungi that form a mutualistic association with the roots of certain herbaceous and woody plants (Harley and Smith, 1983). The hyphae of mycorrhizal fungi augment the root system of plants, promoting water and nutrient, especially P, uptake. In return, the plants provide organic compounds to support the growth of heterotrophic fungi. Mycorrhizal inoculation has been shown to increase growth and nutrient uptake in many terrestrial forbs, woody plants, and C_4 grasses (Hoeksema et al., 2010).

There also is evidence of mycorrhizal association with wetland plants (Cooke and Lefor, 1998; Wetzel and van der Valk, 1996; Turner et al., 2000) though it is less than in terrestrial plants since anaerobic soil conditions limit their colonization (Anderson et al., 1984; Peat and Fitter, 1993; Cornwell et al., 2001). Hoefnagels et al. (1993) reported the presence of mycorrhizae in *S. alterniflora* in saline tidal marshes as did Wang et al. (2011) for mangroves. Freshwater species also exhibit mycorrhizal associations including *Carex* where mycorrhizal infection was reported in 16 of 23 species tested (Miller et al., 1999). Vegetation of peatlands, both woody vegetation and herbaceous emergents, have mycorrhizal associations (Thormann et al., 1999). Mycorrhizae are more common in drier parts of the wetlands and decline with increasing soil saturation and inundation (Hoefnagels et al., 1993; Rickerl et al., 1994; Miller and Bever, 1999). Seasonal drying enhances mycorrhizal colonization of wetlands (Miller, 2000). Because mycorrhizal spores are present in flooded soils, the absence of mycorrhizae is due to the anaerobic conditions rather than dispersal limitation (Miller, 2000). Studies indicate widespread mycorrhizal association with wetland plants, including *Carex*, *Scirpus*, and *Phalaris*, in restored freshwater marshes (Bauer et al., 2003). While the benefits of mycorrhizae to nutrient uptake and enhanced growth by wetland plants have not been demonstrated in the field, there is some evidence that mycorrhizae may aid wetland plants during periods of low soil moisture and low P availability (Khan, 2004).

The Individualistic View—Tolerance and Inhibition

Gleason (1917, 1926) proposed an alternative hypothesis to facilitation based on the *individual* nature of the plant community. According to Gleason (1926),

> *Every species of plant is a law unto itself, the distribution in space depends on its individual peculiarities of migration and environmental requirements. Its disseminules migrate everywhere and grow wherever they find favorable conditions.*

With the individualistic model, the successional sequence depends on the organisms that arrive at the site first and on favorable environmental conditions (light, water, nutrients) for propagules to germinate, grow, and reproduce. The Gleasonian idea of succession is sometimes referred to as allogeneic, that is, succession is controlled by environmental factors and not by organisms.

Connell and Slatyer (1977) refined Gleason's individualistic concept of succession by creating two distinct hypotheses, inhibition and tolerance. Both models begin with Gleason's premise that succession begins with species whose propagules reach the site and become established. According to Connell and Slatyer, in the tolerance model, the early occupants modify the environment so that it becomes less suitable for subsequent recruitment of early successional species but this modification has little or no effect on subsequent recruitment of late successional species (Figure 3.2(b)). According to the inhibition model, early occupants modify the environment so that it becomes less suitable for subsequent recruitment of both early and late successional species (Figure 3.2(c)). Evidence in support of the inhibition model includes shrubs that form a closed canopy and inhibit establishment of trees (Niering and Egler, 1955; Niering and Goodwin, 1974). In wetlands, invasive species such as *Phragmites*, *Typha*, and

Phalaris that possess characteristics of strong competitor species (Grime, 1977) may fit this model. Evidence in support of the tolerance model is scant but it is likely that it exists.

Other Gleasonian Models

Based on studies of old field succession, Egler (1954) developed two hypotheses to describe plant succession. One hypothesis, *relay floristics*, describes the successive appearance and disappearance of groups of species (Figure 3.3(a)). Each group of species invades the site at a certain stage of development. Over time, the coloniz-ing group makes environmental conditions unsuitable for themselves but suitable for invasion by the next group. The second hypothesis, *initial floristic composition*, posits that all potential propagules are present in the soil/seedbank or can readily disperse to the site such that nearly all species are present when succession is initiated. Annuals dominate the site first, followed by grasses, then woody species. During succession, as each new successive species drops out, a new group of species, there from the start, assumes predominance (Figure 3.3(b)). The initial floristic composition hypothesis stresses the importance of the seedbank in guiding succession (Cairns and Heckman, 1996), which may not be practical on sites where human alteration has occurred for many years. This is especially true for wetland restoration sites on agricultural lands that were drained a century or more ago.

Another Gleasonian approach is that of van der Valk (1981) that is based on species life history traits and adaptation to the environment of the site. Known as the *environmental sieve* model, it was developed for freshwater marshes and uses water levels (flooding, drawdown), traits of adult individuals, and seedbank characteristics such as presence or absence of species to qualitatively predict plant communities. Weiher and Keddy (1995) introduced the concept of *assembly rules* to describe how wetland plant communities organize. In many respects, assembly rules is an extension of the environmental sieve model because it focuses on the small number of environmental factors that organize communities. In wetlands, obvious factors, such as hydrology, salinity, fertility, and others (burial, herbivory, disturbance, competition), organize plant communities by filtering out species that are not adapted to these conditions (Keddy, 2010). The combination of these factors (e.g., deep flooding and high salinity) also limits the ultimate size of the species pool.

From a restoration perspective, Hobbs and Norton (2004) describe three types of filters, abiotic, biotic, and socioeconomic, that determine its outcome (Table 3.6). Abiotic filters include climate, substrate, and landscape structure, including patch size and isolation. Legacy effects, especially agriculture that modifies the soil environment both through tillage and fertilizer additions, also is important as is disturbance since so often restoration sites are degraded by land clearing. Biotic filters include species interactions, competition, predation and mutualism, along with propagule dispersal, species arrival, and succession. There are also biological legacy effects such as the presence/absence of remnant species. Last but not least, Hobbs and Norton include socioeconomic filters such as community goals and costs that invariably describe the benefits and cost of the restoration and when, and to what degree, it is implemented, managed, and monitored.

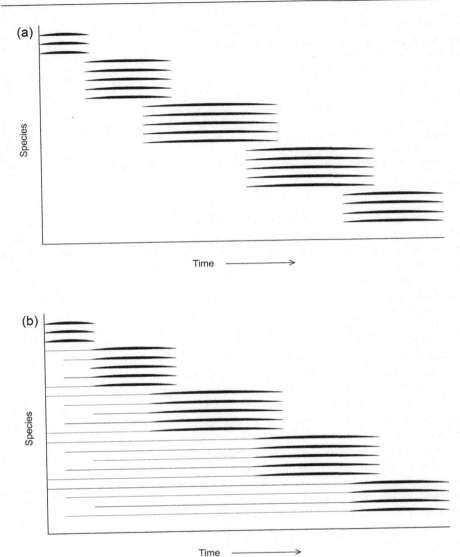

Figure 3.3 Egler's theories of plant community succession: (a) Relay floristics whereby a series of plant communities becomes established, each community making the environment unsuitable for themselves but suitable for the invading community. (b) Initial floristic composition whereby all species are present in the seedbank or are able to disperse, at the beginning. During succession, different species (grasses, shrubs, trees) come to dominate at different times.
Modified from Egler, F.E., January 1, 1954. Vegetation science concepts I. Initial floristic composition, a factor in old-field vegetation development with 2 figs. Vegetatio 4 (6): 412–417, Figures 1 and 2. Reproduced with kind permission from Springer Science and Business Media.

Table 3.6 Abiotic, Biotic, and Socioeconomic Filters That Determine Outcomes of Ecological Restoration

Abiotic	Biotic	Socioeconomic
Climate	Competition	Goals of the community
Substrate	Predation–trophic interactions	Costs to the community
Landscape structure	Propagule availability	
Disturbance[1]	Mutualisms	
Land use legacies[1]	Succession (biological) legacies	

From Hobbs and Norton (2004), except as noted.[1]

Assembly rules and other Gleasonian models imply that the endpoint community following restoration may consist of a number of alternative stable states depending on the importance of a particular species (i.e., a keystone species) and the order in which species arrive (Lockwood, 1997). These models also emphasize the priority effects of early arriving species that become dominant in the community (Young et al., 2001). Gleasonian models further posit that assembly may be deterministic (controlled by the environmental template) or stochastic (controlled by chance events such as an episodic flood) (Trowbridge, 2007). However, assembly rules and traditional succession models share much in common, including the idea that communities develop through time toward relatively stable states and the importance of biotic interactions, especially competition and facilitation, but they differ in that assembly rules explicitly allows for the possibility of multiple stable states (Young et al., 2001).

One might argue that restoration ecology traditionally relies on the facilitation model since restoration practitioners use deliberate means to accelerate succession and ecosystem development. Such activities include introduction of propagules, seeds and plants, nutrient additions (to increase nitrogen capital) and sometimes carbon, and amendments such as topsoil and other sources of organic matter. Even so, understanding the environmental conditions that "filter" species and the life history traits of desirable (and undesirable) species is essential for knowing which species will colonize on their own, which will need help getting there, and, once there, prospering, and, finally, which species may invade.

Ecosystem Development

Succession theory focuses on changes in the structure of the community, species replacement, richness, and niches, over time. Odum (1969) expanded on these ideas to include changes in ecosystem functions, energy flow, and nutrient cycling. His ideas fall squarely in the camp of facilitation. According to Odum, ecosystem development is an orderly process of community development that is reasonably directional and, therefore, predictable. It results from modification of the physical environment by the community. Odum made a number of predictions concerning how ecosystem attributes

Table 3.7 Trends Expected in the Development of Ecosystems

Ecosystem Attribute	Early Stage of Succession	Late Stage of Succession
Energetics		
Net production	High	Low
Biomass	Low	High
Food chains	Linear, grazing	Weblike, detritus
Total organic matter	Low	High
Structure		
Species diversity	Low	High
Life History		
Size of organisms	Small	Large
Life cycles	Short, simple	Long, complex
Niche specialization	Broad	Narrow
Growth form	"r"-selection	"K"-selection
Nutrient Cycling		
Mineral cycles	Open	Closed
Role of detritus	Unimportant	Important

Condensed from Odum, Eugene P., April 18, 1969. The strategy of ecosystem development. Science 164 (3877): 262–270. Reproduced with permission from the American Association for the Advancement of Science.

including energetics, community structure, life history, and nutrient cycling change over time between the developmental and mature stages of succession (Table 3.7).

During the early, developmental stage, net primary production is high, biomass and total organic matter are low. Food webs are simple and grazer-based and species richness is low. Small organisms with short, simple life cycles and broad niches dominate. As the ecosystem matures, net primary production decreases while biomass and organic matter stocks increase. In the mature stage, most of the CO_2 assimilated by the ecosystem is used to maintain existing biomass rather than support new growth. Organisms are larger in size, possess longer, more complex life cycles, and are more niche specialists than during the early stages. During the early stages, organisms trend toward "r"-selection, whereas in the mature stage, "K"-strategists are favored. Nutrient cycles also differ between the early and late stages of ecosystem development. During the early stages, mineral cycles are more open, i.e., the ecosystem receives most of its nutrients from outside the system, and detritus and organic matter are unimportant as a source of nutrients. As the ecosystem matures, mineral cycles become more closed as recycling through detritus and soil organic matter becomes increasingly important.

Studies of ecosystem development during primary succession provide support for some of Odum's ideas. Nitrogen often is the primary nutrient that limits development of the plant community and studies show that a minimum amount of N capital, N stored in litter and soil organic matter, is needed to create a self-sustaining ecosystem. For terrestrial ecosystems, a threshold of about $70–100\,g\,N/m^2$ is needed to do this (Marrs et al., 1983; Bradshaw, 1983). This minimal amount of N capital depends not only on the amount stored in litter and soil but also on the decomposition rate (Marrs et al., 1983; Bradshaw, 1983). For example, decomposition in ecosystems of cold, wet, and waterlogged environments, including many wetlands, is low, so more N capital is needed to sustain the ecosystem. In warmer and (moderately) dry climates, decomposition proceeds faster and less N capital is needed since the N pools turn over more quickly. This is also true for wetlands that dry down during the growing season as, during the dry phase, aerobic decomposition will predominate, oxidizing organic matter in litter and soils, and recycling N more rapidly than when the site is flooded (Weller, 1978). Some of Odum's predictions have been questioned as being incorrect (Smith, 1996). Nevertheless, this framework is useful for predicting trajectories and timelines of ecosystem development following restoration.

Biological Invasions

A pervasive problem in restoration is invasive and aggressive plant species that come to dominate a site, forming monotypic communities. Freshwater marshes, in particular, are susceptible to invasion and dominance of such competitor species. They include cattail (*Typha* spp.), reed canary grass (*P. arundinacea*), common reed (*Phragmites australis*), purple loosestrife (*Lythrum salicaria*), and others (Galatowitsch et al., 1999). In estuarine environments, saline tidal marshes are invaded by nonnative halophytes such as *S. alterniflora* or by hybrid forms of *Spartina* that often were introduced to build land.

The role of humans in the spread of invasive plants and animals cannot be understated (Elton, 1958; Mack et al., 2000). Our ability to travel great distances and in large numbers throughout the world makes humans the greatest agents of dispersal. Seeds of many invaders arrived in the ballast waters of ships (Harper, 1965; Wilcox, 1989). Some species, for example, *P. arundinacea*, reed canary grass, were introduced to North America and elsewhere as a forage grass (Wilkins and Hughes, 1932; Galatowitsch et al., 1999). Likewise, purple loosestrife was introduced as an ornamental, medicinal, or bee-keeping plant (Pellett, 1966). The wetlands of coastal Hawaii are a classic example of the spread of invasive plants into terrestrial and wetland ecosystems. Bantilan-Smith et al. (2009) censused 35 created, restored, and natural coastal wetlands. Of the 85 species identified, only 16 were native and 3 of the 4 most abundant species were exotic. In addition to loss of species diversity, invaders alter ecosystem processes. Invasion of brackish marshes by *Phragmites* has been shown to accelerate N cycling through increased plant uptake, N mineralization, and nitrate reduction (Windham and Ehrenfeld, 2003). It is important to recognize that most introductions of new species do not take hold. In fact, only about 10% of introduced species become established, even with deliberate and repeated introductions (Williamson, 1996). And only 10% of introduced species become a pest, i.e., invasive.

Community Susceptibility to Invasion

Bare and physically disturbed soils are particularly open to invasion (Harper, 1965). These environments offer much open space and light. Invasion of such sites often is

Table 3.8 Community Attributes and Species Traits That Contribute to Plant Invasion of Restoration Sites

Community attributes	Vacant niche	Disturbed ground/bare soil, hydrologic alteration/ditches, nutrient enrichment
	Species richness	Low richness/more empty niches
	Enemy release	Release from competitors, predators, parasites, pathogens on new range
	Connectivity	Roadside ditches, riparian areas
Species traits	r-strategists	Propagule pressure/prodigious seed production, dispersal/wind, water-dispersed
	Broad tolerance	Varying hydrology, nutrient level
	Efficient use	Resources/light, nutrients
	Hybrid vigor	Hybridization of *Spartina*, *Phragmites*, *Lythrum*
	Allelopathy	Toxins in roots, litter

referred to as the vacant niche hypothesis (Mack et al., 2000; Table 3.8). An example is the invasion of western and eastern Pacific saline marshes by *S. alterniflora*, a native of the eastern US coast that colonizes previously unvegetated mudflats (Callaway and Josselyn, 1992; Daehler and Strong, 1996). On many terrestrial and wetland restoration sites, especially created wetlands, bare soil is exposed during site preparation which makes them especially susceptible to invasion.

Wetland restoration sites also are exposed to disturbances that alter hydrology and water chemistry. Nutrient-enriched wetlands often are dominated by competitor species such as *Phragmites*, *Typha*, *Phalaris*, and *Myriophyllum* (water milfoil) (Galatowitsch et al., 1999). Nutrient additions have been shown to stimulate *Phalaris* at the expense of native sedge species (Green and Galatowitsch, 2002). In saline tidal marshes, the spread of the invasive *S. alterniflora* into mudflat and vegetated intertidal habitat is enhanced by N additions (Tyler et al., 2007). And use of deicing salts has led to increased dominance by the salt-tolerant *Phragmites* in freshwater wetlands (Richburg et al., 2001).

Hydrologic alterations also promote invasion. For example, ditches and culverts harbor more invasive species than wetlands without these alterations (Kercher et al., 2004). Road-side ditches and riparian areas are ideal for invasive species as they are strongly connected to other parts of the landscape, either through human dispersal (roads) or water (Zedler and Kercher, 2004). Stabilizing water levels through dam construction may pave the way for invasives (Zedler and Kercher, 2004) as do erosion and sedimentation (Werner and Zedler, 2002; Stiles et al., 2008). In restored vernal pools in California, pools with deeper water levels were more resistant to invasion than shallow pools (Collinge et al., 2011).

Species richness is thought to affect a community's resistance to invasion. Communities with more species are predicted to be more resistant to invasion (Elton, 1958; Mack et al., 2000). For example, in vernal pools, seeding intensity increased biotic resistance to invasion (Collinge et al., 2011). Species sometimes arrive to new sites without their native controls, enabling them to proliferate in the new habitat (i.e., the enemy release hypothesis (Klironomos, 2002; Mitchell and Power, 2003)). They may escape from competitors, predators, parasites, pathogens, and grazers who, on their home range, hold their populations in check (Elton, 1958).

Life History Traits of Invaders

Many invaders possess life history traits that enable them to readily disperse to and colonize sites (Table 3.3). One is the ability to produce large numbers of seeds that are readily dispersed, such as by wind. In wetlands, seeds that disperse by water or waterfowl are carried to distant sites where they colonize (Zedler and Kercher, 2004). Both propagule pressure (the number of offspring or seeds produced per individual) and the intrinsic rate (r) of increase of the population (i.e., generation time) enhance a species' ability to invade (Williamson, 1996). Other predictors include reproductive strategy (i.e., r vs K), common abundance in native habitat, large range in native habitat, climatic matching (i.e., those species from a similar climate are more likely to be successful invaders), and phenotypic plasticity over a wide range of environments (Rejmanek and Richardson, 1996; Williamson, 1996; Rejmanek, 2000). Invasive species often exhibit greater morphological plasticity, such as variation in shoot and root production, in response to disturbance, especially nutrients. For example, *P. arundinacea* allocates more biomass to roots under low nutrient conditions (Green and Galatowitsch, 2001; Maurer and Zedler, 2002), but when nutrients are added, it produces fewer roots and allocates more to aboveground biomass. Over a range of nutrient additions, *Phalaris* exhibited greater plasticity in root–shoot allocation than native species (Green and Galatowitsch, 2001).

There is evidence that mutualism between invasive plant and animal species is not uncommon and facilitates invasion by both (Simberloff and Von Holle, 1999). Studies have shown that exotic herbivores facilitate invasion of exotic plant species (Parker et al., 2006). The result is an invasional meltdown whereby the two invaders increase the likelihood of their survival and impact on native species. Thus, facilitation is perhaps as important as release from an enemy when it comes to species invasion. Hybridization also is known to contribute to species invasion. Many aggressive species in wetlands, including *Typha*, *Spartina*, and *Lythrum*, are hybrids (Galatowitsch et al., 1999; Ellstrand and Schierenbeck, 2000) that, if left unchecked, will dominate freshwater and estuarine wetlands. *Spartina townsendii* (*anglica*), a hybrid between the native *Spartina maritima* and *S. alterniflora* from North America, long ago invaded coastal mudflats of Britain and Western Europe (Elton, 1958; Thompson, 1991). Finally, some invaders may come to dominate through release of alleopathic chemicals (Gallardo et al., 1998a,b). Gallardo et al. (1998b) showed that extracts of cattail (*T. domingensis*) inhibited growth of water fern (*Salvinia minima*).

Once established, successful invaders may maintain their dominance by having greater tolerance to environmental factors, waterlogging and salinity, or more efficient use of resources such as light and nutrients (Zedler and Kercher, 2004). Some invasive species, *Typha* and *Phalaris* for example, tolerate deeper flooding than native species (Kercher and Zedler, 2004). Mechanisms of invasion by *Typha* suggest that the presence of both plants and litter acts to reduce richness of native species (Farrer and Goldberg, 2009; Vaccaro et al., 2009; Larkin et al., 2012). Using mesocosms and field sampling, Larkin et al. (2012) demonstrated that the dense shading by live vegetation and copious litter production alter environmental conditions such as light and temperature and decrease native species richness. The effect of litter was stronger than that of the vegetation. *Typha* litter also increased N cycling (N mineralization, soil ammonium) and thus modified ecosystem processes that further promote its spread (Farrer and Goldberg, 2009). Other competitors such as *Phalaris* and *Phragmites* employ shading to outcompete smaller, native species. Wetzel and van der Valk (1998) demonstrated that a tall, spreading horizontal canopy enabled *Phalaris* to outcompete *Carex* as well as *Typha*.

Invasive Species and Restoration

Limiting and controlling the spread of invasive species will almost always be needed to successfully restore ecosystems, especially wetlands. As Zedler and Kercher (2004) point out, many wetlands are "sinks" on the landscape, occupying the lowest areas, with disturbances, nutrients, sediment, and other pollutants propagating downstream and altering the environmental template. Their low position on the landscape also ensures a steady supply of propagules to the site. The best advice is to be proactive. Ameliorate the stressors such as nutrients, road salt and other contaminants that favor the expansion of invasive species. Restrict entry to the site by ensuring that propagules are not brought in with amendments such as organic mulch. On bare soils, quickly seed or plant a nurse or cover crop to fill vacant niches. When dealing with invasive species, continued vigilance is critical.

References

Abraham, K.F., Jefferies, R.L., Alisauskas, R.T., 2005. The dynamics of landscape change and snow geese in mid-continent North America. Global Change Biology 11, 841–855.

Anderson, R.C., Liberta, A.E., Dickman, L.A., 1984. Interaction of vascular plants and vesicular-arbuscular mycorrhizal fungi across a soil moisture gradient. Oecologia 64, 111–116.

Bantilan-Smith, M., Bruland, G.L., MacKenzie, R.A., Henry, A.R., Ryder, C.R., 2009. A comparison of the vegetation and soils of natural, restored, and created coastal lowland wetlands in Hawaii. Wetlands 29, 1023–1035.

Bauer, C.R., Kellogg, C.H., Bridgham, S.D., Lamberti, G.A., 2003. Mycorrhizal colonization across hydrologic gradients. Wetlands 23, 961–968.

Bertness, M.D., Callaway, R.M., 1994. Positive interactions in communities. Trends in Ecology and Evolution 5, 191–193.

Bertness, M.D., Hacker, S.D., 1994. Physical stress and positive associations among marsh plants. American Naturalist 144, 363–372.

Blossey, B., Skinner, L.C., Taylor, J., 2001. Impact and management of purple loosestrife (*Lythrum salicaria*) in North America. Biodiversity and Conservation 10, 1787–1807.

Boutin, C., Keddy, P.A., 1993. A functional classification of wetland plants. Journal of Vegetation Science 4, 591–600.

Bradshaw, A.D., 1983. The reconstruction of ecosystems. Journal of Applied Ecology 20, 1–17.

Bradshaw, A.D., Chadwick, M.J., 1980. The Ecology and Reclamation of Derelict and Degraded Land. University of California Press, Berkeley, California.

Cairns Jr., J., Heckman, J.R., 1996. Restoration ecology: the state of an emerging field. Annual Review of Energy and the Environment 21, 167–189.

Callaway, J.C., Josselyn, M.N., 1992. The introduction and spread of smooth cordgrass (*Spartina alterniflora*) in South San Francisco Bay. Estuaries 15, 218–226.

Carpenter, S.R., Ludwig, D., Brock, W.A., 1999. Management of eutrophication for lakes subject to potentially irreversible change. Ecological Applications 9, 751–771.

Cavieres, L.A., Bandano, E.I., Sierra-Almeida, A., Gomez-Gonzalez, S., Molina-Montenegro, M.A., 2006. Positive interactions between alpine plant species and the nurse cushion plant *Laretia acaulis* do not increase with elevation in the Andes of central Chile. New Phytologist 169, 59–69.

Chapin III, F.S., Walker, L.R., Fastie, C.L., Sharman, L.C., 1994. Mechanisms of primary succession following deglaciation at Glacier Bay, Alaska. Ecological Monographs 64, 149–174.

Choi, Y.D., 2004. Theories for ecological restoration in a changing environment: toward futuristic restoration. Ecological Research 19, 75–81.

Clements, F.E., 1916. Plant Succession. Publication 242. Carnegie Institute, Washington, DC.

Collinge, S.K., Ray, C., Gerhardt, F., 2011. Long-term dynamics of biotic and abiotic resistance to exotic species invasion in restored vernal pool plant communities. Ecological Applications 21, 2105–2118.

Connell, J.H., Slatyer, R.O., 1977. Mechanisms of succession in natural communities and their role in community stability and organization. American Naturalist 111, 1119–1144.

Cooke, J.C., Lefor, M.W., 1998. The mycorrhizal status of selected plant species from Connecticut wetlands and transition zones. Restoration Ecology 6, 214–222.

Cornwell, W.K., Bedford, B.L., Chapin, C.T., 2001. Occurrence of arbuscular mycorrhizal fungi in a phosphorus-poor wetland and mycorrhizal response to phosphorus fertilization. American Journal of Botany 88, 1824–1829.

Cowles, H.C., 1899. The ecological relations of the vegetation on the sand dunes of Lake Michigan. Botanical Gazette 27, 97–117, 167–202, 281–308, 361–391.

Crocker, R.L., Major, J., 1955. Soil development in relation to vegetation and surface age at Glacier Bay, Alaska. Journal of Ecology 43, 427–448.

Daehler, C.C., Strong, D.R., 1996. Status, prediction, and prevention of introduced cordgrass *Spartina* spp. invasions in Pacific estuaries, USA. Biological Conservation 78, 51–58.

Didham, R.K., Watts, C.H., 2005. Are systems with strong underlying abiotic regimes more likely to exhibit alternative stable states? Oikos 110, 409–416.

Doyle, T.W., O'Neil, C.P., Melder, M.P.V., From, A.S., Palta, M.M., 2007. Tidal freshwater swamps of the southeastern United States: effects of land use, hurricanes, sea-level rise, and climate change. In: Conner, W.H., Doyle, T.W., Krauss, K.W. (Eds.), Ecology of Tidal Freshwater Forested Wetlands of the Southeastern United States. Springer, Dordrecht, The Netherlands, pp. 1–28.

Egerova, J., Proffitt, E., Travis, S.E., 2003. Facilitation of survival and growth of *Baccharis halimifolia* L. by *Spartina alterniflora* Loisel in a created Louisiana salt marsh. Wetlands 23, 250–256.

Egler, F.E., 1954. Vegetation science concepts I. Initial floristic composition, a factor in old-field vegetation development. Vegetatio 4, 412–417.

Ellstrand, N.C., Schierenbeck, K.A., 2000. Hybridization as a stimulus for the evolution of invasiveness in plants. Proceedings of the National Academy of Sciences USA 97, 7043–7050.

Elton, C.S., 1958. The Ecology of Invasions by Plants and Animals. Methuen and Co. Ltd, London.

Farrer, E.C., Goldberg, D.E., 2009. Litter drives ecosystem and plant community changes in cattail invasion. Ecological Applications 19, 398–412.

Faulkner, S.P., Chambers, J.L., Conner, W.H., Keim, R.F., Day, J.W., Gardiner, E.S., Hughes, M.S., King, S.L., McLeod, K.W., Miller, C.A., Nyman, J.A., Shafer, G.P., 2007. Conservation and use of coastal wetland forest in Louisiana. In: Conner, W.H., Doyle, T.W., Krauss, K.W. (Eds.), Ecology of Tidal Freshwater Forested Wetlands of the Southeastern United States. Springer, Dordrecht, The Netherlands, pp. 447–460.

Flores, J., Jurado, E., 2003. Are nurse-protégé interactions more common among plants from arid environments? Journal of Vegetation Science 14, 911–916.

Franco, A.C., Nobel, P.S., 1988. Interactions between seedlings of *Agave deserti* and the nurse plant *Hilaria rigida*. Ecology 69, 1731–1740.

Fukama, T., Lee, W.G., 2006. Alternative stable states, trait dispersion and ecological restoration. Oikos 113, 353–356.

Galatowitsch, S.M., Anderson, N.O., Ascher, P.D., 1999. Invasiveness in wetland plants in temperate North America. Wetlands 19, 733–755.

Gallardo, M.T., Martin, B.B., Martin, D.F., 1998a. An annotated bibliography of allelopathic properties of cattails, *Typha* spp. Florida Scientist 61, 52–58.

Gallardo, M.T., Martin, B.B., Martin, D.F., 1998b. Inhibition of water fern *Salvinia minima* by cattail (*Typha domingensis*) extracts and by 2-chlorophenol and salicylaldehyde. Journal of Chemical Ecology 24, 1483–1490.

Glaser, P.H., 1987. The Ecology of Patterned Boreal Peatlands of Northern Minnesota: A Community Profile. U.S. Fish and Wildlife Service, Washington, DC. Report 85(7.14).

Gleason, H.A., 1917. The structure and development of the plant association. Bulletin of the Torrey Botanical Club 43, 463–481.

Gleason, H.A., 1926. The individualist concept of the plant association. Bulletin of the Torrey Botanical Club 53, 7–26.

Green, E.K., Galatowitsch, S.M., 2001. Differences in wetland plant community establishment with additions of nitrate-N and invasive species (*Phalaris arundinacea* and *Typha* × *glauca*). Canadian Journal of Botany 79, 170–178.

Green, E.K., Galatowitsch, S.M., 2002. Effects of *Phalaris arundinacea* and nitrate-N addition on the establishment of wetland plant communities. Journal of Applied Ecology 39, 134–144.

Grime, J.P., 1977. Evidence for the existence of three primary strategies in plants and its relevance to ecological and evolutionary theory. The American Naturalist 111, 1169–1194.

Groeneveld, E.V.G., Masse, A., Rochefort, L., 2007. *Polytrichum strictum* as a nurse-plant in peatland restoration. Restoration Ecology 15, 709–719.

Hagerthey, S.E., Newman, S., Rutchey, K., Smith, E.P., Godin, J., 2008. Multiple regime shifts in a subtropical peatland: community-specific thresholds to eutrophication. Ecological Monographs 78, 547–565.

Harley, J.L., Smith, S.E., 1983. Mycorrhizal Symbiosis. Academic Press, London.

Harper, J.L., 1965. Establishment, aggression, and cohabitation in weedy species. In: Baker, H.G., Stebbins, G.L. (Eds.), The Genetics of Colonizing Species. Academic Press, New York, pp. 243–265.

Heffernan, J.B., 2008. Wetlands as alternative stable state in desert streams. Ecology 89, 1261–1271.

Hobbs, R.J., Cramer, V.A., 2008. Restoration ecology: interventionist approaches for restoring and maintaining ecosystem function in the face of rapid environmental change. Annual Review of Environment and Resources 33, 39–61.

Hobbs, R.J., Norton, D.A., 1996. Towards a conceptual framework for restoration ecology. Restoration Ecology 4, 93–110.

Hobbs, R.J., Norton, D.A., 2004. Ecological filters, thresholds and gradients in resistance to ecological reassembly. In: Temperton, V.M., Hobbs, R.J., Nuttle, T., Halle, S. (Eds.), Assembly Rules and Ecological Restoration. Island Press, Washington, DC, pp. 72–95.

Hobbs, R.J., Higgs, E., Harris, J.A., 2009. Novel ecosystems: implications for conservation and restoration. Trends in Ecology and Evolution 24, 599–605.

Hobbs, R.J., Hallett, L.M., Ehlich, P.R., Mooney, H.A., 2011. Intervention ecology: applying ecological principles in the twenty-first century. BioScience 61, 442–450.

Hobbs, R.J., Higgs, E., Hall, C.M., Bridgewater, P., Chapin III, F.S., Ellis, E.C., Ewel, J.J., Hallett, L.M., Harris, J., Hulvey, K.B., Jackson, S.T., Kennedy, P.L., Kueffer, C., Lach, L., Lantz, T.C., Lugo, A.E., Mascara, J., Murphy, S.D., Nelson, C.R., Perring, M.P., Richardson, D.M., Seastedt, T.R., Stabdish, R.J., Starzomski, B.M., Suding, K.N., Tognetti, P.M., Yakob, L., Yung, L., 2014. Managing the whole landscape: historical, hybrid and novel ecosystems. Frontiers in Ecology and the Environment 12, 557–564.

Hoeksema, J.D., Chaudhary, V.B., Gehring, C.A., Johnson, N.C., Karst, J., Koide, R.T., Pringle, A., Zabinski, K., Bever, J.D., Moore, J.C., Wilson, G.W.T., Klironomos, J.N., Umbanhowar, J., 2010. A meta-analysis of context-dependency in plant response to inoculation with mycorrhizal fungi. Ecology Letters 13, 394–407.

Hoefnagels, M.H., Broome, S.W., Shafer, S.R., 1993. Vesicular-arbuscular mycorrhizae in salt marshes in North Carolina. Estuaries 16, 851–858.

Huston, M., 1979. A general hypothesis of species diversity. American Naturalist 113, 81–101.

Huston, M., Smith, T., 1987. Plant succession: life history and competition. The American Naturalist 130, 168–198.

Jefferies, R.L., Jano, A.P., Abraham, K.M., 2006. A biotic agent promotes large-scale catastrophic change in the coastal marshes of Hudson Bay. Journal of Ecology 94, 234–242.

Joyce, C.B., 2014. Ecological consequences and restoration potential of abandoned wet grasslands. Ecological Engineering 66, 91–102.

Kassas, M., 1995. Desertification: a general review. Journal of Arid Environments 30, 115–128.

Keddy, P.A., 2010. Wetland Ecology: Principles and Conservation. Oxford University Press, Cambridge, UK.

Kercher, S.M., Zedler, J.B., 2004. Flood tolerance in wetland angiosperms: a comparison of invasive and noninvasive species. Aquatic Botany 80, 89–102.

Kercher, S.M., Carpenter, Q.J., Zedler, J.B., 2004. Interrelationships of hydrologic disturbance, reed canary grass (*Phalaris arundinacea* L.) and native plants in Wisconsin wet meadows. Natural Areas Journal 24, 316–325.

Khan, A.G., 2004. Mycotrophy and its significance in wetland ecology and management. In: Wong, M.H. (Ed.), Developments in Ecosystems, vol. 1. Elsevier B.V, Amsterdam, The Netherlands, pp. 95–114.

Klironomos, J.N., 2002. Feedback with soil biota contributes to plant rarity and invasiveness in communities. Nature 47, 67–70.

Larkin, D.J., Freyman, M.J., Lishawa, S.C., Geddes, P., Tuchman, N.C., 2012. Mechanisms of dominance by the invasive hybrid cattail *Typha × glauca*. Biological Invasions 14, 65–77.

Lewis III, R.R., 2005. Ecological engineering for successful management and restoration of mangrove forests. Ecological Engineering 24, 403–418.

Lewis III, R.R., Gilmore, R.G., 2007. Important considerations to achieve successful mangrove forest restoration with optimum fish habitat. Bulletin of Marine Science 80, 823–837.

Lindeman, R.L., 1941. The developmental history of Cedar Creek bog, Minnesota. American Midland Naturalist 25, 101–112.

Lockwood, J.L., 1997. An alternative to succession: assembly rules offer guide to restoration efforts. Restoration and Management Notes 15, 45–50.

McCook, L.J., 1994. Understanding ecological community succession: causal models and theories, a review. Vegetatio 110, 115–147.

McKee, K.L., Rooth, J.E., Feller, I.C., 2007. Mangrove recruitment after forest disturbance is facilitated by herbaceous species in the Caribbean. Ecological Applications 17, 1678–1693.

Mack, R.N., Simberloff, D., Lonsdale, W.M., Evans, H., Clout, M., Bazzaz, F.A., 2000. Biotic invasions: causes, epidemiology, global consequences and control. Ecological Applications 10, 689–710.

Marrs, R.H., Roberts, R.D., Bradshaw, A.D., 1983. Nitrogen and the development of ecosystems. In: Lee, J.A., McNeill, S., Rorison, I.H. (Eds.), Nitrogen as an Ecological Factor. Blackwell Scientific Publications, Oxford, pp. 113–136.

Maurer, D.A., Zedler, J.B., 2002. Differential invasion of a wetland grass explained by tests of nutrients and light availability on establishment and clonal growth. Oecologia 131, 279–288.

Mayer, A.L., Rietkerk, M., 2004. The dynamic regime concept for ecosystem management and restoration. BioScience 54, 1013–1020.

Miller, R.M., Smith, C.I., Jastrow, J.D., Bever, J.D., 1999. Mycorrhizal status of the genus Carex, Cyperaceae. American Journal of Botany 86, 547–553.

Miller, S.P., 2000. Arbuscular mycorrhizal colonization of semi-aquatic grasses along a wide hydrologic gradient. New Phytologist 145, 145–155.

Miller, S.P., Bever, J.D., 1999. Distribution of arbuscular mycorrhizal fungi in stands of the wetland grass Panicum hemitomon along a wide hydrologic gradient. Oecologia 119, 586–592.

Mitchell, C.E., Power, A., 2003. Release of invasive plants from fungal and viral pathogens. Nature 421, 625–627.

Moore, P.D., Bellamy, D.J., 1974. Peatlands. Elek Science, London.

Niering, W.A., Egler, F.E., 1955. A shrub community of Viburnum lentago, stable for twenty-five years. Ecology 36, 356–360.

Niering, W.A., Goodwin, B.H., 1974. Creation of relatively stable shrublands with herbicides: arresting "succession" on rights-of-way and pastureland. Ecology 55, 784–795.

Niering, W.A., Whittaker, R.H., Lowe, C.W., 1963. The saguaro: a population in relation to the environment. Science 142, 15–23.

Odum, E.P., 1969. The strategy of ecosystem development. Science 164, 262–270.

Odum, E.P., Barrett, G.W., 2005. Fundamentals of Ecology, fifth ed. Thomson, Brooks/Cole, Belmont, California.

Olson, J.S., 1958. Rates of succession and soil changes on southern Lake Michigan sand dunes. Botanical Gazette 119, 125–190.

Padilla, F.M., Pugnaire, F.I., 2006. The role of nurse plant in the restoration of degraded environments. Frontiers in Ecology and the Environment 4, 196–202.

Parker, J.D., Burkpile, D.E., Hay, M.E., 2006. Opposing effects of native and exotic herbivores on plant invasions. Science 311, 1459–1461.

Peat, A.J., Fitter, A.H., 1993. The distribution of arbuscular mycorrhizae in the British flora. New Phytologist 125, 845–854.

Pellett, M., 1966. Purple loosestrife, colorful honey plant. American Bee Journal 106, 134–135.

Peters, D.P.C., Bestelmeyer, B.T., Herrick, J.E., Fredrickson, E.L., Monger, H.C., Havstad, K.M., 2006. Disentangling complex landscapes: new insights into arid and semiarid system dynamics. BioScience 56, 491–501.

Pianka, E.R., 1970. On r- and K-selection. American Naturalist 104, 592–597.

Rejmanek, M., 2000. Invasive plants: approaches and predictions. Austral Ecology 25, 497–506.

Rejmanek, M., Richardson, D.M., 1996. What attributes make some plant species more invasive? Ecology 77, 1655–1661.

Richburg, J.A., Patterson III, W.A., Lowenstein, F., 2001. Effect of road salt and *Phragmites auatralis* invasion on the vegetation of a western Massachusetts calcareous lake-basin fen. Wetlands 21, 247–255.

Rickerl, D.H., Sancho, F.O., Ananth, S., 1994. Vesicular-arbuscular endomycorrhizal colonization of wetland plants. Journal of Environmental Quality 23, 913–916.

Schat, H., 1984. A comparative ecophysiological study on the effects of waterlogging and submergence on dune slack plants: growth, survival, and mineral nutrition in sand culture experiments. Oecologia 62, 279–286.

Scheffer, M., Carpenter, S.R., 2003. Catastrophic regime shifts in ecosystems: linking theory to observation. Trends in Ecology and Evolution 18, 648–656.

Scheffer, M., Hosper, S.H., Meijer, M.L., Moss, B., 1993. Alternative equilibria in shallow lakes. Trends in Ecology and Evolution 8, 275–279.

Scheffer, M., Carpenter, S., Folley, J.A., Folke, C., Walker, B., 2001. Catastrophic shifts in ecosystems. Nature 413, 591–596.

Simberloff, D., Von Holle, B., 1999. Positive interactions of nonindigenous species: invasional meltdown? Biological Invasions 1, 21–32.

Smith, M.D., Knapp, A.K., Collins, S.L., 2009. A framework for assessing ecosystem dynamics in response to chronic resource alterations induced by global change. Ecology 90, 3279–3289.

Smith, R.L., 1996. Ecology and Field Biology. Addison Wesley Longman, Inc, Menlo Park, California.

Stiles, C.A., Bemis, B., Zedler, J.B., 2008. Evaluating edaphic conditions favoring reed canary grass invasion in a restored native prairie. Ecological Restoration 26, 61–70.

Straskrabova, J., Prach, K., 1998. Five years of restoration of alluvial meadows: a case study from central Europe. In: Joyce, C.B., Wade, P.M. (Eds.), European Wet Grasslands: Biodiversity, Management, and Restoration. John Wiley and Sons Ltd, New York, pp. 295–303.

Suding, K.N., Hobbs, R.J., 2009. Threshold models in restoration and conservation: a developing framework. Trends in Ecology and Evolution 24, 271–279.

Thompson, J.D., 1991. The biology of an invasive plant. BioScience 41, 393–401.

Thormann, M.N., Currah, R.S., Bayley, S.E., 1999. The mycorrhizal status of the dominant vegetation along a peatland gradient in southern boreal Alberta. Wetlands 19, 438–450.

Trowbridge, W.B., 2007. The role of stochasticity and priority effects in floodplain restoration. Ecological Applications 17, 1312–1324.

Turner, S.D., Amon, J.P., Schneble, R.M., Friese, C.F., 2000. Mycorrhizal fungi associated with plants in groundwater-fed fen wetlands. Wetlands 20, 200–204.

Tyler, A.C., Lambrinos, J.G., Grosholz, E.D., 2007. Nitrogen inputs promote the spread of an invasive marsh grass. Ecological Applications 17, 1886–1898.

Vaccaro, L.E., Bedford, B.L., Johnston, C.A., 2009. Litter accumulation promotes dominance of invasive species of cattails (*Typha* spp.) in Lake Ontario wetlands. Wetlands 29, 1036–1048.

van der Valk, A.G., 1981. Succession in wetlands: a Gleasonian approach. Ecology 62, 688–696.

Wang, Y., Huang, Y., Qui, Q., Xin, G., Yang, Z., Shi, S., 2011. Flooding greatly affects the diversity of arbuscular mycorrhizal fungi communities in the roots of wetland plants. PLoS One 6 (9), e24512.

Weiher, E., Keddy, P.A., 1995. The assembly of experimental wetland plant communities. Oikos 73, 323–335.

Weller, M.W., 1978. Management of freshwater marshes for wildlife. In: Good, R.E., Whigham, D.F., Simpson, R.L. (Eds.), Freshwater Wetlands: Ecological Processes and Management Potential. Academic Press, New York, pp. 267–284.

Werner, K.J., Zedler, J.B., 2002. How sedge meadow soils, microtopography, and vegetation respond to sedimentation. Wetlands 22, 451–466.

Wetzel, P.R., van der Valk, A.G., 1996. Vesicular-arbuscular mycorrhizae in prairie pothole wetland vegetation in Iowa and North Dakota. Canadian Journal of Botany 74, 883–890.

Wetzel, P.R., van der Valk, A.G., 1998. Effects of nutrient and soil moisture on competition between *Carex stricta, Phalaris arundinacea*, and *Typha latifolia*. Plant Ecology 138, 179–190.

Wilcox, D.A., 1989. Mitigation and control of purple loosestrife (*Lythrum salicaria* L.) along highway corridors. Environmental Management 13, 365–370.

Wilkins, F.S., Hughes, H.D., 1932. Agronomic trials with reed canary grass. Journal of the American Society of Agronomy 24, 18–28.

Williamson, M., 1996. Biological Invasions. Chapman and Hall, London.

Windham, L., Ehrenfeld, J.G., 2003. Net impact of plant invasion on nitrogen-cycling processes within a brackish tidal marsh. Ecological Applications 13, 883–897.

Young, T.P., Chase, J.M., Huddleston, R.T., 2001. Community succession and assembly: comparing, contrasting and combining paradigms in the context of ecological restoration. Ecological Restoration 19, 5–18.

Zedler, J.B., 2000. Progress in wetland restoration ecology. Trends in Ecology and Evolution 15, 402–407.

Zedler, J.B., Kercher, S., 2004. Causes and consequences of invasive plants in wetlands: opportunities, opportunists, and outcomes. Critical Reviews in Plant Sciences 23, 431–452.

Consideration of the Landscape

4

Chapter Outline

Introduction

Worldwide, there is widespread cumulative loss of wetlands from catchments and watersheds, leading to loss of landscape-level functions such as water quality improvement, flood attenuation, and biodiversity (Brinson, 1988; Hemond and Benoit, 1988; Stein and Ambrose, 2001). There is increasing recognition that a landscape approach is needed to restore these functions. The 1992 National Research Council report, Restoration of Aquatic Ecosystems, described *the need to consider (proposed) restoration projects in their landscape and watershed contexts on a scale appropriate to the needs of affected plant and animal species.* Early watershed-based approaches to wetland restoration involved restoring them to protect water quality (Shabman, 1995; NRC, 2001; USEPA, 2008, 2013). The value of a watershed-based approach is that it allows for prioritization of wetlands for restoration (an important economic consideration), establishes where they will be placed in the landscape, and determines how (by what techniques) they will be restored (Zedler, 2006). Watershed-based approaches tend to be employed more in developed countries such as the US and Australia (Brierly and Fryirs, 2005; USEPA, 2008, 2013) where agricultural and urbanization is most intensive and where the legal tools and economic resources are available.

In the US, the impetus for a watershed-based approach came from the realization that, as wetlands were lost to development, they were being replaced by mitigation wetlands of a different type (often a pond) and in different numbers and locations, sometimes outside of the watershed (Bedford, 1996). Essentially, the cumulative loss of wetlands and wetland-dependent functions was not being replaced through creating and restoring them. One idea behind a watershed-based approach is to restore regional wetland diversity by restoring different types of wetlands in different locations based on wetland templates (see, for example, the hydrogeomorphic classification in Chapter 2) determined from the past and in regional assessments (Bedford, 1999).

At the watershed scale, wetland restoration may involve passive approaches, ameliorating or removing stressors that impact the site (McIver and Starr, 2001). This type of approach might include reducing nutrient loads through better management practices such as more efficient off-site fertilizer use or reintroducing flood flows and sediment to riparian and floodplain wetlands. A classic example is the one-time managed flood on the Colorado River (USA) that was used to restore riparian habitat by scouring, transporting, and depositing sand (Patten et al., 2001; Stevens et al., 2001). Active approaches also are needed and consist of restoring hydrology by plugging drainage ditches or removing fill, replanting vegetation, and removing invasive species. A watershed-based approach may involve a combination of passive and active approaches to restore a diversity of wetland types (riparian, in-stream, created) and prioritize their placement in the watershed (Mitsch, 1992).

Stressors of Natural, Agricultural, and Urban Landscapes

Successful wetland restoration depends on eliminating or ameliorating the effects of stressors associated with altered hydrology and pollutants, sediment, nutrients, and others that originate off-site. In almost all situations, wetland hydrology—the depth, duration, frequency, and timing of flow—must be restored. Altered hydrology usually is an on-site problem; however, delivery of water to wetlands along streams and rivers may be modified by dams upstream that dampen the annual cycle of flood pulses and reduce sediment transport needed to maintain characteristic wetland communities (Stevens et al., 2001). Other stressors that originate off-site include excess sediment and especially nutrients that may affect wetlands far downstream. Restoring wetlands in natural, agricultural, and urban landscapes, respectively, requires increasing attention to stressors, both on-site and off-site (Figure 4.1). The intensity and degree of alterations to the wetlands increases with human development of the watershed, reaching its peak in urban environments where both the wetland and watershed are highly altered (Figure 4.1).

In relatively intact (natural) landscapes, most stressors, if they are present, originate on site. They usually involve hydrologic alterations such as levees, impoundments, or tide gates that restrict flow and movement of water (Table 4.1). More often, they are structures, ditches, or tile drains that accelerate the conveyance of water off the site. Wetlands in these natural landscapes should be easiest to restore since most activities are focused on-site.

With increasing development of the landscape, greater attention and resources must be focused off-site on stressors that affect water quality and species movement that depend on connectivity. For example, wetlands in agricultural watersheds are exposed to excess sediment, nutrients, herbicides, and pesticides that are linked to production of food crops (Table 4.1). It is difficult to remediate stressors such as nutrients because of the diffuse pathways of flow, including groundwater and overland flow, off the cropland. Alterations to the site itself are not inconsequential either. Wetlands in

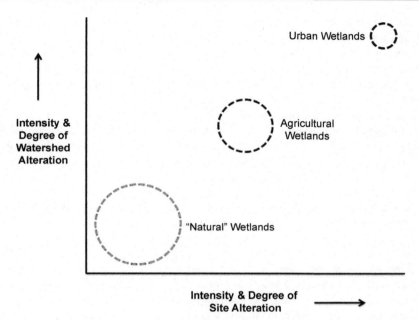

Figure 4.1 Relationship between intensity and degree of site alteration and watershed alteration and ease of restoring wetlands in natural, agricultural, and urban landscapes. The larger the "circle," the greater the probability of success.

Table 4.1 Common On- and Off-site Stressors That Impact Wetlands in Agricultural and Urban Watersheds

	Agricultural	Urban
Hydrological	Ditching, tile drainage, levees	Ditching, fill, mosquito ditches, levees, tide gates, flashy (high peak) flows
Chemical	Sediment, nutrients (N, P), herbicides, pesticides	Sediment, nutrients (N, P), heavy metals, hydrocarbons (oils, grease), deicing salts, pharmaceutical and personal care products
Biological	Absence of ecotones, invasive plants and animals	Absence of ecotones, invasive plants and animals

agricultural landscapes are invariably ditched or drained so that wetland hydrology is altered or completely absent. Restoration of biodiversity is severely constrained by the near-elimination of vegetated buffers and the homogenization of the terrestrial landscape. Agricultural activities lead to loss of woody riparian vegetation and diversity of birds and fish (Mensing et al., 1998). The need to take a watershed approach to restoring wetlands in agricultural landscapes is especially important because of their near-elimination in these regions and its adverse impacts to biodiversity, flood abatement, and water quality (Zedler, 2003).

Restoring wetlands in urban landscapes, however, represents the greatest challenge because of the sheer number and magnitude of stressors. Hydrologic alterations are widespread and intense and include levees, ditches, mosquito ditches, and tide gates in coastal areas, or placement of fill atop them. The intensity and timing of flow also is dramatically altered because of the prevalence of impervious surface in the watershed. Furthermore, impervious surfaces efficiently convey pollutants downstream that degrade water quality and affect health of aquatic organisms. In urban landscapes, pollutants include not only sediment and nutrients but also deicing salts, heavy metals, and other artificial pollutants such as oils, grease, and other hydrocarbons and chemicals from pharmaceutical and personal care products. Edges and ecotones around urban wetlands are practically nonexistent and, other than the water flowing through the wetland, there is no connectivity to other intact natural elements of the landscape.

At larger spatial scales, pollution from *airsheds* may be a stressor on wetlands and other ecosystems. Nutrient-poor wetlands such as fens or bogs may be particularly susceptible to atmospheric deposition of N from fossil fuel combustion from automobiles and coal-fired electrical generating plants. Atmospheric N deposition, such as ammonium and nitrogen oxides, has been implicated in declining plant species diversity in peatlands, bogs, and fens (Bobbink et al., 1998; Bobbink and Lamers, 2002) and eutrophication of estuaries and shallow seas (Paerl, 1993; Paerl and Whitall, 1999; Howarth et al., 2002). In some parts of the world, power plants and metal ore smelters emit large amounts of acidity and sulfur dioxide that reduce biodiversity, contribute to wetland eutrophication (Lamers et al., 2002), acidify soils, and stress and kill vegetation in the local area (Winterhalder, 1996).

Another constraint on restoring wetlands, especially in agricultural and urban landscapes, is the prevalence of invasive species, plants and animals (see Chapter 3, Theory). Invasive species are a problem in almost all restoration projects: terrestrial, wetland, and aquatic ecosystems. In most cases, the effort to eradicate such species results in a stalemate where constant vigilance, including herbicides and other methods of control, is necessary to keep the invaders in check.

Wetland Functions and Placement on the Landscape

When restoring wetlands, their placement in the landscape determines the type and intensity of functions—hydrologic, water quality, biodiversity, and habitat—they deliver. A wetland's ability to deliver certain ecosystem functions depends on both internal properties such as size, morphometry, hydrology, vegetation, and soils (Preston and Bedford, 1988) and landscape properties. Landscape or off-site properties such as watershed size, position in the watershed, and land use determine if, and to what extent, a wetland provides a given function, as do factors like climate, topography, geology, and soils. Several approaches exist to quantify wetland functions, recognizing that landscape position determines if and how much of a given function the wetland will provide. Probably the best developed one is the hydrogeomorphic approach described in Chapter 2 (Definitions). Other methods include hydrogeological (Winter, 1986, 2001), riparian geomorphology (Harris, 1988), and hydrologic equivalency approaches (Bedford, 1996, 1999).

Hydrologic functions such as water detention and storage are determined by size of the wetland and by their placement in the landscape. Headwater wetlands and wetlands along streams and rivers serve this function very well because they are positioned to intercept and detain water immediately flowing off the terrestrial landscape (NRC, 2001). However, headwater wetlands are small in size, so their ability to detain water may be limited. Riparian and streamside wetlands are especially effective for water quality improvement (NRC, 2001) because they serve as the *edge* or ecotone between terrestrial lands and aquatic ecosystems. They have large surface area relative to their size that enhances sediment trapping and nutrient removal (see Chapter 6, Forested Wetlands). Riparian wetlands may remove as much as 80% of overland P flux to streams and nearly 90% of N, mostly through denitrification (Peterjohn and Correll, 1984).

Water quality improvement is a major concern for the Mississippi River watershed (USA) where nutrient (N) export from the Corn Belt of the Midwestern US contributes to long-term hypoxia, or low oxygen, in waters of the northern Gulf of Mexico. A scientific panel convened by the U.S. National Oceanic and Atmospheric Administration recommended restoring and creating more than $99,000 \, km^2$ of riparian, floodplain, and created wetlands to reduce N inputs to the Gulf by 40% (Mitsch et al., 2001). It was recognized that not only the area but the placement of these wetlands in the landscape was critical to achieve these reductions (NRC, 2001). Mitsch and Day (2006) recommended a more compact arrangement of created and restored wetlands ($22,000 \, km^2$) to address the issue using a combination of in-field wetlands to intercept agricultural runoff and diversion wetlands to retain floodwaters along major tributaries and rivers.

Riparian and streamside wetlands also provide other functions such as shading to moderate stream temperatures, adding detritus to support aquatic food webs, and stabilizing stream banks and channels (Vannote et al., 1980; Naiman and Decamps, 1997; NRC, 2001). They are corridors for dispersal of plant and animal species and movement of animals throughout watersheds (Meffe and Carroll, 1997; Naiman and Decamps, 1997; NRC, 2001). The width and spatial pattern of riparian vegetation is important because it affects not only nutrient removal but diversity of riparian birds (Mensing et al., 1998; Gergel et al., 2002) and carbon export to streams (Gergel et al., 1999). Lack of continuity and varying width of riparian buffers reduces water and pollutant retention, necessitating the need to create wider buffers to meet management goals to reduce pollutants (Weller et al., 1998).

Endangered species such as salmon depend on the configuration and connectivity of wetlands adjacent to rivers and streams. In the U.S. Pacific Northwest, juvenile salmon depend on interconnected wetlands during their migration to the sea (Simenstad et al., 2000, 2006). Their habitats include dendritic tidal channels where they forage as well as tidal freshwater sloughs where they stay for extended periods. Restoration of a single habitat is insufficient to foster the recovery of these endangered fish species. Rather, restoration projects must be designed and implemented to include the diversity of habitats that comprise the estuarine landscape including wetlands (Simenstad et al., 2000).

Wetlands that lack strong connections to aquatic ecosystems also are important for animals, especially amphibians. The classic example is geographically isolated or depressional wetlands that are critical for herpetofauna (amphibians and some reptiles). These organisms breed in isolated wetlands (Semlitsch et al., 1996) that provide favorable breeding habitat because they often lack predators (fish) (Snodgrass et al., 1996, 2000). Herpetofauna typically disperse only a short distance, sometimes less than 1 km during their lifetime, and so removal of these wetlands from the landscape leads to loss of habitat and insurmountable dispersal distances to breed and maintain metapopulations and overall biodiversity (Semlitsch and Bodie, 1998). Both depressional wetlands and the intact terrestrial buffers around them are essential for the survival of many amphibian species (Burke and Gibbons, 1995; Semlitsch, 1998).

The Need for a Watershed Approach

A watershed-based approach to wetland restoration developed from our recognition that many wetlands are connected to aquatic ecosystems that serve as conduits for pollutants and dispersal of organisms. In countries such as the US where restoration is used as a tool for mitigation (see Chapter 2, Definitions), a watershed approach is needed because wetland loss often is offset by replacement wetlands that lack the diversity of wetlands in the landscape (Bedford, 1996) and their placement in appropriate locations.

Wetland restoration at the watershed scale typically focuses on water quality improvement though other benefits such as detention/alteration of flood flows and facilitating species movement and dispersal may be realized as well (Cedfelt et al., 2000; NRC, 2001; Hoffman and Baattrup-Pederson, 2007). Candidate restoration sites may be prioritized based on hydrology (i.e., retention time) and load (Almendinger, 1999; Newbold, 2005). Sites may be placed in a catchment of known water quality problems or that receive large amounts of water and nutrients relative to other catchments (Almendinger, 1999). On-site properties such as hydraulic residence time, hydraulic flux, and wetland characteristics (area, volume, average depth) also are important for optimized nutrient removal.

Sometimes, a combination of restoration techniques is employed to remediate water quality. Richardson et al. (2011) used a three-phase approach to improve water quality in Sandy Creek, a 600-ha urban and suburban watershed in North Carolina (USA). Restoration involved recontouring and planting 600 m of degraded stream, building a 1.6-ha stormwater wetland pond, and constructing a 0.6-ha treatment wetland. Significant reductions in sediment, phosphorus, nitrogen, and fecal coliform bacteria concentrations were observed following completion of the three-phase project (see Chapter 11, Case Studies).

Efforts to improve water quality at landscape and watershed scales require creating and restoring large amounts of wetlands. The amount needed will depend on the nutrient load as well as their strategic placement in the landscape. For example, Mitsch et al. (2001) suggested that restoration of riparian zones and wetlands on low-order streams could reduce N loads to the Mississippi River (USA) by 20–50%

Table 4.2 Estimates of Riparian/Wetland Area Needed to Improve the Water Quality of Receiving Waters

Activity	Percentage of Watershed Restored	Reduction in Load (%)	Source
Riparian restoration (Mississippi river basin, USA)	0.7–1.8	19–50 (N)	Mitsch et al. (2001)
Wetland creation and restoration (Mississippi river basin, USA)	2.7–6.6	19–50 (N)	Mitsch et al. (2001)
Wetland creation and restoration (Mississippi river basin, USA)	3–4	46 (N)	Kovacic et al. (2006)
Wetland creation and restoration (Sweden)[a]	5–10	25–50 (N) 85–100 (P)	Tonderski et al. (2005)
	5	75 (N)	Arheimer and Wittgren (2002)
Wetland creation (NE Spain)	1.4	50 (N)	Moreno et al. (2007)
Wetlands (China)	~5	>90 (P)	Yin et al. (1993)

Note: N = nitrogen, P = phosphorus.
[a]Involves conversion of arable land to wetland.

(Table 4.2). The amount of land needed to do this would be only 1–2% of the entire watershed. The amount of N removed could be doubled if floodplain forests were restored across 3–7% of the watershed. Similarly, Kovacic et al. (2006) calculated that creating wetlands on 3–4% of the watershed could reduce N loads by 46% (Table 4.2).

Modeling studies in Sweden suggest that river nitrate concentrations could be reduced by 6% if 0.4% of the watershed was restored to wetlands (Arheimer and Wittgren, 2002). Scaling up, if 5% of the watershed was restored to wetlands, then N loads could be reduced by 75%. Work by Arheimer and Wittgren (1994) estimated that conversion of 1% of the watershed to wetlands could reduce N transport by 10–16% to the Baltic Sea. Finally, Tonderski et al. (2005) suggested that converting 2% of the watershed's arable land to wetlands could reduce anthropogenic N loads by 10% and P loads by 34%. Scaling this to 5–10% of the watershed yields removal rates of 25–50% for N and 85–100% for P.

In China, conversion of 5% of the watershed to wetlands and ponds can measurably reduce nutrient concentrations and the effects of eutrophication (Yan et al., 1998; Yin et al., 1993). These multipond systems are an ancient creation, consisting of a mixture of wetlands, ditches, and open water that provide drinking water for livestock and humans, irrigation of crops, production of fish, and wetland-dependent functions such as water detention and purification (Yin and Shan, 2001). In northeast Spain, creating and restoring wetlands on 3–6% of the watershed could remove most of the nitrogen from agricultural wastewater (Moreno et al., 2007). In their study, constructed wetlands covered 1.4% of the water and removed up to 50% of the total N load from agricultural wastewater. In a review paper, Verhoeven et al. (2006) suggested that restoring 2–7% of the watershed area to wetlands *can contribute significantly to water quality improvement at the watershed scale.*

Restoring wetlands for biodiversity is more difficult than for other functions like water quality improvement and flood attenuation (Zedler, 2003) that are determined mostly by on-site (wetland) characteristics such as hydraulic and nutrient loading rate, wetland size/area/depth, and hydraulic retention time. Some animals require large areas and connections to enable migration from one patch to another. Plants also require connections for seed dispersal either by water or by birds that migrate through vegetated and riparian corridors.

Wetland restoration at the watershed scale should strive to integrate multiple functions and benefits (Almendinger, 1999). An example of a multifunction restoration project is abandoned rice fields in Spain that were restored to marsh for water quality improvement and avian habitat (Comin et al., 2001). Restoration efforts also should focus on the benefits that accrue to society (Teal and Peterson, 2005). Such benefits include timber harvesting and fisheries (see, for example, Chapter 9, Mangroves), recreational fishing, hunting, bird watching, activities such as canoeing and kayaking, and education. Teal and Peterson (2005) and others argue that support and "buy-in" from society is critical if we are to restore wetlands at large spatial scales.

The Nature Conservancy's Kankakee Sands Preserve in northwestern Indiana (USA) is a large-scale restoration of the terrestrial (prairie, oak savanna) and wetland (marsh) landscape that existed prior to European settlement. Covering more than 3100 ha, the restoration involves plugging ditches to restore hydrology, removing invasive species, and reintroducing fire. The primary goal is to restore biodiversity to this predominantly agricultural landscape by recreating a large (12,000 ha) contiguous patch of restored and intact natural habitats. Located only 100 km from the city of Chicago, Kankakee Sands provides recreational (bird watching) and environmental educational opportunities for the people of the region. See Chapter 11, Case Studies, to learn more about wetland restoration at Kankakee Sands.

Selecting and Prioritizing Sites for Restoration

A watershed-based approach to wetland restoration requires identifying candidate sites and prioritizing their placement. On-site properties such as topography and connectivity to aquatic habitats are needed to restore sites where they will be successful, that is, the site can support wetland hydrology, soils, and vegetation. Off-site properties are needed to describe the landscape characteristics, land use, including source landscapes such as catchments with high nutrient loads or intact landscape patches that are sources of biodiversity. Once sites are identified, they are prioritized according to their ability to intercept nutrients, attenuate floodwaters, connect habitats to enhance biodiversity, or for multiple benefits.

Watershed and Wetland Characteristics to Identify Suitable Sites

Landscape and watershed characteristics can be used to predict the potential benefits of restoration for water quality and biodiversity, including avian and aquatic species. A number of studies show that terrestrial land use—amount of and distance to a given type, such as agriculture, is the major predictor of nutrient loading to streams and rivers (Osborne and Wiley, 1988; Lenat and Crawford, 1994; Detenbeck et al., 1993; Johnson et al., 1997; Herlihy et al., 1998; Johnson et al., 2001; Jones et al., 2001; Gergel et al., 2002). Percentage of arable/cultivated land and number of stock ponds (Moreno-Mateos et al., 2008) also is related to nutrient load. Agricultural landscapes, in particular, produce large nutrient loads to aquatic ecosystems, so incorporating heterogeneity into the landscape can provide beneficial ecosystem services. Moreno-Mateos et al. (2008) suggest that landscape heterogeneity can be increased by reducing the sizes of catchments dominated by agriculture, increasing the area covered by natural vegetation, and creating numerous wetlands throughout the catchment.

Agricultural and urban land use, including impervious surface, contribute other pollutants such as salinity, sediment, and metals that impair water quality (Gergel et al., 2002; Moreno-Mateos et al., 2008). Percentage of impervious surface is an important predictor of flow that affects water quality. Impervious surface alters hydrology by increasing surface flows, shortening the time to flood peaks, and increasing nutrient loads (Paul and Meyer, 2001), at the same time reducing the detention/retention time of water in the wetland.

Biodiversity is enhanced by increasing landscape heterogeneity, especially the amount of natural vegetation. Diversity of aquatic invertebrate and fish are related to percentage of forest in the watershed, especially the amount of and continuity of riparian forest (Wang et al., 1997; Harding et al., 1998; Jones et al., 1999). Abundance and diversity of riparian bird species is related to width of riparian buffers (Stauffer and Best, 1980; Dickson et al., 1995). Studies suggest that a buffer of at least 25–100 m is needed to support healthy riparian avian communities (Croonquist and Brooks, 1993; Keller et al., 1993; Darveau et al., 1995; Hodges and Krementz, 1996).

Many aquatic organisms exhibit a nonlinear response to land conversion (Gergel et al., 2002). For agricultural landscapes, declines in biodiversity were observed once 50% of the watershed was converted. Modeling studies suggest that similar thresholds exist for nutrient loading where loads do not increase significantly until a certain (40%) percentage of landscape is converted to a source of nutrients (Gergel, 2005). In urban watersheds, diversity of aquatic species was evident once ~20% of the watershed was urbanized (Wang et al., 1997).

Gergel (2005) evaluated the effects of landscape area versus heterogeneity on nutrient loading to aquatic ecosystems. She found that, in homogeneous landscapes such as agriculture and undisturbed landscapes, area was a good predictor of nutrient loading. However, in more heterogeneous landscapes such as mixed agriculture and natural areas, the arrangement of land cover types and point sources of nutrients—ditches, drainage and sewer pipes, and roads—were better predictors. Moreno-Mateos et al. (2008) employed a number of landscape characteristics, both for an agricultural catchment and for wetlands within it, to predict the effect of wetland creation and restoration on water quality. At the catchment scale, characteristics included catchment

Table 4.3 **Landscape/Watershed and Wetland Characteristics Useful for Determining, Identifying, and Prioritizing Placement of Wetland Restoration Sites on the Landscape**

Landscape/Watershed Characteristics	Wetland Characteristics
Land use	Topography
Land use ratio	Hydric soils/saturation index
Distance from "source"	Size, area, volume
Connectivity (streams, ditches)	Edge area, edge volume
Stream order	Hydraulic flux and retention time
	Nutrient concentration and load
Patch number, area, density	
Aggregation index	Soil properties (e.g., organic matter), vegetation

area, percentage of arable land, stock ponds, natural vegetation and wetlands, patch number, density, mean area, edge density, and landscape diversity. For wetlands, similar measures were calculated, including area, patch properties, perimeter/area ratio, and aggregation index (see Table 1 of Moreno-Mateos et al., 2008).

Table 4.3 lists watershed and wetland characteristics that can inform placement of wetlands for water quality improvement. Watershed characteristics include land use/land use ratio, distance from "source" ecosystems, and connectivity, including streams and ditches. Stream order also is important when it comes to prioritizing their placement, especially for water quality improvement. For biodiversity, characteristics describing landscape and wetland patches and aggregation are important. Patch size, area, density, and aggregation are important considerations when it comes to creating large habitat blocks and connectivity. Wetland characteristics such as topography and presence of hydric soils are important for placing wetlands in locations where they will develop and persist. Characteristics such as size/area, volume, hydraulic retention time, and nutrient load are important for optimizing nutrient removal as are soil properties such as organic matter content that promote processes like denitrification.

Prioritizing Sites for Restoration

A number of methods exist for identifying and prioritizing wetland restoration sites some of which are presented in Table 4.4. Most methods tend to focus on a single function such as water quality (Almendinger, 1999; White and Fennessy, 2005; Flanagan and Richardson, 2010), flood attenuation (McAllister et al., 2000), or habitat/biodiversity (Llewellyn et al., 1996; Twedt et al., 2006) though some incorporate a suite of functions (Russell et al., 1997; Olson and Harris, 1997; Verhoeven et al., 2008). Almendinger (1999) developed a method to prioritize sites for water quality improvement based on loading, path, and processing factors. Loading factors, derived from our understanding of how constructed wetlands for wastewater treatment, known as *treatment wetlands*, work include inflow concentration and hydraulic residence

Table 4.4 **Methods for Identifying and Prioritizing Sites for Wetland Restoration**

Location	Goal	Methodology
Minnesota river (USA)[a] (wetlands)	Water quality	Select and prioritize sites based on loading (inflow concentration, hydraulic residence time, wetland area, and depth), path (topography, soils, land use), and processing (water level stability, vegetation cover) factors
Ohio (USA)[b] (wetlands)	Water quality	1. Identify suitable sites based on topography, hydric soils, stream order saturation index/soil permeability, land use 2. Prioritize based on contribution to water resource integrity once restored
Northern plains (USA)[c] (prairie pothole wetlands)	Flood attenuation	Select and prioritize based on land use, soils, stream and ditch density, precipitation, property value
Southern California (USA)[d] (riparian wetlands)	Multiple	Select and prioritize based on wetness index (based on upslope contributing area, surface slope, soil permeability) and size of prospective sites, land cover, and proximity to existing riparian vegetation
Southern California (USA)[e] (riparian wetlands)	Multiple	1. Identify and classify riparian reaches based on channel slope and floodplain width 2. Prioritize based on land cover, degree of fragmentation and connectivity
Netherlands[f] (wetlands as operational landscape units (OLUs))	Multiple	1. Define restoration target (e.g., plant or animal species, wetland-dependent function). 2. Identify spatial configurations of hydrology and dispersal crucial for (1). 3. Determine the size, shape, and location of OLUs needed as building blocks for (1).

[a]Almendinger (1999).
[b]White and Fennessy (2005).
[c]McAllister et al. (2000).
[d]Russell et al. (1997).
[e]Olson and Harris (1997).
[f]Verhoeven et al. (2008).

time (Newbold, 2005; Kadlec and Wallace, 2009). Path factors consist of watershed characteristics and include topography, soils, and (agricultural) land use. Processing factors include water level stability and vegetative cover in the wetland.

White and Fennessy (2005) describe a two-phase approach for restoring wetlands in Ohio (USA) for water quality improvement. In Phase 1, the total population of sites suitable for restoration is identified. Watershed and wetland characteristics such as hydric soils, topography, stream order, saturation index, and land use are used to identify sites. In Phase 2, available sites are prioritized according to their contribution to water resource integrity once restored. This might depend on the surrounding land use or the nutrient load that the wetland receives. Flanagan and Richardson (2010) prioritized sites for phosphorus removal based on land use (i.e., the location of prior converted wetlands) and natural land cover within stream buffers. An example of successful prioritization is the Glaciated Interior Plains of the Midwest US, also known as the Corn Belt, where $1100\,km^2$ of wetland and riparian habitat were restored, mostly in catchments and watershed with high nutrient loads (Fennessy and Craft, 2011).

Restoration sites also are prioritized to optimize for flood attenuation or for multiple benefits. McAllister et al. (2000) used climate (precipitation) and landscape characteristics such as land use, soils, stream and ditch density, and property values to estimate flood attenuation by prairie pothole wetlands. Russell et al. (1997) describe an approach for selecting sites for riparian restoration in southern California. Sites determined to have the greatest potential were those with high relative potential wetness (based on elevation), large size, and proximity to existing riparian vegetation. Agricultural or barren sites with high wetness potential also were targeted, but at lower priority. Olson and Harris (1997) prioritized sites for riparian restoration by classifying riparian reaches based on land cover, degree of fragmentation, and connectivity. High-quality riparian wetlands were those with >90% riparian vegetation and were targeted for preservation. Sites with 60–90% riparian vegetation were candidate sites for restoration whereas sites with greater than 10% of irreversible (impervious) surface were not considered for restoration.

Verhoeven et al. (2008) described an approach based on operational landscape units (OLUs) to identify landscape connections and apply them to conceptualize wetland restoration in watersheds. OLUs are defined as "combinations of landscape patches with their hydrogeological and biotic connections" that take into account hydrological connections (flooding and groundwater flow paths) and biotic connections (dispersal and organism transport). An example of an OLU is a stream valley, the streambed and associated riparian zone and ranges in size from tens to hundreds of hectares. Verhoeven et al. (2008) describe a three-step approach to identify OLUs: (1) Define the restoration *target* (the plant or animal species) or wetlands function(s) to be restored; (2) Identify and analyze the spatial mechanisms in the landscape of hydrology and dispersal crucial for successful restoration; and (3) Define the location and extent (size and shape) of the OLUs needed as building blocks for reestablishment of desired plant and animal species and/or wetland functions. The OLU approach is designed to identify and prioritize restoration sites in watersheds with strongly altered hydrology and high degree of fragmentation. A key component is the identification of source areas (e.g., intact species-rich nature preserves), receptor areas suitable for restoration, and connecting pathways (hydrologic flow paths and intact riparian buffers). Such an approach, when implemented and used properly by water and natural resource managers, may go a long way to improve protection and stewardship of endangered species, biodiversity, and clean water.

A hypothetical example for prioritizing restoration in the watershed is shown in Figure 4.2. In Figure 4.2(a), the watershed is shown as it existed before restoration but contains some remnant high-quality streamside and isolated wetlands. In Figure 4.2(b), prioritization of restoration sites is focused on improving water quality. Since N and P transport from the land occurs by different mechanisms, reducing N and P

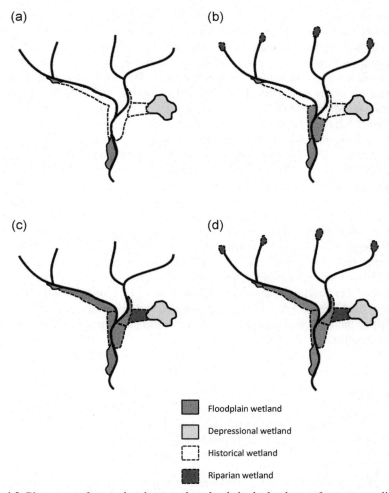

(a)   (b)

(c)   (d)

Floodplain wetland

Depressional wetland

Historical wetland

Riparian wetland

Figure 4.2 Placement of created and restored wetlands in the landscape for water quality improvement and biodiversity enhancement. (a) The existing watershed with unconnected streamside and isolated wetland fragments. (b) Creating wetlands along headwater streams for P removal and restoring floodplain wetlands lower in the watershed for N removal. (c) Creating and restoring wetlands to connect existing wetlands—streamside and isolated —to enhance biodiversity. (d) Creating and restoring wetlands for the dual benefits of water quality improvement and biodiversity enhancement.

loads requires simultaneous yet differential placement of wetlands within the watershed. Phosphorus is transported from uplands to streams largely bound to sediment, especially clay-size particles that travel limited distances (Cooper and Gilliam, 1987), so small wetlands, such as riparian buffers, are placed along headwater streams to trap sediment and P. Nitrogen movement is more diffuse and is transported largely in dissolved form such as nitrate or NH_3 and NO_x from atmospheric deposition

(Howarth et al., 1996; Goolsby et al., 1999). To remove N, larger wetlands are placed lower in the watershed, where retention times are longer, and N is removed by denitrification. Studies in Sweden support the idea of differential restoration for N versus P by focusing on P removal near the source, in headwater streams, watershed outlets, and in areas upstream of lakes (Tonderski et al., 2005) while restoring wetlands in lower reaches of the watershed for N removal (Arheimer and Wittgren, 1994, 2002; Montreuil and Merot, 2006).

In Figure 4.2(c), efforts focus on restoring wildlife habitat and biodiversity. Here wetlands are restored along stream and river corridors, connecting existing patches of intact vegetation or by enlarging remnant wetland habitat. In Figure 4.2(d), the goals of wetland restoration are twofold, improve water quality, and enhance biodiversity. Riparian wetlands are placed along headwaters to remove P. Floodplain wetlands are placed in upper and lower reaches of the watershed to provide corridors for dispersal and sites for N removal. Large wetlands are restored and connected to existing wetlands to provide larger "island" habitats for refuge, feeding, and breeding.

In summary, a landscape approach requires to: (1) take stock of the historical loss of wetlands, their types, numbers, size, and placement in the watershed; (2) determine the cumulative impacts of these losses on plants and animals and ecosystem functions and services; (3) develop a plan to restore them in types, numbers, and places where they are most likely to succeed based on on-site and landscape characteristics; and (4) prioritize their placement based on preestablished goals, potential for success, and optimum benefits.

References

Almendinger, J.E., 1999. A method to prioritize and monitor wetland restoration for water quality improvement. Wetlands Ecology and Management 6, 241–251.

Arheimer, B., Wittgren, H.B., 1994. Modeling the effects of wetlands on regional nitrogen transport. Ambio 23, 378–386.

Arheimer, B., Wittgren, H.B., 2002. Modeling nitrogen removal in potential wetlands at the catchment scale. Ecological Engineering 19, 63–80.

Bedford, B.L., 1996. The need to define hydrologic equivalence at the landscape scale for freshwater wetland mitigation. Ecological Applications 6, 57–68.

Bedford, B.L., 1999. Cumulative effects on wetland landscapes: links to wetland restoration in the United States and Canada. Wetlands 19, 775–788.

Bobbink, R., Lamers, L.P.M., 2002. Effects of increased nitrogen deposition. In: Bell, J.N.B., Treshow, M. (Eds.), Air Pollution and Plant Life. John Wiley and Sons, Chichester, England, pp. 201–235.

Bobbink, R., Hornung, M., Roelofs, J.G.M., 1998. The effects of air borne nitrogen pollutants on species diversity in natural and semi-natural European vegetation. Journal of Ecology 86, 717–738.

Brierley, G.J., Fryirs, K.A., 2005. Geomorphology and River Management. Blackwell Publishing, Malden, Massachusetts.

Brinson, M.M., 1988. Strategies for assessing the cumulative effects of wetland alteration on water quality. Environmental Management 12, 655–662.

Burke, V.J., Gibbons, J.W., 1995. Terrestrial buffer zones and wetland conservation: a case study of freshwater turtles in a Carolina Bay. Conservation Biology 9, 1365–1369.

Cedfelt, P.T., Watzin, M.C., Richardson, B.D., 2000. Using GIS to identify functionally significant wetlands in the northeastern United States. Environmental Management 26, 13–24.

Comin, F.A., Romero, J.A., Hernadez, O., Menendez, M., 2001. Restoration of wetlands from abandoned rice fields for nutrient removal, and biological community and landscape diversity. Restoration Ecology 9, 201–208.

Cooper, J.R., Gilliam, J.W., 1987. Phosphorus redistribution from cultivated fields into riparian areas. Soil Science Society of America Journal 51, 1600–1604.

Croonquist, M.J., Brooks, R.P., 1993. Effects of habitat disturbance on bird communities in riparian corridors. Journal of Soil and Water Conservation 48, 65–70.

Darveau, M., Beauchesne, P., Belanger, L., Huot, J., Larue, P., 1995. Riparian forest strips as habitat for breeding birds in boreal forest. Journal of Wildlife Management 59, 67–78.

Detenbeck, N.E., Johnston, C.A., Niemi, G.J., 1993. Wetland effects on lake water quality in the Minneapolis/St. Paul metropolitan area. Landscape Ecology 8, 39–61.

Dickson, J.G., Williamson, J.H., Conner, R.N., Ortego, B., 1995. Streamside zones and breeding birds in eastern Texas. Wildlife Society Bulletin 23, 750–755.

Fennessy, S., Craft, C., 2011. Agricultural conservation practices increase wetland ecosystem services in the glaciated Interior Plains. Ecological Applications 21, S49–S64.

Flanagan, N.E., Richardson, C.J., 2010. A multi-scale approach to prioritize wetland restoration for watershed-level water quality improvement. Wetlands Ecology and Management 18, 695–706.

Gergel, S.E., Turner, M.G., Kratz, T.K., 1999. Dissolved organic carbon as an indicator of the scale of watershed influence on lakes and rivers. Ecological Applications 9, 1377–1390.

Gergel, S.E., Turner, M.G., Miller, J.R., Melack, J.M., Stanley, E.H., 2002. Landscape indicators of human impacts to riverine systems. Aquatic Sciences 64, 118–128.

Gergel, S.E., 2005. Spatial and non-spatial factors: when do they affect landscape indicators of watershed loading? Landscape Ecology 20, 177–189.

Goolsby, D.A., Battaglin, W.A., Lawrence, G.B., Artz, R.S., Aulenbach, B.T., Hooper, R.P., Keeney, D.R., Stensland, G.J., 1999. Flux and Sources of Nutrients in the Mississippi-Atchafalaya River Basin: Topic 3 for the Integrated Assessment on Hypoxia in the Gulf of Mexico. NOAA Coastal Ocean Program Decision Analysis Series No. 17. NOAA Coastal Ocean Program, Silver Spring, MD.

Harding, J.S., Benfield, E.F., Bolstad, P.V., Helfman, G.S., Jones III, E.D.B., 1998. Stream biodiversity: the ghost of land use past. Proceedings of the National Academy of Sciences 95, 14843–14847.

Harris, R.R., 1988. Associations between stream valley geomorphology and riparian vegetation as a basis for landscape analysis in the eastern Sierra Nevada, California, USA. Environmental Management 12, 219–228.

Hemond, H.F., Benoit, J., 1988. Cumulative impacts on water quality functions of wetlands. Environmental Management 12, 639–653.

Herlihy, A.T., Stoddard, J.L., Johnson, C.B., 1998. The relationship between stream chemistry and watershed land-cover data in the mid-Atlantic region, U.S. Water, Air, and Soil Pollution 105, 377–386.

Hodges Jr., M.F., Krementz, D.G., 1996. Neotropical migratory breeding bird communities in riparian forests of different widths along the Altamaha River, Georgia. Wilson Bulletin 108, 496–506.

Hoffman, C.C., Baattrup-Pederson, A., 2007. Re-establishing freshwater wetlands in Denmark. Ecological Engineering 30, 157–166.

Howarth, R.W., Sharpley, A., Walker, D., 2002. Sources of nutrient pollution to coastal waters in the United States: implications for achieving coastal water quality goals. Estuaries 25, 656–676.

Howarth, R.W., Billen, G., Swaney, D., Townsend, A., Jaworski, N., Lathja, K., Downing, J.A., Elmgren, R., Caraco, N., Jordan, T., Berendse, F., Freney, J., Kudeyarov, V., Murdoch, P., Zaho-Lina, Z., 1996. Regional nitrogen budget and riverine N and P fluxes for drainages to the North Atlantic Ocean: natural and human influences. Biogeochemistry 35, 75–139.

Johnson, G.D., Meyers, W.L., Patil, G.P., 2001. Predictability of surface water pollution loading in Pennsylvania using watershed-based landscape measurements. Journal of the American Water Resources Association 37, 821–835.

Johnson, L.B., Richards, C., Host, G.E., Arther, J.W., 1997. Landscape influences on water chemistry in midwestern stream ecosystems. Freshwater Biology 37, 193–208.

Jones III, E.B.D., Helfman, G.S., Harper, J.O., Bolstad, P.V., 1999. Effects of riparian forest removal on fish assemblages in southern Appalachian streams. Conservation Biology 13, 1454–1465.

Jones, K.B., Neale, A.C., Nash, M.S., Van Remortel, R.D., Wickham, J.D., Riitters, K.H., O'Neill, R.V., 2001. Predicting nutrient and sediment loading to streams from landscape metrics: a multiple watershed study from the Mid-Atlantic region. Landscape Ecology 16, 301–312.

Kadlec, R.H., Wallace, S.D., 2009. Treatment Wetlands. CRC Press, Boca Raton, Florida.

Keller, C.M.E., Robbins, C.S., Hatfield, J.S., 1993. Avian communities in riparian forests of different widths in Maryland and Delaware. Wetlands 13, 137–144.

Kovacic, D.A., Twait, R.M., Wallace, M., Bowling, J., 2006. Use of created wetlands to improve water quality in the Midwest – Lake Bloomington case study. Ecological Engineering 28, 258–270.

Lamers, L.P.M., Falla, S.J., Samborska, E.M., van Dulken, I.A.R., van Hengstrum, G., Roelofs, J.G.M., 2002. Factors controlling the extent of eutrophication and toxicity in sulfate-polluted freshwater wetlands. Limnology and Oceanography 47, 585–593.

Lenat, D.R., Crawford, J.K., 1994. Effects of land use on water quality and aquatic biota of three North Carolina Piedmont streams. Hydrobiologia 294, 185–199.

Llewellyn, D.W., Shaffer, G.P., Craig, N.J., Creasman, L., Pashley, D., Swan, M., Brown, C., 1996. A decision support system for prioritization restoration sites on the Mississippi alluvial plain. Conservation Biology 10, 1446–1455.

McAllister, L.S., Peniston, B.E., Leibowitz, S.G., Abbruzzese, B., Hyman, J.B., 2000. A synoptic assessment for prioritizing wetland restoration efforts to optimize flood attenuation. Wetlands 20, 70–83.

McIver, J., Starr, L., 2001. Restoration of degraded lands in the interior Columbia River basin: passive versus active approaches. Forest Ecology and Management 153, 15–28.

Meffe, G.K., Carroll, G.R., 1997. Principles of Conservation Biology. Sinauer Associates Ltd, Sunderland, Massachusetts.

Mensing, D.M., Galatowitsch, S.M., Tester, J.R., 1998. Anthropogenic effects on the biodiversity of riparian wetlands of a northern temperate landscape. Journal of Environmental Management 53, 349–377.

Mitsch, W.J., 1992. Landscape design and the role of created, restored and natural riparian wetlands in controlling nonpoint source pollution. Ecological Engineering 1, 27–47.

Mitsch, W.J., Day Jr., J.W., 2006. Restoration of Wetlands in the Mississippi-Ohio-Missouri (MOM) River basin: experience and needed research. Ecological Engineering 26, 55–69.

Mitsch, W.J., Day Jr., J.W., Gilliam, J.W., Groffman, P.M., Hey, D.L., Randall, G.W., Wang, N., 2001. Reducing nitrogen loading to the Gulf of Mexico from the Mississippi River basin: strategies to counter a persistent ecological problem. BioScience 51, 373–388.

Montreuil, O., Merot, P., 2006. Nitrogen removal in valley bottom wetlands: assessment in headwater catchments distributed throughout a large basin. Journal of Environmental Quality 35, 2113–2122.

Moreno, D., Pedrocchi, C., Comin, F.A., Garcia, M., Gabezas, A., 2007. Creating wetlands for improvement of water quality and landscape restoration in semi-arid zones degraded by intensive agricultural use. Ecological Engineering 30, 103–111.

Moreno-Mateos, D., Mander, U., Comin, F.A., Pedrocchi, C., Uuemaa, E., 2008. Relationships between landscape pattern, wetland characteristics, and water quality in agricultural catchments. Journal of Environmental Quality 37, 2170–2180.

Naiman, R.J., Decamps, H., 1997. The ecology of interfaces: riparian zones. Annual Review of Ecology and Systematics 28, 621–658.

National Research Council (NRC), 2001. Compensating for Wetland Losses under the Clean Water Act. National Academy Press, Washington, D.C.

National Research Council (NRC), 1992. Restoration of Aquatic Ecosystems. National Academy Press, Washington, D.C.

Newbold, S.C., 2005. A combined hydrologic simulation and landscape design model to prioritize sites for wetlands restoration. Environmental Modeling and Assessment 10, 251–263.

Olson, C., Harris, R., 1997. Applying a two-stage system to prioritize riparian restoration at the San Luis Rey River, San Diego County, California. Restoration Ecology 5, 43–55.

Osborne, L.L., Wiley, M.J., 1988. Empirical relationships between land use/cover and stream water quality in an agricultural watershed. Journal of Environmental Management 26, 9–27.

Paerl, H.W., 1993. Emerging role of atmospheric nitrogen deposition in coastal eutrophication: biogeochemical and trophic perspectives. Canadian Journal of Fisheries and Aquatic Sciences 50, 2254–2269.

Paerl, H.W., Whitall, D.R., 1999. Anthropogenically-derived atmospheric nitrogen deposition, marine eutrophication and harmful algal bloom expansion: Is there a link? Ambio 28, 307–311.

Patten, D.T., Harpman, D.A., Voita, M.J., Randle, T.J., 2001. A managed flood on the Colorado River: background, objectives, design, and implementation. Ecological Applications 11, 635–643.

Paul, M.J., Meyer, J.L., 2001. Streams in urban landscapes. Annual Review of Ecology and Systematics 32, 333–365.

Peterjohn, W.T., Correll, D.L., 1984. Nutrient dynamics in an agricultural watershed: observations on the role of a riparian forest. Ecology 65, 1466–1475.

Preston, E.M., Bedford, B.L., 1988. Evaluating cumulative effects on wetland functions: a conceptual overview and genetic framework. Environmental Management 12, 565–583.

Richardson, C.J., Flanagan, N.E., Ho, M., Pahl, J.W., 2011. Integrated stream and wetland restoration: a watershed approach to improved water quality on the landscape. Ecological Engineering 37, 25–39.

Russell, G.D., Hawkins, C.P., O'Neill, M.P., 1997. The role of GIS in selecting sites for riparian restoration based on hydrology and land use. Restoration Ecology 5, 56–68.

Semlitsch, R.D., 1998. Biological delineation of terrestrial buffer zones for pond-breeding salamanders. Conservation Biology 12, 1113–1119.

Semlitsch, R.D., Bodie, J.R., 1998. Are small, isolated wetlands expendable? Conservation Biology 12, 1129–1133.

Semlitsch, R.D., Scott, D.E., Pechmann, J.H.K., Gibbons, J.W., 1996. Structure and dynamics of an amphibian community: evidence from a 16-year study of a natural pond. In: Cody, M.L., Smallwood, J.A. (Eds.), Long-term Studies of Vertebrate Communities. Academic Press, New York, pp. 217–247.

Shabman, L.A., 1995. Making watershed restoration happen: what does economics offer? In: Cairns Jr., J. (Ed.), Rehabilitating Damaged Ecosystems. Lewis Publishers, Boca Raton, Florida, pp. 35–47.

Simenstad, C.A., Reed, D., Ford, M., 2006. When is restoration not? Incorporating landscape-scale processes to restore self-sustaining ecosystems in coastal wetland restoration. Ecological Engineering 26, 27–39.

Simenstad, C.A., Hood, W.G., Thom, R.M., Levy, D.A., Bottom, D.L., 2000. Landscape structure and scale constraints on restoring estuarine wetlands for Pacific coast juvenile salmon. In: Weinstein, M.P., Kreeger, D.A. (Eds.), Concepts and Controversies in Tidal Marsh Ecology. Kluwer Academic Publishers, Dordrecht, The Netherlands, pp. 597–630.

Snodgrass, J.W., Bryan, A.L., Lide, R.F., Smith, G.M., 1996. Factors affecting the occurrence and structure of fish assemblages in isolated wetlands of the upper coastal plain, USA. Canadian Journal of Fisheries and Aquatic Sciences 53, 443–454.

Snodgrass, J.W., Komoroski, M.J., Bryan Jr., A.L., Burger, J., 2000. Relationships among isolated wetland size, hydroperiod and amphibian species richness: implications for wetland regulation. Conservation Biology 14, 414–419.

Stauffer, D.F., Best, L.B., 1980. Habitat selection by birds of riparian communities: evaluating effects of habitat alterations. Journal of Wildlife Management 44, 1–15.

Stein, E.D., Ambrose, R.F., 2001. Landscape-scale analysis and management of cumulative impacts to riparian ecosystems: past, present, and future. Journal of the American Water Resources Association 37, 1597–1614.

Stevens, L.E., Ayers, T.J., Bennett, J.B., Christensen, K., Kearsley, M.J.C., Meretsky, V.J., Phillips III., A.M., Parnell, R.A., Spence, J., Sogge, M.K., Springer, A.E., Wegner, D.L., 2001. Planned flooding and Colorado river trade-offs downstream from Glen Canyon dam, Arizona. Ecological Applications 11, 701–710.

Teal, J.M., Peterson, S., 2005. Restoration benefits in a watershed context. Journal of Coastal Research 40, 132–140.

Tonderski, K.S., Arheimer, B., Pers, C.B., 2005. Modeling the impact of potential wetlands on phosphorus retention in a Swedish catchment. Ambio 34, 544–551.

Twedt, D.J., Uihlein III., W.B., Elliot, A.B., 2006. A spatially explicit decision support model for reforestation of bird habitat. Conservation Biology 20, 100–110.

US EPA (U.S. Environmental Protection Agency), 2008. Handbook for Developing Plans to Restore and Protect Our Waters. EPA 841-B-08–002. U.S. EPA, Washington, D.C. http://water. epa.gov/polwaste/nps/upload/2008_04_18_NPS_watershed_handbook_handbook-2.pdf (accessed 26.03.15.).

USEPA (U.S. Environmental Protection Agency), 2013. EPA Region 5 Wetlands Supplement: Incorporating Wetlands into Watershed Planning. U.S. EPA, Washington, D.C.

Vannote, R.L., Minshall, G.W., Cummins, K.W., Sedell, J.R., Cushing, C.E., 1980. The river continuum concept. Canadian Journal of Fisheries and Aquatic Sciences 37, 130–137.

Verhoeven, J.T.A., Yin, B.C., Hefting, M.M., 2006. Regional and global concerns over wetlands and water quality. Trends in Ecology and Evolution 21, 96–103.

Verhoeven, J.T.A., Soobs, M.B., Jannsen, R., Omtzigt, N., 2008. An operation landscape unit approach for identifying key landscape connections in wetland restoration. Journal of Applied Ecology 45, 1496–1503.

Wang, L., Lyons, J., Kanehl, P., Gatti, R., 1997. Influences of watershed land use on habitat quality and biotic integrity in Wisconsin streams. Fisheries 22, 6–12.

Weller, D.E., Jordan, T.E., Correll, D.L., 1998. Heuristic models for material discharge from landscapes with riparian buffers. Ecological Applications 8, 1156–1169.

White, D., Fennessy, S., 2005. Modeling the suitability of wetland restoration potential at the watershed scale. Ecological Engineering 24, 359–377.

Winter, T.C., 1986. Effects of ground-water recharge on configuration of the water table beneath sand dunes and on seepage in lakes in the sandhills of Nebraska, USA. Journal of Hydrology 86, 221–237.

Winter, T.C., 2001. The concept of hydrologic landscapes. Journal of the American Water Resources Association 37, 335–349.

Winterhalder, K., 1996. Environmental degradation and rehabilitation of the landscape around Sudbury, a major mining and smelting area. Environmental Reviews 4, 185–224.

Yan, W., Yin, C., Tang, H., 1998. Nutrient retention by multipond systems: mechanisms for the control of nonpoint source pollution. Journal of Environmental Quality 27, 1009–1017.

Yin, C., Shan, B., 2001. Multipond systems: a sustainable way to control diffuse phosphorus pollution. Ambio 30, 369–375.

Yin, C., Zhao, M., Jin, W., Lan, Z., 1993. A multi-pond system as a protective zone for the management of lakes in China. Hydrobiologia 251, 321–329.

Zedler, J.B., 2003. Wetlands at your service: reducing impacts of agriculture at the watershed scale. Frontiers in Ecology and the Environment 1, 65–72.

Zedler, J., 2006. Taking a Watershed Approach in the Absence of a Watershed Plan. Presentation at the Fifth Stakeholder's Forum of Federal Wetlands Mitigation. Environmental Law Institute, Washington, D.C.

Part Two

Restoration of Freshwater Wetlands

Inland Marshes

Chapter Outline

Introduction

Inland marshes are a wide-ranging group of wetlands but their common element is dominance of herbaceous emergent vegetation, mostly grasses (Poaceae), sedges (Cyperaceae), and rushes (Juncaceae) in varying mixtures and proportions. Soils are highly variable, ranging from mineral soils to soils rich in organic matter and even peat. Landscape position and the source of water, precipitation, surface water, and groundwater, also vary. As a result, hydroperiod (depth, duration, and frequency of inundation) and water chemistry vary among them. In North America, inland marshes include prairie potholes of the north central US and Canada, playas of the southern plains, limesink wetlands of the southeastern US, marshes of the Great Lakes, and near coast marshes of the Mississippi River delta. Worldwide, the Pantanal of South America, Okavango Delta of southern Africa, and reed (*Phragmites*) marshes of the Danube River and Volga River deltas of Europe are well known examples of inland marshes.

Many inland marshes are found in climates where rainfall occurs seasonally, i.e., they have a distinct wet season and dry season. In drier climates where grasslands dominate the terrestrial landscape and where forest vegetation is sparse to absent, fire is important in maintaining marsh vegetation. Inland marshes also are common along lakeshores where water level fluctuations limit colonization by trees. On the Great Lakes of the US and Canadian border, marshes are found in protected bays and other sheltered areas. In many regions, marshes persist by artificial disturbance such as grazing or mowing. Essentially, inland marshes can be found just about anywhere and they are maintained in an intermediate stage of succession by disturbances such as fluctuating water levels, fire, grazing, or mowing.

Creating and Restoring Wetlands. http://dx.doi.org/10.1016/B978-0-12-407232-9.00005-1

Inland marshes occupy a variety of landscape positions and receive water from a number of sources. Lakeshore and riverine marshes have strong surface water connections and their hydroperiod is controlled by these connections. Hydrology of riverine marshes is driven by the seasonality of river pulsing whereas hydroperiod of lakeshore marshes is controlled by lake levels that vary seasonally with precipitation, including snowmelt and temperature/growing season (Figure 5.1). In contrast, marshes in depressions have weak surface water connections and receive most of their water from precipitation that varies seasonally or from groundwater whose supply is more constant.

Water source and soil determine the chemistry and fertility of inland marshes. Riverine marshes often receive substantial amounts of sediment, base cations (Ca, Mg, K), and

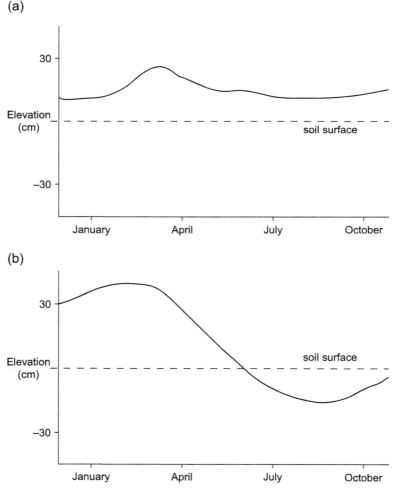

Figure 5.1 Hydroperiod—depth, duration, frequency of inundation—of (a) lakeshore marsh (Great Lakes) and (b) depressional marsh (northwestern Indiana).

phosphorus (P) during overbank flooding. Marshes that receive most of their water from precipitation are relatively low in nutrients (Table 5.1). In particular, base cations are considerably lower in precipitation than in stream or river water. The concentration of bicarbonate in floodwaters depends on underlying soils and geology with high concentrations in limestone and karst landscapes. In arid and semiarid regions, marshes may be inundated with water of varying salinity and ionic composition including Na, Cl, and SO_4.

Like all wetland plants, vegetation of inland marshes is adapted to periodic to continuous inundation or saturation of soils with water. Plants, like animals, need oxygen to support respiration to produce energy for cell growth and maintenance and oxygen needs to be supplied to the roots to sustain belowground respiration processes. Adaptations of herbaceous vegetation to waterlogging include stems and leaves containing aerenchyma, large intercellular spaces that transport oxygen-rich air to belowground tissues (Burdick and Mendelssohn, 1990), pressurized ventilation whereby differences in temperature between leaf stomata and roots force fresh air down leaf petioles into the roots and waste gases out (Dacey, 1980; Brix et al., 1992), adventitious roots to take oxygen directly out of the water column (Wample and Reid, 1979; Jackson, 1985), and radial oxygen loss that oxygenates the soils in the immediate vicinity of the root (Teal and Kanwisher, 1966; Mendelssohn and Postek, 1982). Marsh plants also have metabolic adaptations that enable them to survive periods of anoxia. These adaptations involve a switch from aerobic to anaerobic respiration with the production of acetaldehyde and ethanol as intermediate and end products, respectively, as opposed to CO_2 (Mendelssohn et al., 1981). Acetaldehyde and ethanol are toxic if they accumulate in the plant tissues, but many wetland plants, including rice (*Oryza sativa*) and *Spartina*, possess the enzyme, alcohol dehydrogenase that catalyzes the conversion to ethanol which diffuses out of the roots (Bertani et al., 1980).

Table 5.1 **Concentration of Selected Ions (mg/L) in Precipitation versus River Water**

	Precipitation[a]	Stream/Fresh River Water[b]
Sodium	0.1–5	1–6
Potassium	<0.1–0.3	0.2–2
Calcium	0.2–1.2	1.6–15
Magnesium	<0.1–0.3	0.4–4
Chloride	0.1–9	0.5–8
Sulfate	1.3–2.9	6.3–11
Bicarbonate	<0.1	0.9–58
PO_4^{3+}	<0.01–0.02	0.02
NH_4^+	0.2–0.22	0.04
NO_3^-	0.3–1.5	1–2

[a]From Gorham (1961), Likens and Bormann (1977), and Schlesinger (1978).
[b]From Livingstone (1963), Likens and Bormann (1977), and Meybeck (1979).

Though inland marshes are found in a diversity of climates, landscapes, and soil types, vegetation is surprisingly uniform and dominated by cosmopolitan genera such as *Typha, Phragmites, Juncus, Phalaris, Carex, Cyperus* and other grasses, sedges and rushes (Table 5.2). Marshes, especially those that dry down during the growing season, also possess high diversity of flowering plants, including those in the Asteraceae (*Bidens* sp.), Polygonaceae (*Polygonum* sp.), and Onagraceae (*Ludwigia* sp.) among others. These species flower at different times of the growing season and produce brightly colored flowers that attract pollinators such as butterflies, bees, and other insects.

Vegetation of inland marshes exhibits strong patterns of zonation corresponding to differences in hydroperiod and timing of flooding that is determined by elevation relative to the changing water levels (Figure 5.2). These zones often are visible as rings or bands around water bodies such as lakes and ponds. *Typha, Phragmites,* and *Juncus* grow in standing water and often are found in the deepest portion of the wetland. Species of *Scirpus, Schoenoplectus,* and *Carex* are found at higher elevations in the marsh, in saturated but not flooded soils. At even higher elevations that dry down longer, *Leersia* sp. (cutgrass) and asters may predominate. At the marsh edge, soils may not be flooded or saturated long enough to support hydrophytic vegetation, and here upland grasses, shrubs, or trees may dominate. Marsh vegetation also varies temporally through the growing season. Phenology, the timing of plant transformations such as leaf out, maximum growth, flowering, and seed production, vary among species. For example, species such as *Pontedaria, Sagittaria,* and *Schoenoplectus* flower in early summer and set seed and senesce by midsummer. Other species, often annuals (e.g., *Bidens*), flower, set seed, and senesce toward the end of the growing season. Thus, the inland marsh is a mosaic of species and plant communities with distinct spatial and temporal properties.

Table 5.2 Common Plants of Inland Freshwater Marshes

Family	Genus
Alismataceae	*Alisma, Echinodorus, Sagittaria*
Amaranthaceae	*Amaranthus*
Cyperaceae	*Carex, Cladium, Cyperus, Scirpus, Schoenoplectus, Eleocharis, Glyceria*
Juncaceae	*Juncus* (needlerush)
Nymphaeaceae	*Nymphaea, Nuphar*
Nelumbonaceae	*Nelumbo* (lotus)
Onagraceae	*Ludwigia* (primrose)
Poaceae	*Calamagrostis, Leersia, Oryza sativa* (domesticated rice), *Panicum, Paspalum, Phalaris, Phragmites, Spartina, Sporobolus, Zizania* (wild rice), *Zizaniopsis, Distichlis,* many others
Pontedariaceae	*Eichhornia, Pontedaria*
Polygonaceae	*Polygonum* (knotweed)
Sparganiaceae	*Sparganium*
Typhaceae	*Typha* (cattail)

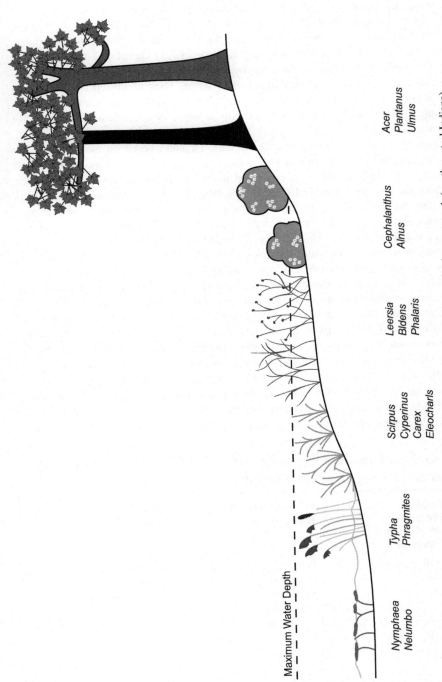

Figure 5.2 Zonation of plant communities along an elevation/hydroperiod gradient in a freshwater marsh (south central Indiana).

Several classification systems have been developed for inland marshes based primarily on hydroperiod (Table 5.3). Shaw and Fredine (1956) delineated six types of marshes. Five were based on hydrology, seasonally flooded, meadows, shallow marsh, deep marsh, and open water. Stewart and Kantrud (1971) identified correspondingly similar marshes in their classification scheme: wetland-low prairie, wet meadow, shallow marsh, deep marsh, and permanent open water. They identified two additional communities, intermittent alkali and fens (alkaline bogs), that are brackish to saline in nature. Both classification systems employ a saline or alkali class to account for salinity.

Euliss et al. (2004) developed a classification system based on two attributes of hydrology, relation to groundwater (discharge vs recharge) and atmospheric water (drought vs deluge), to describe plant and animal communities of inland marshes. The *Wetland Continuum* of Euliss et al. describes 16 plant communities that occur along the two axes. At the deluge end of the continuum, marsh vegetation consists of open water or submerged aquatic habitat. At the drought end, terrestrial and wetland annuals predominate. Along the groundwater axis, discharge wetlands, i.e., those wetlands in which groundwater discharges into, are characterized by open water or submerged wetland perennials. During drought, wetland annuals dominate the marsh. Recharge wetlands, those wetlands whose water recharges groundwater, are drier than discharge wetlands. During drought, recharge wetlands are dominated by terrestrial annuals. During deluge, they consist of open water or submersed wetland perennials. Euliss et al. (2004) further describes how invertebrate, amphibian, and bird communities respond to changing water levels. During drought, terrestrial invertebrates, passive dispersers with drought-resistant eggs or cysts, and organisms with r-selected traits for reproduction such as short life cycle and large number of offspring with low survivorship, dominate. During deluge, organisms with aquatic traits such as algae, amphipods, and amphibians, and active dispersers and K-strategists that have longer life cycles and produce fewer offspring with greater survivorship, dominate the marsh.

Table 5.3 Two Classification Schemes for "Fresh" Water Marshes, Based on Hydroperiod and Salinity

Shaw and Fredine (1956)	Stewart and Kantrud (1971)	Dominant Vegetation
Hydroperiod		
Seasonally flooded	Low prairie	Annuals, grasses
Meadows	Wet meadow	Sedges, grasses
Shallow	Shallow marsh	Cutgrass, knotweed
Deep	Deep marsh	Cattails, rushes
Open	Permanent	Open water with littoral zone
Salinity		
Saline	Alkali	Widgeon grass (*Ruppia*)
	Fen (alkaline bog)	

Many inland marshes exhibit cycles of drought and flood that occur over a period of years to decades. Voigts (1976) followed changes in invertebrate communities over a 5- to 20-year cycle of drought and flood in marshes of the prairie pothole region. During drought, invertebrate communities were dominated by isopods. As the climate moistened and water levels increased, emergent vegetation colonized the marsh along with amphipods, chironomid larvae, and other insect larvae. In the flood or deluge stage, floating aquatic plants replaced emergent vegetation and copepods, that swim, were dominant (Figure 5.3).

In arid and semiarid regions such as western North America, eastern Europe–Mediterranean, Middle East, and inland China (Chapman, 1974), inland marshes are inundated or saturated with brackish or saline water. The chemical composition of inland saline waters is highly variable but, overall, it differs from seawater by containing less Na and Cl and more SO_4^{2-}, Ca, and Mg (Chapman, 1974). Like seawater, pH of floodwaters of inland saline marshes is higher than that of freshwater marshes and often is greater than 7. Stewart and Kantrud (1972) classified six types of saline marshes in the prairie pothole region: fresh, slightly brackish, moderately brackish, brackish, subsaline, and saline. In their classification system (see Chapter 2, Definitions), the U.S.

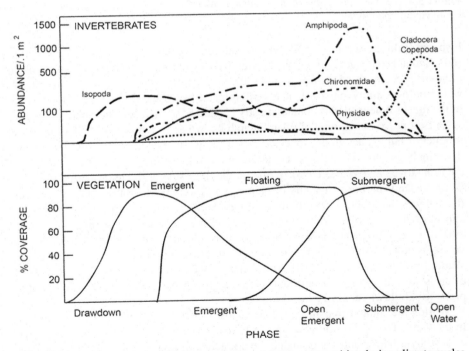

Figure 5.3 Changes in inland marsh plant and invertebrate communities during climate cycles of drought and flood.
Redrawn from Voigts, D.K., April 1976. Aquatic invertebrate abundance in relation to changing marsh vegetation. American Midland Naturalist 95 (2), 313–322. Reproduced with permission of American Midland Naturalist.

Figure 5.4 Drainage tile emptying into a ditch, Benton County, Indiana.
Photo credit: Todd Royer.

Fish and Wildlife Service (USFWS) uses salinity to categorize freshwater (and estuarine) wetlands (Cowardin et al., 1979). Wetlands are classified as fresh (<0 ppt salinity),
oligo- (0.5–5), meso- (5–18), poly- (18–30), eu- (30–40), and hypersaline (>40). Floodwaters of low salinity marshes tend to be dominated by Ca and CO_3 whereas floodwaters
of the most saline wetlands contain mostly Na and SO_4^{2-} (LaBaugh, 1989).

In Europe, a special type of inland marsh known as wet grassland exists that is the
product of human activities (Joyce, 2014). They were created by clearing forests and
draining bogs and wet marshes and are maintained by grazing or mowing for hay production (Joyce and Wade, 1998). While on the drier side of the hydroperiod gradient,
wet grasslands harbor a diversity of plants, birds, and invertebrates, especially butterflies (Joyce and Wade, 1998). In the past half century, many wet grasslands have disappeared due to agricultural intensification and others to abandonment (Luoto et al.,
2003; Joyce, 2014). On abandoned sites, restoration of these habitats and their biodiversity is done by reintroducing grazing and/or mowing (Luoto et al., 2003; Joyce and
Wade, 1998; Schaich et al., 2010; Joyce, 2014). The pace of restoration of the systems
often is slowed by residual nutrients in the soil, especially P, from past fertilization
practices (Tallowin et al., 1998).

Worldwide, inland marshes have been lost mostly by conversion to agriculture.
In the Corn Belt of the Midwest US, vast areas of marsh and forested wetlands were
drained by ditching and tile drainage—the practice of laying subsurface tiles or pipes
at shallow (1–1.5 m) depths to convey water off the land (Figure 5.4). In Indiana,
for example, more than 19,000 km^2 (87%) of wetlands were lost since the arrival of
European colonists (Dahl, 1990). In the wheat growing regions of the north central
US and central Canada, the most important waterfowl breeding areas for waterfowl
in North America (Batt et al., 1989), losses were less but still considerable. In the

Figure 5.5 (a) A recently plugged ditch in an agricultural field to restore hydrology. (b) The same site 2 years later.

prairie pothole region of North Dakota, nearly 10,000 km² (49%) of marsh wetlands were drained for agriculture during the past 200 years (Dahl, 1990). Today, the rate of wetland loss in the US has slowed (Dahl, 2011) but elsewhere drainage of wetlands continues as human population grows and global demand for food increases. For example, large freshwater marshes in the Sanjiang Plain of northeastern China have been converted to rice, corn, soybeans, and other agricultural commodities in the past 50 years (Figure 5.5), similar to what occurred in the US more than a century ago.

Theory

Succession and ecosystem development of most marshes is driven by allogeneic factors such as hydroperiod, salinity, in the case of inland saline marshes, and disturbance (fire, grazing, mowing) that maintain the herbaceous vegetation against encroachment by woody vegetation. Successional dynamics is best described by the Gleasonian rather than the Clementsian approach (van der Valk, 1982). According to van der Valk (1981), life history traits such as lifespan, propagule longevity, and environmental requirements of the seedbank determine the replacement of species during the cycles of flooding and drying with no climax community evident (van der Valk, 1982).

Hydroperiod is the primary driver of marsh structure and this is especially true in regions with unstable (drought, deluge) climates (Weller, 1978). In these climates, over a period of years, the marsh may transition from vegetated wetland to open water to dry down and is accompanied by a shift from emergent vegetation to floating and submersed vegetation, and eventually to an ecosystem dominated by terrestrial annuals (see Figure 5.3). The dry down phase is critical for germination of wetland seeds (Weller, 1978). It also enhances nutrient cycling processes by oxidizing organic matter and detritus that built up. In marshes restored for wildlife habitat, water levels are manipulated to simulate the effects of natural drawdown cycles. Weller (1978) states that "stability seems deadly to a marsh system" and that fluctuating water levels lead to more productive marshes. Birds also benefit from the ever-changing water levels and habitats as the wet–dry cycles produce an array of habitats and food sources for avian species (Murkin et al., 1997).

Fire also is an important agent of disturbance, maintaining herbaceous vegetation against encroachment by shrubs and trees, especially in regions where precipitation is low or falls seasonally or where drought is a frequent occurrence. Furthermore, herbaceous vegetation when dry is highly combustible, yet, after a burn, resprouts quickly. Fire-adapted marshes include depressional wetlands of the southeastern US, wetlands in prairie landscapes, freshwater marshes in coastal Louisiana, and even peat-forming marshes such as the Florida Everglades. Fire favors establishment and growth of marsh vegetation by reducing competition from woody species and encourages flowering. The benefits of fire include increased light and nutrient availability, especially base cations and P, and warmer soil temperatures that promote seed germination (Christensen, 1977, 1993; Whelan, 1995).

Like fire, grazing promotes growth of herbaceous vegetation. Domesticated animals such as goats, cows, sheep, and horses grazed in marshes for millennia. Some studies suggest that grazing may increase species diversity and the formation of complex distribution patterns and sharper boundaries between vegetation zones (Bakker and Ruyter, 1981). In many parts of the world, overgrazing has led to degradation and denudation of marsh vegetation. Native herbivores, including geese and muskrat (*Ondatra zibethicus*), also create disturbances by feeding (Lynch et al., 1947; O'Neil, 1949; Weller, 1994).

When food is abundant, muskrats reproduce quickly and may consume large areas of vegetation for food and to build lodges. Nutria (*Myocastor coypus*), introduced from South America and similar to muskrat but larger (Kinler et al., 1987), also may denude large areas of freshwater and brackish marsh when populations are large and environmental conditions are favorable (Odum et al., 1984; Weller, 1994). Nutria is common in Gulf Coast states such as Louisiana, but populations also exist in the Chesapeake Bay, Pacific Northwest (Kinler et al., 1987), and in at least 20 states and parts of Canada (Evans, 1970). Furthermore, they were introduced to the Union of Soviet Socialist Republics (USSR) and are established in reed (*Phragmites*) marshes of the Caspian Sea (Aliev, 1965). Similar to muskrat and nutria, grazing by lesser snow geese (*Chen caerulescens carulescens*) has led to widespread destruction of marshes in Hudson Bay, Canada (Jefferies et al., 2006). Furbearers and geese feed preferentially on growing shoots, rootstocks, and tubers that contain large stores of nutrients and carbohydrates needed by the plants to recover from disturbances like herbivory and fire (Lynch et al.,

1947; Weller, 1994). They prefer to feed on oligohaline and brackish plant species such as three-square (*Scirpus olneyi*), cattail (*Typha*) (Lynch et al., 1947; Palmisano, 1972; Kinler et al., 1987). O'Neil (1949) reported that muskrat "eat-outs" were more likely to occur in marshes dominated by three-square than in marshes dominated by other plant species. Eat-outs can cause the root mat to disintegrate, creating unconsolidated, mucky soils (O'Neil, 1949), and threaten the long-term stability of coastal marshes.

Mowing is used to restore and maintain wet grasslands of Europe (Joyce and Wade, 1998) and elsewhere. These habitats are hotspots of biodiversity in the old and highly managed landscapes of the European continent. Periodic mowing increases plant species richness and aboveground biomass (Straskrabova and Prach, 1998) and nurse plants sown with the introduced seed mix have been shown to enhance establishment of target species (Manchester et al., 1998).

Practice

Restoration of inland marshes typically consists of (1) creating shallow water or moist soil impoundments for waterfowl, (2) restoring wetland hydrology on agricultural land by plugging drainage ditches or breaking tile drains, and (3) offsetting environmental damages to existing wetlands through mitigation. Many early restoration projects involved the creation of impoundments for waterfowl resting, breeding, and feeding. In the US, the USFWS constructed moist soil impoundments on many national wildlife refuges beginning in the 1930s. The rationale behind these projects was to support healthy duck populations for hunting. Likewise, Ducks Unlimited, a nongovernmental agency based in Canada, has worked with farmers and other land owners in the prairie pothole region and elsewhere for many years to restore marshes (Lietch, 1989). In 2013, Ducks Unlimited restored or enhanced more than 50,000 acres of wetlands in North America (fact sheet www.ducks.org).

Many inland marshes in the US are restored through government-sponsored programs of the U.S. Department of Agriculture (USDA). Large areas of cultivated land in the US historically were wetlands but were drained for agriculture in the nineteenth and twentieth centuries. The highly productive agricultural lands in the Corn Belt of the Midwest and northern plains were ditched and tiled and the organic rich soils were tilled for corn, soybeans, and wheat. Peatlands in the San Francisco Bay delta and the Florida Everglades were drained and today provide winter vegetables for much of North America. With the passage of the Clean Water Act and other wetland protection measures in the 1970s and 1980s, monies from the USDA Conservation Reserve program (CRP) and Agricultural Wetland Reserve (WRP) program were available to restore wetlands on these lands.

Restoring Hydrology

Restoring the natural patterns of inundation is the critical first step to reintroduce marsh vegetation. Many wetlands restored for waterfowl habitat have control structures to regulate water levels. Water levels may be adjusted to promote desired vegetation, facilitate germination of wetland species, oxidize accumulated organic matter and plant detritus, and control nuisance plant and animal species. Winter drawdown is used in temperate and cool climates to kill undesirable submerged aquatic vegetation such as milfoil (*Myriophyllum* sp.) and reduce populations of animals such as muskrat and carp (*Cyprinus carpio*) (Weller, 1994). Ideally, drawdown of water levels should take place in the same way and at the same time as natural marshes in the region. In natural marshes, drawdown typically occurs later in the growing season when the combination of warm temperatures and high evapotranspiration leads to lowering of water levels and drying of soils that promotes germination of seeds of wetland and other plant species.

Typically, marsh restoration on agricultural lands is achieved by plugging ditches (Figure 5.5) or breaking tile drains (see Figure 5.4). Sometime ditches are entirely filled but the more common and cost-effective method is to plug the ditch at its outlet. In the Midwest US and elsewhere, drainage tiles, usually made of perforated plastic pipe, are buried in the soil in a grid or network to quickly collect and carry surface water off site. Sometimes, tile drains are not buried in a systematic way, so it can be difficult to find (and break) them.

Reestablishing Vegetation

Reestablishing marsh vegetation may be passive, letting natural colonization or self-design occur (Galatowitsch and van der Valk, 1996b; Mitsch and Wilson, 1996; Mitsch et al., 1998), or it may be active, involving site preparation, direct seeding, spreading stockpiled fresh topsoil containing propagules (Erwin et al., 1994; Parikh and Gale, 1998), or by planting. On former agricultural land, site preparation may involve removal of existing vegetation and seedbank by herbicidal treatment, topsoil removal, or solarization (i.e., covering with black sheeting) (Pfeifer-Meister et al., 2012a,b) followed by seeding. Sowing seed is less common in Europe, where the focus mostly is on restoring the natural hydrology and water chemistry (Boers et al., 2006). On large sites, it may not be practical to seed or plant and one must rely on natural colonization. However, natural colonization is highly stochastic and may lead to plant communities that are less desirable than using active methods. Streever and Zedler (2000) state that until we better understand the dynamics of plant colonization, which species will establish naturally and which ones will not, planting should be required unless strong evidence suggests that it is not necessary for a particular project.

For natural colonization, species whose seeds are dispersed by wind will make it to the site in large numbers and many will germinate. Propagules of many annual species are dispersed by wind and they are among the first to colonize. Noon (1996) sampled eight created marshes that ranged in age from 2 to 11 years and found that, during the first 3 years, annuals dominated the marsh. Older marshes were increasingly dominated by perennials and percent cover of vegetation increased with age (Noon, 1996). Wind dispersal is common in many aggressive species, including the perennials *Typha*, *Phalaris*, *Phragmites*, and *Lythrum* (purple loosestrife) (Galatowitsch et al., 1999; Zedler and Kercher, 2004) that often come to dominate created and restored marshes (Campbell et al., 2002; Fink and Mitsch, 2007), especially those that colonize naturally (Reinartz and Warne, 1993; Combroux et al., 2002). Seeding or especially planting has been shown to increase plant diversity and slow invasion (Reinartz and Warne, 1993; Streever and Zedler, 2000; Mahaney et al., 2004). This is not surprising in light of the *initial floristics theory* of Egler (1954) (see Chapter 3, Ecological Theory and Restoration) that hypothesizes that the development of the mature plant community is determined in large part by the species that colonize first.

Many desirable species including *Carex*, *Cyperus* and other sedges do not readily disperse and generally will not colonize the site without deliberate introduction (Budelsky and Galatowitsch, 2000). Kettenring and Galatowitsch (2011b) reported that the seed rain of restored and natural prairie pothole marshes was dominated by annuals and invasive perennial species (Figure 5.6), with *Carex* species contributing very little to the overall seed rain. Furthermore, van der Valk et al. (1999) reported that *Carex* seeds do not remain viable for very long, only a matter of months (Figure 5.7(a)), and that germination is very low, less than 1% (Figure 5.7(b)). They suggested that establishing *Carex* from seed is maximized by using fresh seed, maintaining high soil moisture levels, and increasing soil organic matter (SOM) content to levels found in natural sedge meadows. Experimental studies indicate that seeds of different *Carex* species have varying germination requirements for light (Kettenring et al., 2006), temperature (Kettenring and Galatowitsch, 2007a), and stratification (Kettenring and Galatowitsch, 2007b, 2011a). Seed germination of many *Carex* species is the greatest following wet-cold or moist-cold stratification (Budelsky and Galatowitsch, 1999). Collectively, these studies suggest that *Carex* colonization of restored marshes is limited by the seed rain and germination requirements and that seeding and/or planting is required to reintroduce them.

Mowing often is used to increase plant species diversity and growth of desirable species. This practice is widely used in Europe to restore wetland grasslands (Joyce and Wade, 1998; Joyce, 2014). Berg et al. (2012) evaluated the effects of mowing on

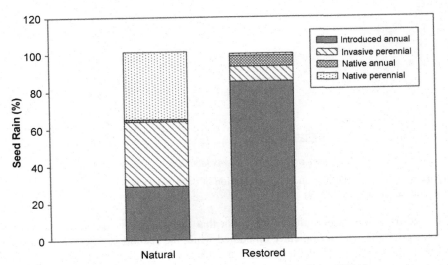

Figure 5.6 Seed rain of natural and restored prairie wetlands in Iowa (USA). From Kettenring and Galatowitsch (2011a).

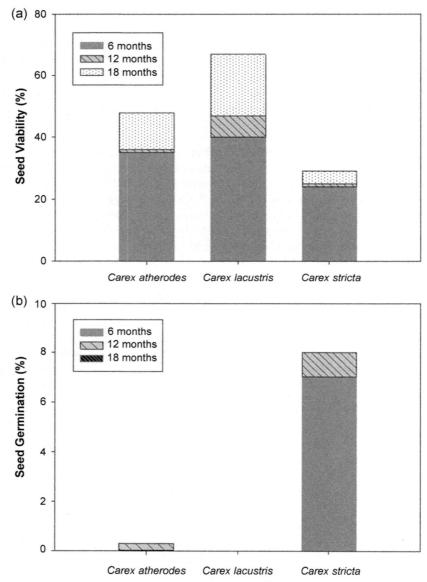

Figure 5.7 (a) Seed viability and (b) germination of three *Carex* species.
From van der Valk et al. (1999).

two wet grassland communities in Estonia that were abandoned 40 years ago. One species-poor site was periodically inundated while the second site, with high species richness, was infrequently flooded. During the 5-year study, the periodically inundated site exhibited greater change in species composition as mowing favored expansion of short-statured species (*Festuca* spp.) at the expense of taller species such as

Phragmites. No increase in overall species richness was evident during the 5 years at either site and the authors suggested that more time, perhaps 10 years, was needed to detect a measurable increase in diversity. Joyce (2014) suggested that for wet grasslands, restoration efforts should be directed toward sites that are treeless, low in soil nutrients, and that have been abandoned less than 20 years.

Restoring and Recreating Soils

A common feature of restored and especially created marsh soils is that they contain less organic matter and N and have higher bulk density and more coarse particles (gravel, sand) than natural marshes (Bischel-Machung et al., 1996; Galatowitsch and van der Valk, 1996c; Noon, 1996; Whittecar and Daniels, 1999; Nair et al., 2001; Brooks et al., 2005; Lu et al., 2007; Gutrich et al., 2009; Hossler and Bouchard, 2010; Marton et al., 2013a,b). Nair et al. (2001) suggested that, for wetlands created on upland sites, wetland soil development could be accelerated by minimizing soil compaction during construction/excavation and incorporating fertilizer and organic matter. Excavation activities also tend to homogenize the spatial variability in soil properties. Several studies report that microtopography is absent from many restoration sites (Whittecar and Daniels, 1999; Hoeltje and Cole, 2009). Microtopography improves wetland functions such as water retention and biogeochemical cycling (e.g., denitrification) by creating areas of aerobic and anaerobic conditions that are in close proximity to each other (Moser et al., 2009). Disking, the practice of mixing the topsoil layer to remove existing vegetation, also enhances species diversity and abundance of hydrophytic species relative to nondisked sites (Moser et al., 2007).

Amendments

Transplanting soil from natural to restored marshes has been used to accelerate the development of the plant community by improving the soil's physical and nutritional properties (Ashworth, 1997; Stauffer and Brooks, 1997; Anderson and Cowell, 2004) and by introducing seeds and propagules (Erwin and Best, 1985; Erwin et al., 1985; Brown and Bedford, 1997; Parikh and Gale, 1998). Typically, donor soil is removed from a wetland impacted by mining or development activities, stockpiled, and then spread over the mitigation marsh. Erwin and Best (1985) applied stockpiled topsoil from a natural wetland to a created marsh in central Florida. After 2 years, areas where topsoil was applied had greater species richness, especially late-successional species and fewer aggressive weedy species including cattail (*Typha*) than areas that did not receive topsoil. Stauffer and Brooks (1997) tested the effects of salvaged marsh surface soil from a donor wetland in Pennsylvania. Similar to the Florida study, marshes receiving salvaged soil had greater species richness, vegetation coverage, and fewer undesirable species. In New York, Brown and Bedford (1997) transplanted wetland soil from small remnant wetlands in drainage ditches to restored freshwater marshes. Plots receiving transplanted soil contained more wetland plant species and more cover of wetland plants than untreated plots.

Transplanted topsoil also enhances nutritional properties by adding inorganic nutrients to promote plant growth and organic matter content that support the heterotrophic community. Stauffer and Brooks (1997) added leaf litter compost to facilitate the establishment of *Carex* in a created marsh. SOM and total inorganic N were greater in plots receiving leaf litter as was survival of hand-planted *Carex*. Anderson and Cowell (2004) compared 17 marshes mulched with topsoil with 16 unmulched marshes, ages 5–11 years old, in Florida. Mulched marshes had greater SOM and base cations (Ca, Mg, K) than unmulched marshes. Sutton-Grier et al. (2009) added compost containing topsoil, wood chips, and biosolids to a restored freshwater marsh in North Carolina. Compost additions increased SOM and plant-available N and P. Microbial biomass and denitrification potential also increased with the addition of compost though there was no consistent effect of compost on plant growth. Ballantine et al. (2015) added straw, topsoil, and/or biochar to restored marshes in New York. Amendments, especially biochar, increased SOM content (Figure 5.8(a)). Straw alone and in combination with biochar stimulated microbial respiration, even 3 years following organic matter additions (Figure 5.8(b)). Topsoil addition also produced a marked increase in cation exchange capacity. Alsfeld et al. (2008) evaluated the effects of adding coarse woody debris (CWD) and organic matter (straw) and incorporating microtopography (ridges and furrows) to enhance biodiversity of created marshes in Delaware. Additions of CWD increased insect diversity and biomass of insects. Amendments did not increase SOM though total and native plant richness increased in amended versus unamended plots. From these studies, the cumulative benefits of organic mulch include greater (1) SOM content, (2) nitrogen, (3) phosphorus, and (4) base cations that enhance microbial activity and plant species richness.

Ecosystem Development

Ecosystem development often is hindered by the difficulty in restoring proper hydrology to the site and often water control structures are used to maintain the proper depth, duration, and seasonality of inundation. In the absence of hydrologic manipulation, created and restored marshes may be too wet, resulting in the creation of open water habitats or ponds (Confer and Niering, 1992; Bedford, 1996; Gwin et al., 1999; Magee et al., 1999; Whittecar and Daniels, 1999; Cole and Shafer, 2002) or they may be too dry, the result being a plant community containing more terrestrial species (Whittecar and Daniels, 1999; Hopple and Craft, 2013). When wetlands are created, there sometimes is a tendency to exaggerate their hydrology, making them deeper and wetter for longer periods (Shaffer et al., 1999; Cole and Brooks, 2000; Hartzell et al., 2007). These created marshes also have steeper slopes (Shulse et al., 2012). As a result, they are likely to have less N retention and trophic complexity (e.g., benthic invertebrates, avifauna, herpetofauna) than natural wetlands which tend to be shallower and have more shoreline complexity (Hansson et al., 2005; Shulse et al., 2012).

Successful establishment of marsh vegetation depends on the methods used. Active restoration techniques such as seeding and planting generally yield benefits sooner than passive methods where revegetation takes place from dispersal and from germination of seeds in the seedbank. Passive restoration has been successful in reestablishing vegetation in depressional marshes of the Midwest (Kellogg and Bridgham, 2002)

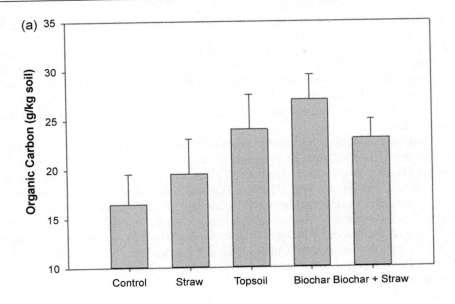

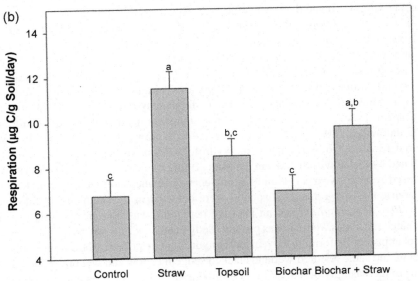

Figure 5.8 (a) Soil organic carbon in control and amended plots 3 years after treatment application. (b) Microbial respiration, averaged across years 1 and 3, in control and amended plots. From Ballantine et al. (2015).

and southeastern US (De Steven et al., 2006, 2010) but not in prairie pothole wetlands. De Steven et al. (2010) suggested that differences in the intensity of regional land use may explain this discrepancy as the southeastern US retains a diverse mix of forests, agriculture, and interspersed wetlands so that propagules are available to disperse to the site. Even for this region, depressional wetlands restored by passive revegetation lack certain characteristic species such as *Panicum hemitomon* (maidencane), *Leersia hexandra* (cutgrass), and the sedge, *Carex striata* (De Steven et al., 2006). In contrast, the prairie pothole region is characterized by widespread intensive agriculture with very few remaining wetlands. O'Connell et al. (2013) suggested that, in agricultural landscapes, distance to the nearest intact reference wetland was a good predictor of whether a restored wetland would recolonize naturally, by self-design, or would require planting.

Restoring prairie potholes usually consists of reintroducing hydrology (Seabloom and van der Valk, 2003), and development of the plant community relies on dispersal and germination of seeds in the seedbank (i.e., the efficient community hypothesis of Galatowitsch and van der Valk (1996b)). Ecosystem development of prairie pothole marsh vegetation is slow and hindered by the lack of propagules dispersing to the site and in the seedbank. Studies indicate that emergent and submersed vegetation develop with the proper hydrology but drier vegetation zones, wet prairie and sedge meadow, often do not develop (Galatowitsch and van der Valk, 1996a,b; Meyer et al., 2008b). Both dispersal constraints and biotic factors such as competition from invasive species may slow or inhibit colonization by native species (Donath et al., 2003; Galatowitsch, 2006). Altered hydrology is another factor that limits development of the plant communities in agricultural landscapes. Row crop agriculture (i.e., corn, soybeans) requires drainage that lowers the regional water table, and created and restored marshes in these regions often are characterized by drier, facultative, and facultative upland species than natural wetlands which contain more facultative wet and obligate wetland species (Heaven et al., 2003; Hopple and Craft, 2013).

In some restorations, reintroduction of hydrology leads to diverse and productive plant communities within 2–3 years (Kellogg and Bridgham, 2002). In this study, low density seeding and planting performed no better than hydrologic restoration alone and sedge species such as *Calamagrostis* were not present in any restorations. Mulhouse and Galatowitsch (2003) evaluated the vegetation of prairie pothole wetlands that were restored 12 years ago by reintroducing hydrology. Many of these were the same sites evaluated by Galatowitsch and van der Valk (1996a,b) 9 years earlier. Species richness increased over the 9-year period and distinct vegetation zones had begun to form. However, wet prairie and sedge meadow species were detected in low numbers. Reed canary grass (*Phalaris arundinacea*), an invasive perennial, was present on every site and dominated the zones where it was present. Sedge meadows did not exist at any site and did not appear to be forming 12 years after hydrologic restoration. Nineteen years after flooding, the same restored wetlands lacked sedge meadow and wet prairie perennials and were increasingly dominated by *Phalaris* and *Typha* (Aronson and Galatowitsch, 2008). Matthews and Spyreas (2010) reported similar findings in Illinois where relative cover of *Phalaris* increased with time in both seeded and planted marshes.

The absence of viable propagules in restored marsh soils often limits natural colonization. Wang et al. (2013) compared belowground propagules in low-intensity farmed soybean fields, wetlands restored by hydrologic restoration on agricultural land, and natural sedge meadows in the Sanjiang Plain of Northeastern China. Compared to natural meadows, farmed and restored sites lacked propagules of key species, *Carex* and *Calamagrostis*, even 14 years after hydrologic restoration. Mulhouse and Galatowitsch (2003) and Aronson and Galatowitsch (2008) recommended that, in agricultural landscapes, seeding, planting, and adaptive management (i.e., invasive species removal) are necessary to recreate wetland plant communities that resemble historical communities.

Plant communities of restored marshes elsewhere also are slow to achieve similarity to natural marshes. Confer and Niering (1992) compared five 3- and 4-year-old freshwater marshes created by the Connecticut Department of Transportation with five natural marshes of comparable size and type. Plant diversity was greater in natural marshes whereas the created marshes contained more aggressive species such as *Typha latifolia* (cattail), *Phragmites australis* (common reed), and *Lythrum salicaria* (purple loosestrife). Natural marshes also had greater wildlife usage. In a follow-up study 8 years later, Moore et al. (1999) reported increased species richness at both restored and reference sites but an increase in the invasive *P. australis* as well. Balcombe et al. (2005a) compared plant communities of 11 mitigation marshes, 4–21 years old, and four natural reference marshes in West Virginia. Restored marshes contained more pioneer species and nonnative species than reference marshes. Comparison of plant communities of 11 emergent wetland restoration projects in Wisconsin (USA) over a 12-year period revealed an increase in hydrophytic (obligate, facultative wet) species (Nedland et al., 2007) but an overall decrease in species diversity and richness. Gutrich et al. (2009) followed plant community development of created depressional marshes over time in Ohio and Colorado (USA). The sites, ranging in age from 5 to 19 years old, were sampled multiple times and compared with reference marshes in each region. In Ohio, floristic composition (species richness, number of native species, number of hydrophytic species, percent cover) of the created sites achieved equivalence to reference marshes within a decade, whereas in Colorado, the plant community was slower to develop. Vegetation of sites with greater initial restoration efforts, including planting of nonwoody species and contouring to create shallow banks, developed faster than sites where these techniques were not used. Spieles (2005) compared vegetation development in 45 mitigation wetlands, including 17 created and 19 restored wetlands in 21 states. The sites were established between 1993 and 2000. There was a trend of increasing prevalence of wetland vegetation and native species with age. Depressional marshes contained a greater proportion of nonnative species whereas marshes in riverine settings contained more native species, indicating the importance of flowing water in dispersal. Matthews and Endress (2010) sampled vegetation in 24 restored marshes in agricultural landscapes of Illinois to characterize the rate of vegetation succession. Over four consecutive years of sampling, annual species were replaced by clonal perennials. Compared to marsh age, site characteristics such as distance from naturally occurring wetlands and row crop agriculture were more important than site age in determining species composition.

A diverse assemblage of vegetated habitats is important for wetland fauna. Three years after hydrology restoration dike removal, restored marshes in northern New York had comparable overall densities and numbers of birds but fewer wetland-dependent and wetland-associated birds relative to natural marshes (Brown and Smith, 1998).

Delphey and Dinsmore (1993) compared breeding bird communities in 1- to 3-year-old restored prairie pothole marshes and natural marshes. Restored wetlands had lower species richness with fewer common yellowthroat (*Geothlypis trichas*), red-winged blackbird (*Agelaius phoeniceus*), marsh wren (*Cistothorus palustris*), and swamp sparrow (*Melospiza georgiana*). However, avian use of restored marshes increases with time. For example, VanRees-Siewert and Dinsmore (1996) reported higher densities of breeding birds in restored marshes in Iowa that were 4 years old as compared to marshes that were 1 year old. Likewise, on Prince Edward Island, Canada, breeding ducks used small restored wetlands more than reference wetlands (Stevens et al., 2003). At the restored sites, waterfowl density and species richness was positively correlated with wetland:vegetated (cattail) area and percent vegetation cover. O'Neal et al. (2008) evaluated waterbird response to fresh marshes restored through the USDA Conservation Reserve Enhancement Program in Illinois (USA). Waterbird use was positively related to hydrology and vegetation structure with greater use where hydrology was restored and where the ratio of open water to vegetation cover was 70% or greater.

Finfish utilization of inland marshes depends on their connectivity to aquatic ecosystems and they quickly use restored marshes once hydrology is reintroduced (Langston and Kent, 1997). Streever and Crisman (1993) compared finfish use in 8- to 20-year-old constructed marshes and natural marshes in Florida. Whereas many species were present in equivalent numbers in the two marsh types, the Everglades pygmy sunfish (*Elassoma evergladei*) was consistently found in lower numbers in the created marshes.

Use of restored marshes by amphibians also depends on the presence and structure of vegetation. Lehtinen and Galatowitsch (2001) surveyed seven recently restored wetlands in Minnesota with five reference wetlands. Amphibians rapidly colonized restored sites such that, within 2 years, restored wetlands contained eight species as compared to 12 in reference wetlands. In Wisconsin, Nedland et al. (2007) reported no difference in anuran and waterfowl use of restored marshes that were sampled 3 and 15 years after restoration, indicating relatively rapid utilization by wetland fauna. Likewise, in West Virginia, amphibian and waterfowl use of 4- to 21-year-old mitigation wetlands was comparable to or greater than in reference wetlands (Balcombe et al., 2005b). Stevens et al. (2002) used call surveys to assess anuran (frog and toad) populations in 2- to 7-year-old restored wetlands created by dredging on Prince Edward Island, Canada and found greater use of restored wetlands than natural wetlands. Rowe and Garcia (2014) evaluated 26 wetlands, 5–15 years old, restored through the USDA Wetlands Reserve Program. Native anurans were positively correlated with site age but negatively associated with invasive plant cover. These studies collectively suggest that amphibian use of created and restored marshes develops quickly once hydrology is restored and vegetation becomes established. Experimental studies of mitigation wetlands designed to test factors that promote amphibian communities indicate that presence of vegetation, upland edges containing gentle slopes, and absence of fish produce the greatest level of species richness and reproductive success (Shulse et al., 2012).

Bat populations also respond positively to marsh restoration. Menzel et al. (2005) compared bat response to restoration of a Carolina Bay marsh (South Carolina, USA) that had been drained. Following restoration, bat activity increased over the restored bay and was comparable to or greater than in reference bays. Menzel et al. suggested that reestablishment of hydrology—water—increased food resources for bats.

Whereas a diverse plant community is slow to develop following restoration, reintroduction of hydrology can quickly lead to development of ecosystem function, especially primary production. For example, a created marsh in Virginia had levels of aboveground biomass that were comparable to an adjacent natural marsh within 1 year (DeBerry and Perry, 2004). Whigham et al. (2002) measured aboveground biomass in 12 marshes restored on former agricultural land over a 3-year period. Hydrology of the sites was restored between 1964 and 1993 by plugging ditches. There was no clear relationship between biomass production and marsh age as aboveground biomass varied considerably both spatially and among years. Productivity of woody and submersed vegetation also develops quickly once hydrology is restored. McKenna (2003) found that primary production by aquatic, herbaceous, and woody wetland species in a 1-year-old marsh restored by retaining water using a control structure was comparable to a nearby reference marsh.

Whereas primary productivity develops quickly following restoration, aboveground biomass stocks develop more slowly. Hossler and Bouchard (2010) compared five created mitigation marshes (3–8 years old) with four natural marshes in Ohio. They found lower aboveground biomass as well as soil organic C, and mineralizable C in the restored marshes. Other studies reported less aboveground biomass in young (1–9 years old) restored marshes than in natural marshes (Fennessy et al., 2008) though there is evidence that biomass increases with site age (Meyer et al., 2008a).

Similar to primary productivity, aquatic and benthic food webs of restored marshes also develop relatively quickly. Mayer and Galatowitsch (1999) compared diatom communities on artificial substrates placed in eight restored prairie pothole marshes (ages 3–15 years) and eight reference marshes and found that species richness, diversity, and equitability were comparable. Likewise, Dodson and Lillie (2001) found no difference in taxa richness of zooplankton (e.g., copepods, cladocerans, *Daphnia*) among 1- to 21-year-old restored marshes and natural marshes. Gleason et al. (2004) compared "egg banks" of invertebrates in restored wetlands and undrained natural wetlands in the prairie pothole region and found no difference in abundance and diversity of invertebrates incubated from them. Stanczak and Keiper (2004) compared benthic invertebrates in created and natural marshes in Ohio. Within 4 years, the created marsh contained invertebrate communities that were comparable to natural wetlands though clams were slow to colonize the site. In Florida, Streever et al. (1996) reported no difference in dipteran communities in 10 1- to 11-year-old created marshes and 10 natural marshes. Likewise, Meyer and Whiles (2008) found similar macroinvertebrate communities in 5- to 16-year-old restored slough wetlands (Nebraska, USA) and reference marshes. Not all studies show rapid colonization by benthic organisms though. In seasonal wetlands in California, aquatic macroinvertebrate communities did not converge to reference wetlands for 10 years following restoration (Marchetti et al., 2010), suggesting that short hydroperiod and lack of connectivity slow their development.

Ecosystem functions related to C and nutrient cycling often differ between restored and natural marshes. Fennessy et al. (2008) compared litter decomposition and soil C, N, and P pools in 10 1- to 9-year-old marshes created for mitigation with nine natural marshes in Ohio. The rate of decomposition was lower and soil organic C, N, and labile P was higher in the natural marshes. Fennessy et al. hypothesized that low nutrient stocks in restored marsh soils led to lower rates of carbon and nutrient cycling. Other studies, however, report higher rates of decomposition in created and restored marshes. Taylor and Middleton (2004) compared decomposition in a natural marsh and a 5-year-old reclaimed marsh in southern Illinois and found that decomposition rates were higher in the reclaimed marsh. Atkinson and Cairns (2001) measured litter decomposition of *Scirpus cyperinus* and *T. latifolia* in 2-year-old and 20-year-old created marshes. Similar to Taylor and Middleton, they found that decomposition was faster in the older marsh, with 85% of the mass remaining after 507 days in the 2-year-old marsh and 76% in the 20-year-old marsh. It is likely that differences in hydrology drive decomposition rates such that decomposition will be lower in wet (anaerobic) sites and faster in dry (aerobic) sites, so getting the hydrology right is critical to support C and nutrient cycles characteristic of natural marshes.

Hydrology is key to the development of a facultative anaerobic microbial community that is characteristic of flooded and saturated soils. Bossio et al. (2006) compared microbial communities of 3-year-old restored marshes in the Sacramento–San Joaquin Delta of California. The marshes were reflooded following decades of drainage for agriculture. Phospholipid fatty acid analysis indicated a rapid shift from a community dominated by aerobic microorganisms to one dominated by anaerobes. With restoration of hydrology, other measures of microbial activity develop quickly. For example, Duncan and Groffman (1994) reported no difference in microbial processes related to C cycling and N cycling in constructed marshes and natural wetlands in New England as respiration, microbial biomass, N mineralization, nitrification, and denitrification did not differ between 3-year-old constructed marshes and natural wetlands. They attributed the rapid development of the microbial community to aggressive establishment of vegetation, 10 species densely planted on the sites.

Other studies indicate that C and nutrient cycles may be slow to develop following restoration and, in fact, require considerable time to become comparable to natural marshes. Hossler et al. (2011) compared C, N, and P cycling in 10 created and restored freshwater marshes, ranging in age from 1 to 39 years old with five natural depressional marshes, also in Ohio. Created and restored marshes contained 90% less C in litter and 80% less C in soil than natural marshes. Denitrification, on average, was 60% less. Meyer et al. (2008a) measured a number of ecosystem functions related to C and N cycling along a chronosequence of 1- to 10-year-old restored wet meadows and three natural wet meadows in Nebraska. Similar to other studies, soil organic C and N were lower in restored meadows though there was evidence of increasing C and N with age. This was most evident in the sloughs where hydroperiod was longer relative to the wetland margin. Soil extractable and microbial N also increased with site age. Marton et al. (2013a) compared denitrification and phosphorus sorption in 10-year-old depressional restored marshes with natural marshes in Indiana. Restored marshes had lower denitrification and P sorption than natural marshes.

Studies of soil development of created and restored marshes yield variable outcomes depending on the time since restoration. Bishel-Machung et al. (1996) compared SOM along a chronosequence of 1- to 8-year-old created freshwater wetlands in Pennsylvania. They found that the created wetlands contained less SOM with no increase over the 8-year period. Created marshes as old as 20 years contained considerably less SOM (4.2–6.5%) than natural marshes (12–21%) in the region (Cole et al., 2001). Shaffer and Ernst (1999) also reported no increase in SOM in created marshes in Oregon that were sampled then resampled 6 years later. Longer term studies, however, document increases in SOM as created marshes age. SOM (and N) stocks increased with age in marshes as old as 16 years but they still contained less C and N than natural marshes (Nair et al., 2001; Lu et al., 2007; Wolf et al., 2011). Microbial processes related to N cycling, mineralization, and denitrification also increased with marsh age (Wolf et al., 2011). Card et al. (2010) and Card and Quideau (2010) measured soil C and microbial community structure in restored prairie pothole wetlands of different ages in Canada. Soil organic C was 4.8% in 1- to 3-year-old marshes and increased with time following restoration such that C content of 7- to 11-year-old sites (7.3%) was approaching that of mature reference marshes (10.7%). Microbial biomass, evenness, and diversity also was less in the younger (<7 years) restored wetlands than in reference wetlands but it increased with time. Similarly, Jinbo et al. (2007) reported an increase in soil organic C following abandonment of agricultural land and colonization by sedges in the Sanjiang Plain of northeastern China. Soil organic C (0–10 cm) averaged 3.1%, 4.4%, and 10.7% in sites abandoned 1, 6, and 13 years ago, respectively, compared to 2.8% in cultivated soil. Anderson et al. (2005) measured temporal and spatial changes in soil properties in Olentangy River (Ohio) freshwater marshes, one planted, one unplanted, 2 and 10 years after the wetlands were established. SOM and P increased as the marshes aged and spatial variability of soil properties increased. Ballantine and Schneider (2009) compared soil properties along a chronosequence of 35 depressional marshes ranging in age from 3 to 55 years old in New York. Surface SOM (0–5, 5–10 cm) increased with marsh age, but even after 55 years, restored marshes contained less SOM than natural marshes in the region. These studies and studies of other created and restored wetlands suggest that microbial processes may develop relatively quickly but that bulk soil properties such as SOM take decades to develop to levels found in natural marshes.

Overall, ecosystem development of inland freshwater and saline marshes is highly variable and is linked to the techniques employed, planting versus natural colonization, and degree of connectivity to aquatic ecosystems as marshes that lack strong connections develop more slowly than marshes with strong connections. Functions such as primary productivity develop relatively quickly, within a matter of years, plant species composition develops more slowly (Table 5.4). Development of species rich communities is especially slow in depressional marshes in agricultural landscapes such as prairie potholes because of the depauperate seedbank and isolation from natural marshes. Fauna such as aquatic macroinvertebrates, amphibians, and fish use created and restored marshes once hydrology is reestablished and vegetation becomes established. Wetland-dependent birds feed in restored marshes once good cover develops but use of marshes for breeding takes longer and depends on development of a diverse plant community and their structural characteristics. Ecosystem functions such as decomposition and nutrient retention also take time, and, of course, development of organic-rich soil characteristic of many marshes is slow, on the order of decades or more.

Table 5.4 **Development of Inland Freshwater Marsh Ecosystem Services Following Creation or Restoration**

	Time (years) to Equivalence to Mature Natural Marshes
Productivity and Habitat Functions	
Primary production	1–3
Plant diversity	3 to >20
Aquatic invertebrates	1–10
Amphibians	2–20
Finfish	5–10
Wetland-dependent birds	
• Feeding	3–5
• Breeding	>3 to >20
Regulation Functions	
Decomposition[a]	>5 to >20
Nutrient (N, P) retention	>10
Soil formation	10s to 100s

[a]Depends on reestablishing the proper hydrology.

Keys to Ensure Success

Successful restoration of inland marshes depends in large part on site history and landscape context. In agricultural landscapes, the absence of propagules in the seed-bank and via dispersal limits natural colonization. Freshwater marshes, in particular, are highly susceptible to invasion by aggressive species such as *Phalaris*, *Typha*, *Phragmites*, and *Lythrum*, so it is important to select sites that are resistant to invasion (see Chapter 3, Ecological Theory and Restoration). Avoid sites with disturbance that exposes bare soil to potential colonizers (Table 5.5). Also avoid sites that receive runoff containing large amounts of nutrients (N, P), excess sediment, and deicing salts that are known to promote invasion (Zedler and Kercher, 2004; Miklovic and Galatowitsch, 2005). Ensure that the site is devoid of propagules of these species and make sure that any added amendments such as SOM do not contain them.

Grade the site to create a depression that eventually will produce the concentric zones of vegetation surrounding a central area of open water. Create a gentle slope that allows easy access for terrestrial fauna, especially amphibians. Marshes created on excavated soils should be plowed using a rip or chisel plow to reduce compaction. Amendments such as organic matter or topsoil are added to improve soil physical and nutritional properties (Whittecar and Daniels, 1999). Incorporate microtopographic features into the soil surface as appropriate.

When seeding or planting, select seeds and young transplants appropriate for the geographic region, inundation zone, and, in arid and semiarid environments, salinity. Transplants, while more expensive, are more likely to resist invasion by aggressive

Table 5.5 Key Characteristics to Ensure Success

Site Selection
- Avoid sites with soil disturbance (cleared land) or stressors (nutrients, road salt)
- Avoid sites with uncontrolled stressors such as runoff containing nutrients, sediment, and road salt
- Reintroduce disturbance as appropriate (wet grasslands)

Construction Considerations
- Configure the site to create concentric rings of vegetation surrounding a central core of open water
- Gentle slope between open water/wetland and upland
- Rip and chisel plow surface/subsurface soil to reduce compaction (excavated sites)
- Incorporate microtopographic features
- Return stockpiled topsoil containing seedbank and soil organic matter or amend with organic matter

Planting Considerations
- Seeds and plants selected for appropriate geographic region, inundation regime, salinity (if applicable)
- Protect young transplants from herbivores (e.g., waterfowl)

Maintenance
- Invasive species monitoring and removal
- Monitoring and control of herbivores

species better than seeding. Keystone marsh species such as *Carex* that dominate wet sedge meadows and other marshes will need to be planted. Newly established seedlings may need to be protected from herbivory using nets or temporary fences. Maintenance of the site will require continual vigilance to identify and remove invaders before they become established. In the early stages, removal by hand may be sufficient. However, herbicides will be necessary once invasive species become established. In some situations, activities to control herbivore populations such as muskrat and nutria and waterfowl such as geese may be needed.

References

Aliev, F.F., 1965. Dispersal of nutria in the USSR. Journal of Mammalogy 46, 101–102.

Alsfeld, A.J., Bowman, J.L., Deller-Jacobs, A., 2008. Effects of woody debris, microtopography, and organic matter amendments on the biotic community of constructed depressional wetlands. Biological Conservation 142, 247–255.

Anderson, C.J., Cowell, B.C., 2004. Mulching effects on the seasonally flooded zone of west-central Florida, USA, wetlands. Wetlands 24, 811–819.

Anderson, C.J., Mitsch, W.J., Nairn, R.W., 2005. Temporal and spatial development of surface soil conditions at two created riverine marshes. Journal of Environmental Quality 34, 272–284.

Aronson, M.F., Galatowitsch, S.M., 2008. Long-term vegetation development of restored prairie pothole wetlands. Wetlands 28, 883–895.

Ashworth, S.M., 1997. Comparison between restored and reference sedge meadow wetlands in south-central Wisconsin. Wetlands 17, 518–527.

Atkinson, R.B., Cairns Jr., J., 2001. Plant decomposition and litter accumulation in depressional wetlands: functional performance of two wetland age classes that were created via excavation. Wetlands 21, 354–362.

Bakker, J.P., Ruyter, J.C., 1981. Effects of five years of grazing on salt-marsh vegetation. Vegetatio 44, 81–100.

Balcombe, C.K., Anderson, J.T., Fortney, R.H., Rentch, J.S., Grafton, W.N., Kordek, W.S., 2005a. A comparison of plant communities of mitigation and reference wetlands in the mid-Appalachians. Wetlands 25, 130–142.

Balcombe, C.K., Anderson, J.T., Fortney, R.H., Kordek, W.S., 2005b. Wildlife use of mitigation and reference wetlands in West Virginia. Ecological Engineering 25, 85–99.

Ballantine, K., Schneider, R., 2009. Fifty five years of soil development in restored freshwater depressional wetlands. Ecological Applications 19, 1467–1480.

Ballantine, K.A., Lehmann, J., Schneider, R.L., Groffman, P.M., 2015. Trade-offs between soil-based functions in wetlands restored with soil amendments of differing lability. Ecological Applications 25, 215–225.

Batt, B.J., Anderson, M.G., Anderson, C.D., Casell, F.J., 1989. The use of prairie potholes by North American ducks. In: van der Valk, A. (Ed.), Northern Prairie Wetlands. Iowa State University, Ames, IA, pp. 204–227.

Bedford, B.L., 1996. The need to define hydrologic equivalence at the landscape scale for freshwater wetland mitigation. Ecological Applications 6, 57–68.

Berg, M., Joyce, C., Burnside, N., 2012. Differential responses of abandoned wet grassland plant communities to reinstated cutting management. Hydrobiologia 692, 83–97.

Bertani, A., Bramblila, I., Menegus, F., 1980. Effect of anaerobiosis on rice seedlings: growth, metabolic rate, and rate of fermentation products. Journal of Experimental Botany 3, 325–331.

Bishel-Machung, L., Brooks, R.P., Yates, S.S., Hoover, K.L., 1996. Soil properties of reference wetlands and wetland creation projects in Pennsylvania. Wetlands 16, 532–544.

Boers, A.M., Frieswyk, C.B., Verhoeven, J.T.A., Zedler, J.B., 2006. Contrasting approaches to the restoration of diverse vegetation in herbaceous wetlands. In: Bobbink, R., Beltman, B., Verhoeven, J.T.A., Whigham, D.F. (Eds.), Wetlands: Functioning, Biodiversity Conservation, and Restoration. Ecological Studies, vol. 191. Springer-Verlag, Berlin, Germany, pp. 225–246.

Bossio, D.A., Fleck, J.A., Scow, K.M., Fujii, R., 2006. Alteration of soil microbial communities and water quality in restored wetlands. Soil Biology and Biochemistry 38, 1223–1233.

Brown, S.C., Bedford, B.L., 1997. Restoration of wetland vegetation with transplanted wetland soil: an experimental study. Wetlands 17, 424–437.

Brown, S.C., Smith, C.R., 1998. Breeding season bird use of recently restored versus natural wetlands in New York. Journal of Wildlife Management 62, 1480–1491.

Brix, H., Sorrel, B.K., Orr, P.T., 1992. Internal pressurization and convective gas flow in some emergent freshwater macrophytes. Limnology and Oceanography 37, 1420–1433.

Brooks, R.P., Wardrop, D.H., Cole, C.A., Campbell, D.A., 2005. Are we purveyors of wetland homogeneity? A model of degradation and restoration to improve wetland mitigation performance. Ecological Engineering 24, 331–340.

Budelsky, R.A., Galatowitsch, S.M., 1999. Effects of moisture, temperature, and time on seed germination of five wetland *Carices*: implications for restoration. Restoration Ecology 7, 86–97.

Budelsky, R.A., Galatowitsch, S.M., 2000. Effects of water regime and competition on the establishment of a native sedge in restored wetlands. Journal of Applied Ecology 37, 971–985.

Burdick, D.M., Mendelssohn, I.A., 1990. Relation between anatomical and metabolic responses to soil waterlogging in the coastal grass *Spartina patens*. Journal of Experimental Botany 41, 223–228.

Campbell, D.A., Cole, C.A., Brooks, R.P., 2002. A comparison of created and natural wetlands in Pennsylvania, USA. Wetlands Ecology and Management 10, 41–49.

Card, S.M., Quideau, S.A., 2010. Microbial community structure in restored riparian soils of the Canadian prairie pothole region. Soil Biology and Biochemistry 42, 1463–1471.

Card, S.M., Quideau, S.A., Oh, S.W., 2010. Carbon characteristics in restored and reference riparian soils. Soil Science Society of America Journal 74, 1834–1843.

Chapman, V.J., 1974. Salt Marshes and Salt Deserts of the World. Interscience Press, New York.

Christensen, N.L., 1977. Fire and soil plant nutrient relations in a pine-wiregrass savanna on the coastal plain of North Carolina. Oecologia 31, 27–44.

Christensen, N.L., 1993. The effect of fire on nutrient cycles in longleaf pine ecosystems. Proceedings of the Tall Timbers Fire Ecology Conference 18, 205–214.

Cole, C.A., Brooks, R.P., 2000. A comparison of the hydrologic characteristics of natural and created mainstem floodplain wetlands in Pennsylvania. Ecological Engineering 14, 221–231.

Cole, C.A., Shaffer, D., 2002. Section 404 wetland mitigation and permit success criteria in Pennsylvania, USA, 1986–1999. Environmental Management 30, 508–515.

Cole, C.A., Brooks, R.P., Wardrop, D.H., 2001. Assessing the relationship between biomass and soil organic matter in created wetlands of central Pennsylvania, USA. Ecological Engineering 17, 423–428.

Combroux, I.C.S., Bornette, G., Amoros, C., 2002. Plant regenerative strategies after a major disturbance: the case of a riverine wetland restoration. Wetlands 22, 234–246.

Confer, S.R., Niering, W.A., 1992. Comparison of created and natural freshwater emergent wetlands in Connecticut (USA). Wetlands Ecology and Management 2, 143–156.

Cowardin, L.M., Carter, V., Golet, F.C., LaRoe, E.T., 1979. Classification of Wetlands and Deepwater Habitats of the United States. United States Fish and Wildlife Service. Biological Services Program. FWS/OBS-79/31.

Dacey, J.W.H., 1980. Internal winds in water lilies: an adaptation for life in anaerobic sediments. Science 210, 1017–1019.

Dahl, T.E., 1990. Wetlands Losses in the United States 1780's to 1980's. U.S. Department of the Interior, Fish and Wildlife Service, Washington, DC.

Dahl, T.E., 2011. Status and Trends of Wetlands in the Conterminous United States 2004 to 2009. U.S. Department of the Interior, Fish and Wildlife Service, Washington, DC.

DeBerry, D.A., Perry, J.E., 2004. Primary succession in a created freshwater marsh. Castanea 69, 185–193.

Delphey, P.J., Dinsmore, J.J., 1993. Breeding bird communities of recently restored and natural prairie potholes. Wetlands 13, 200–206.

De Steven, D., Sharitz, R.R., Barton, C.D., 2010. Ecological outcomes and evaluation of success in passively restored southeastern depressional wetlands. Wetlands 30, 1129–1140.

De Steven, D., Sharitz, R.R., Singer, J.H., Barton, C.D., 2006. Testing a passive revegetation approach for restoring coastal plain depression wetlands. Restoration Ecology 14, 452–460.

Dodson, S.I., Lillie, R.A., 2001. Zooplankton communities of restored depressional wetlands in Wisconsin, USA. Wetlands 21, 292–300.

Donath, T.W., Hozel, N., Otte, A., 2003. The impact of site conditions and seed dispersal on restoration success in alluvial meadows. Applied Vegetation Science 6, 13–22.

Duncan, C.P., Groffman, P.M., 1994. Comparing microbial parameters in natural and constructed wetlands. Journal of Environmental Quality 23, 298–305.

Egler, F.E., 1954. Vegetation science concepts I. Initial floristic composition. A factor in old field vegetation development. Vegetatio 4, 412–417.

Erwin, K.L., Best, G.R., 1985. Marsh community development in a central Florida phosphate surface-mined reclaimed wetland. Wetlands 5, 155–166.

Erwin, K.L., Best, G.R., Dunn, W.J., Wallace, P.M., 1985. Marsh and forest wetland reclamation of a central Florida phosphate mine. Wetlands 4, 87–104.

Erwin, K.L., Smith, C.M., Cox, W.R., Rutter, R.P., 1994. Successful construction of a freshwater herbaceous marsh in south Florida, USA. In: Mitsch, W.J. (Ed.), Global Wetlands: Old World and New. Elsevier, New York, pp. 493–508.

Euliss Jr., N.H., LaBaugh, J.W., Fredrickson, L.H., Mushet, D.M., Laubhan, M.K., Swanson, G.A., Winter, T.C., Rosenberry, D.O., Nelson, R.D., 2004. The wetland continuum: a conceptual framework for interpreting biological studies. Wetlands 24, 448–458.

Evans, J., 1970. About Nutria and Their Control. U.S. Department of the Interior. Bureau of Sport Fisheries and Wildlife. Resource Publication 86. U.S. Government Printing Office, Washington, DC.

Fennessy, M.S., Rokosch, A., Mack, J.J., 2008. Patterns of plant decomposition and nutrient cycling in natural and created wetlands. Wetlands 28, 300–310.

Fink, D.F., Mitsch, W.J., 2007. Hydrology and nutrient biogeochemistry in a created river diversion oxbow wetland. Ecological Engineering 30, 93–102.

Galatowitsch, S.M., 2006. Restoring prairie pothole wetlands: does the species pool concept offer decision-making guidance for re-vegetation? Applied Vegetation Science 9, 261–270.

Galatowitsch, S.M., van der Valk, A.G., 1996a. Characteristics of recently restored wetlands in the prairie pothole region. Wetlands 16, 75–83.

Galatowitsch, S.M., van der Valk, A.G., 1996b. The vegetation of restored and natural prairie wetlands. Ecological Applications 6, 102–112.

Galatowitsch, S.M., van der Valk, A.G., 1996c. Vegetation and environmental conditions in recently restored wetlands in the prairie pothole region of the USA. Vegetatio 126, 89–99.

Galatowitsch, S.M., Anderson, N.O., Ascher, P.D., 1999. Invasiveness in wetland plants in temperate North America. Wetlands 19, 733–755.

Gleason, R.A., Euliss Jr., N.H., Hubbard, D.E., Duffy, W.G., 2004. Invertebrate egg banks of restored, natural and drained wetlands in the prairie pothole region of the United States. Wetlands 24, 562–572.

Gorham, E., 1961. Factors influencing supply of major ions to inland waters, with special reference to the atmosphere. Geological Society of America Bulletin 72, 795–840.

Gutrich, J.J., Taylor, K.J., Fennessy, M.S., 2009. Restoration of vegetation communities of created depressional wetlands in Ohio and Colorado (USA): the importance of initial effort for mitigation success. Ecological Engineering 35, 351–368.

Gwin, S.E., Kentula, M.E., Schaffer, P.W., 1999. Evaluating the effects of wetland regulation through hydrogeomorphic classification and landscape profiles. Wetlands 19, 477–489.

Hansson, L., Bronmark, C., Nilsson, P.A., Abjornsson, K., 2005. Conflicting demands on wetland ecosystem services: nutrient retention, biodiversity or both? Freshwater Biology 50, 705–714.

Hartzell, D., Bidwell, J.R., Davis, C.A., 2007. A comparison of natural and created depressional wetlands in central Oklahoma using metrics from indices of biological integrity. Wetlands 27, 794–805.

Heaven, J.B., Gross, F.E., Gannon, A.T., 2003. Vegetation comparison of a natural and a created emergent marsh wetlands. Southeastern Naturalist 2, 195–206.

Hoeltje, S.M., Cole, C.A., 2009. Comparison of functions of created wetlands of two age classes in central Pennsylvania. Environmental Management 43, 597–608.

Hopple, A., Craft, C., 2013. Managed disturbance enhances biodiversity of restored wetlands in the agricultural Midwest. Ecological Engineering 61, 505–510.

Hossler, K., Bouchard, V., 2010. Soil development and establishment of carbon-based properties in created freshwater marshes. Ecological Applications 20, 539–553.

Hossler, K., Bouchard, V., Fennessy, M.S., Frey, S.D., Anemaet, E., Herbert, E., 2011. No-net-loss not met for nutrient function in freshwater marshes: recommendations for wetland mitigation policies. Ecosphere 2, 1–36.

Jackson, M.B., 1985. Ethylene and responses of plants to soil waterlogging and submergence. Annual Review of Plant Physiology 36, 145–174.

Jefferies, R.L., Jano, A.P., Abraham, K.M., 2006. A biotic agent promotes large-scale catastrophic change in the coastal marshes of Hudson Bay. Journal of Ecology 94, 234–242.

Jinbo, Z., Changchun, S., Shenmin, W., 2007. Dynamics of soil organic carbon and its fractions after abandonment of cultivated wetlands in northeast China. Soil and Tillage Research 96, 350–360.

Joyce, C.B., 2014. Ecological consequences and restoration potential of abandoned wet grasslands. Ecological Engineering 66, 91–102.

Joyce, C.B., Wade, P.M., 1998. European Wet Grasslands: Biodiversity, Management, and Restoration. John Wiley and Sons, Chichester, UK.

Kellogg, C.H., Bridgham, S.D., 2002. Colonization during early succession of restored freshwater marshes. Canadian Journal of Botany 80, 176–185.

Kettenring, K.M., Galatowitsch, S.M., 2007a. Temperature requirements for dormancy break and seed germination vary greatly among 14 wetland *Carex* species. Aquatic Botany 87, 209–220.

Kettenring, K.M., Galatowitsch, S.M., 2007b. Tools for *Carex* revegetation in freshwater wetlands: understanding dormancy loss and germination temperature requirements. Plant Ecology 193, 157–169.

Kettenring, K.M., Galatowitsch, S.M., 2011a. *Carex* seedling emergence in restored and natural prairie wetlands. Wetlands 31, 273–281.

Kettenring, K.M., Galatowitsch, S.M., 2011b. Seed rain of restored and natural prairie wetlands. Wetlands 31, 283–294.

Kettenring, K.M., Gardner, G., Galatowitsch, S.M., 2006. Effect of light on seed germination of eight wetland *Carex* species. Annals of Botany 98, 869–874.

Kinler, N.W., Linscombe, G., Ramsey, P.R., 1987. Nutria. In: Movak, M., Obbard, M.E., Malloch, B. (Eds.), Wild Furbearer Management and Conservation in North America. Ontario Ministry of Natural Resources, Ontario, Canada, pp. 327–343.

LaBaugh, J.W., 1989. Chemical characteristics of water in northern prairie pothole wetlands. In: van der Valk, A. (Ed.), Northern Prairie Wetlands. Iowa State University, Ames, IA, pp. 56–90.

Langston, M.A., Kent, D.M., 1997. Fish recruitment to a constructed wetland. Journal of Freshwater Ecology 12, 123–129.

Lehtinen, R.M., Galatowitsch, S.M., 2001. Colonization of restored wetlands by amphibians in Minnesota. American Midland Naturalist 145, 388–396.

Lietch, J.A., 1989. Politicoeconomic overview of prairie potholes. In: van de Valk, A. (Ed.), Northern Prairie Wetlands. Iowa State University Press, Ames, IA, pp. 2–14.

Likens, G.E., Bormann, F.H., 1977. Biogeochemistry of a Forested Ecosystem. Springer-Verlag, New York.

Livingstone, D.A., 1963. Chapter G. Chemical composition of rivers and lakes. In: Fleischer, M. (Ed.), Data of Geochemistry. Geological Survey Professional Paper 440-G. United States Government Printing Office, Washington, DC.

Lu, J., Wang, H., Wang, W., Yin, C., 2007. Vegetation and soil properties in restored wetlands near Lake Taihu, China. Hydrobiologia 581, 151–159.

Luoto, M., Pykala, J., Kuussaari, M., 2003. Decline of landscape-scale habitat and species diversity after the end of cattle grazing. Journal of Nature Conservation 11, 171–178.

Lynch, J.J., O'Neil, T., Lay, D.W., 1947. Management significance of damage by geese and muskrats to Gulf coast marshes. Journal of Wildlife Management 11, 50–76.

McKenna Jr., J.E., 2003. Community metabolism during early development of a restored wetland. Wetlands 23, 35–50.

Manchester, S., Treweek, J., Mountford, O., Pywell, R., Sparks, T., 1998. Restoration of a target wet grassland community on ex-arable land. In: Joyce, C.B., Wade, P.M. (Eds.), European Wet Grasslands: Biodiversity, Management, and Restoration. John Wiley and Sons, Chichester, UK, pp. 277–294.

Magee, T.K., Ernest, T.L., Kentula, M.E., Dwire, K.A., 1999. Floristic comparison of freshwater wetlands in an urbanizing environment. Wetlands 19, 517–534.

Mahaney, W.M., Wardrop, D.H., Brooks, R.H., 2004. Impacts of sedimentation and nitrogen enrichment on wetland plant community development. Plant Ecology 175, 227–243.

Marchetti, M.P., Garr, M., Smith, A.D.H., 2010. Evaluating wetland restoration success using aquatic macroinvertebrate assemblages in the Sacramento Valley, California. Restoration Ecology 18, 457–466.

Marton, J.M., Fennessy, M.S., Craft, C.B., 2013a. Restoring nutrient removal and carbon sequestration in the Glaciated Interior Plains. Journal of Environmental Quality 43, 409–417.

Marton, J.M., Fennessy, M.S., Craft, C.B., 2013b. The restoration of wetland ecosystem services by USDA conservation practices: an example from Ohio. Restoration Ecology 22, 117–124.

Matthews, J.W., Endress, A.G., 2010. Rate of succession in restored wetlands and the role of site context. Applied Vegetation Science 13, 346–355.

Matthews, J.W., Spyreas, G., 2010. Convergence and divergence in plant community trajectories as a framework for monitoring wetland restoration progress. Journal of Applied Ecology 47, 1128–1136.

Mayer, P.M., Galatowitsch, S.M., 1999. Diatom communities as ecological indicators of recovery in restored prairie wetlands. Wetlands 19, 765–774.

Mendelssohn, I.A., Postek, J.W., 1982. Elemental analysis of deposits on the roots of *Spartina alterniflora* Loisel. American Journal of Botany 69, 904–912.

Mendelssohn, I.A., McKee, K.L., Patrick Jr., W.H., 1981. Oxygen deficiency in *Spartina alterniflora* roots: metabolic adaptation to anoxia. Science 214, 439–441.

Menzel, J.M., Menzel, M.A., Kilgo, J.C., Ford, W.M., Edwards, J.W., 2005. Bat response to Carolina Bays and wetland restoration in the southeastern U.S. Coastal Plain. Wetlands 25, 542–550.

Meybeck, M., 1979. Concentrations des eaux fluviales en elements majeurs et apports en solution aux oceans. Review de Geologie Dynamique et de Geographie Physique 21, 215–246.

Meyer, C.K., Whiles, M.R., 2008. Macroinvertebrate communities in restored and natural Platte River slough wetlands. Journal of the North American Benthological Society 27, 626–639.

Meyer, C.K., Bauer, S.G., Whiles, M.R., 2008a. Ecosystem recovery across a chronosequence of restored wetlands in the Platte River valley. Ecosystems 11, 193–208.

Meyer, C.K., Whiles, M.R., Baer, S.G., 2008b. Plant community recovery following restoration in temporally variable riparian wetlands. Restoration Ecology 18, 52–64.

Miklovic, S., Galatowitsch, S.M., 2005. Effect of NaCl and *Typha angustifolia* L. on marsh community establishment: a greenhouse study. Wetlands 25, 420–429.

Mitsch, W.J., Wilson, R.F., 1996. Improving the success of wetland creation and restoration with know-how, time, and self-design. Ecological Applications 6, 77–83.

Mitsch, W.J., Wu, X., Nairn, R.W., Weihe, P.E., Wang, N., Deal, R., Boucher, C.E., 1998. Creating and restoring wetlands: a whole-ecosystem experiment in self-design. BioScience 48, 1019–1030.

Moore, H.H., Niering, W.A., Marsicano, L.J., Dowdell, M., 1999. Vegetation change in created emergent wetlands (1988–1996) in Connecticut (USA). Wetlands Ecology and Management 7, 177–191.

Moser, K.F., Ahn, C., Noe, G.B., 2007. Characterization of microtopography and its influence on vegetation patterns in created wetlands. Wetlands 27, 1081–1097.

Moser, K.F., Ahn, C., Noe, G.B., 2009. The influence of microtopography on soil nutrients in created mitigation wetlands. Restoration Ecology 17, 641–651.

Mulhouse, J.M., Galatowitsch, S.M., 2003. Revegetation of prairie pothole wetlands in the mid-continental US: twelve years post-flooding. Plant Ecology 169, 143–159.

Murkin, H.R., Murkin, E.J., Ball, J.P., 1997. Avian habitat selection and wetland dynamics: a 10-year experiment. Ecological Applications 7, 1144–1159.

Nair, V.D., Graetz, D.A., Reddy, K.R., Olila, O.G., 2001. Soil development in phosphate-mined created wetlands of Florida, USA. Wetlands 21, 232–239.

Nedland, T.S., Wolf, A., Reed, T., 2007. A reexamination of restored wetlands in Manitowoc County, Wisconsin. Wetlands 27, 999–1015.

Noon, K.F., 1996. A model of created wetland primary succession. Landscape and Urban Planning 34, 97–123.

Odum, W.E., Smith III, T.J., Hoover, J.K., McIvor, C.C., 1984. The Ecology of Tidal Freshwater Marshes of the United States East Coast: A Community Profile. U.S. Fish and Wildlife Service. FWS/OBS-83/17.

O'Connell, J.L., Johnson, L.A., Beas, B.A., Smith, L.M., McMurray, S.T., Haukos, D.A., 2013. Predicting dispersal limitation in plants: optimizing planting decisions for isolated wetland restoration in agricultural landscapes. Biological Conservation 159, 343–354.

O'Neal, B.J., Heske, E.J., Stafford, J.D., 2008. Waterbird response to wetlands restored through the Conservation Reserve Enhancement Program. Journal of Wildlife Management 72, 654–664.

O'Neil, T., 1949. The Muskrat in the Louisiana Coastal Marshes. Louisiana Department of Wildlife and Fisheries, New Orleans, LA.

Palmisano, A.W., 1972. Habitat preference of waterfowl and fur animals in the northern Gulf coast marshes. In: Proceedings of the Coastal Marsh and Estuary Management Symposium. Louisiana State University, Baton Rouge, LA, pp. 163–190.

Parikh, A., Gale, N., 1998. Vegetation monitoring of created dune swale wetlands, Vandenberg Air Force Base, California. Restoration Ecology 6, 83–93.

Pfeifer-Meister, L., Roy, B.A., Johnson, B.R., Krueger, J., Bridgham, S.D., 2012a. Dominance of native grasses leads to community convergence in wetland restoration. Plant Ecology 213, 637–647.

Pfeifer-Meister, L., Johnson, B.R., Roy, B.A., Carreno, S., Stewart, J.L., Bridgham, S.D., 2012b. Restoring wetland prairies: tradeoffs among native plant cover, community composition, and ecosystem functioning. Ecosphere 3, 1–19. Article 121.

Reinartz, J.A., Warne, E.L., 1993. Development of vegetation in small created wetlands in southeastern Wisconsin. Wetlands 13, 153–164.

Rowe, J.C., Garcia, T.S., 2014. Impacts of wetland restoration efforts on an amphibian assemblage in a multi-invader community. Wetlands 34, 141–153.

Schaich, H., Rudner, M., Konold, W., 2010. Short-term impact of river restoration and grazing on floodplain vegetation in Luxembourg. Agriculture, Ecosystems, and Environment 139, 142–149.

Schlesinger, W.H., 1978. Community structure, dynamics and nutrient cycling in the Okefenokee swamp-forest. Ecological Monographs 48, 43–65.

Seabloom, E.W., van der Valk, A.G., 2003. Plant diversity, composition and invasion of restored and natural prairie pothole wetlands: implications for restoration. Wetlands 23, 1–12.

Shaffer, P.W., Ernst, T.L., 1999. Distribution of soil organic matter in freshwater emergent/open water wetlands in the Portland, Oregon metropolitan area. Wetlands 19, 505–516.

Shaffer, P.W., Kentula, M.E., Gwin, S.E., 1999. Characterization of wetland hydrology using hydrogeomorphic classification. Wetlands 19, 490–504.

Shaw, S.P., Fredine, C.G., 1956. Wetlands of the United States: Their Extent and Their Value to Other Wildlife. Circular 39. United States Department of the Interior, Fish and Wildlife Service, Washington, DC.

Shulse, C.D., Semlitsch, R.D., Trauth, K.M., Gardner, J.E., 2012. Testing wetland features to increase amphibian success and species richness for mitigation and restoration. Ecological Applications 22, 1675–1688.

Spieles, D.J., 2005. Vegetation development in created, restored, and enhanced mitigation wetland banks of the United States. Wetlands 25, 51–63.

Stanczak, M., Keiper, J.B., 2004. Benthic invertebrates in created and adjacent natural wetlands in northeastern Ohio, USA. Wetlands 24, 212–218.

Stauffer, A.L., Brooks, R.P., 1997. Plant and soil responses to salvaged marsh surface and organic matter amendments at a created wetland in central Pennsylvania. Wetlands 17, 90–105.

Stevens, C.E., Diamond, A.W., Gabor, T.S., 2002. Anuran call surveys on small wetlands in Prince Edward Island, Canada restored by dredging of sediments. Wetlands 22, 90–99.

Stevens, C.E., Gabor, T.S., Diamond, A.W., 2003. Use of restored small wetlands by breeding waterfowl in Prince Edward Island, Canada. Restoration Ecology 11, 3–12.

Stewart, R.E., Kantrud, H.A., 1971. Classification of Natural Ponds and Lakes in the Glaciated Prairie Region. U.S. Department of the Interior. Fish and Wildlife Service. Bureau of Sport Fisheries and Wildlife, Washington, DC.

Stewart, R.E., Kantrud, H.A., 1972. Vegetation of Prairie Potholes, North Dakota, in Relation to Quality of Water and Other Environmental Factors. U.S. Geological Survey Professional Paper 585-D.

Straskrabova, J., Prach, K., 1998. Five years of restoration of alluvial meadows: a case study from Central Europe. In: Joyce, C.B., Wade, P.M. (Eds.), European Wet Grasslands: Biodiversity, Management, and Restoration. John Wiley and Sons, Chichester, UK, pp. 295–303.

Streever, W.J., Crisman, T.L., 1993. A comparison of fish populations from natural and constructed freshwater marshes in central Florida. Journal of Freshwater Ecology 8, 149–153.

Streever, B., Zedler, J., 2000. To plant or not to plant. BioScience 50, 188–189.

Streever, W.J., Portier, K.M., Crisman, T.L., 1996. A comparison of dipterans from ten created and ten natural wetlands. Wetlands 16, 416–428.

Sutton-Grier, A., Ho, M., Richardson, C.J., 2009. Organic amendments improve soil conditions and denitrification in a restored riparian wetland. Wetlands 29, 343–352.

Tallowin, J., Kirkham, F., Smith, R., Mountford, O., 1998. Residual effects of phosphorus fertilization on the restoration of floristic diversity to wet grassland. In: Joyce, C.B., Wade, P.M. (Eds.), European Wet Grasslands: Biodiversity, Management, and Restoration. John Wiley and Sons, Chichester, UK, pp. 249–263.

Taylor, J., Middleton, B.A., 2004. Comparison of litter decomposition in a natural versus coal-slurry pond reclaimed as wetland. Land Degradation and Development 15, 439–446.

Teal, J.M., Kanwisher, J.W., 1966. Gas transport in the marsh grass, *Spartina alterniflora*. Journal of Experimental Botany 17, 355–361.

van der Valk, A.G., 1981. Succession in wetlands: a Gleasonian approach. Ecology 62, 688–696.

van der Valk, A.G., 1982. Succession in temperate North America wetlands. In: Gopal, B., Turner, R.E., Wetzel, R.G., Whigham, D.F. (Eds.), Wetlands Ecology and Management. National Institute of Ecology and International Scientific Publications, Jaipur, India, pp. 169–179.

van der Valk, A.G., Bremholm, T.L., Gordon, E., 1999. The restoration of sedge meadows: seed viability, seed germination requirements, and seedling growth of *Carex* species. Wetlands 19, 756–764.

VanRees-Siewert, K.L., Dinsmore, J.J., 1996. Influence of wetland age on bird use of restored wetlands in Iowa. Wetlands 16, 577–582.

Voigts, D.K., 1976. Aquatic invertebrate abundance in relation to changing marsh vegetation. American Midland Naturalist 95, 312–322.

Wang, G., Middleton, B., Jiang, M., 2013. Restoration potential of sedge meadows in hand-cultivated fields in northeastern China. Restoration Ecology 21, 801–808.

Wample, R.L., Reid, D.M., 1979. The role of endogenous auxins and ethylene in the formation of adventitious roots and hypocotyls hypertrophy in flooded sunflower plants (*Helianthus annuus* L.). Physiologia Plantarum 45, 219–226.

Weller, M.W., 1978. Management of freshwater marshes for wildlife. In: Good, R.E., Whigham, D.F., Simpson, R.L. (Eds.), Freshwater Wetlands: Ecological Processes and Management Potentia. Academic Press, New York, pp. 267–284.

Weller, W.W., 1994. Freshwater Marshes: Ecology and Wildlife Management. University of Minnesota Press, Minneapolis, MN.

Whelan, R.J., 1995. The Ecology of Fire. Cambridge University Press, Cambridge, UK.

Whigham, D.F., Pittek, M., Hoffmockel, K.H., Jordan, T., Pepin, A.L., 2002. Biomass and nutrient dynamics in restored wetlands on the outer coastal plain of Maryland, USA. Wetlands 22, 562–574.

Whittecar, G.R., Daniels, W.L., 1999. Use of hydrogeomorphic concepts to design created wetlands in southeastern Virginia. Geomorphology 31, 355–371.

Wolf, K.L., Ahn, C., Noe, G.B., 2011. Development of soil properties and nitrogen cycling in created wetlands. Wetlands 31, 699–712.

Zedler, J.B., Kercher, S., 2004. Causes and consequences of invasive plants in wetlands: opportunities, opportunists, and outcomes. Critical Reviews in Plant Sciences 23, 431–452.

Forested Wetlands

Chapter Outline

Introduction

Forested wetlands, like inland marshes, consist of a number of different plant communities and are characterized by a number of different hydroperiods and water sources. They are found in a number of geomorphic or landscape positions, along streams and rivers, in topographic depressions, and on flats. Riparian forests are located in headwaters and on lower-order streams. They receive much of their water from overland flow from adjacent terrestrial lands (Brinson, 1993). Their hydroperiod is characterized by brief, infrequent inundation, but sometimes to great depth (Figure 6.1). Depending on the frequency and duration of inundation, they may or may not possess the characteristic wetland hydrology, hydrophytic vegetation and hydric soils described in Chapter 2, Definitions. Floodplain forests are farther located downstream on larger, higher-order streams and rivers. They receive most of their water from overbank flow (Brinson, 1993). Flooding is of longer duration than in riparian forests.

Riparian and floodplain forests span the longitudinal continuum from the headwaters downstream to the receiving waters, lakes, and estuaries (Figure 6.2(a)). Riparian forests represent a transient sink for materials transported from higher portions of the watershed (Figure 6.2(b)). This is especially true for sediment and phosphorus (P). Floodplain wetlands, being lower in watershed, serve as longer-term sinks for material exported from the terrestrial landscape on their way to the sea (Figure 6.2(b)).

Floodplain forests, and to a lesser extent riparian areas, are exposed to the periodic pulsing of river flooding. The *flood pulse concept* of Junk et al. (1989) posits that

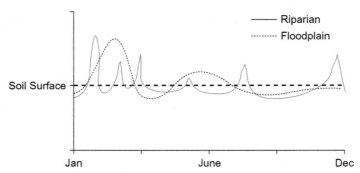

Figure 6.1 Hydrograph showing the depth, duration, frequency, and seasonality of flooding of riparian and floodplain wetlands.

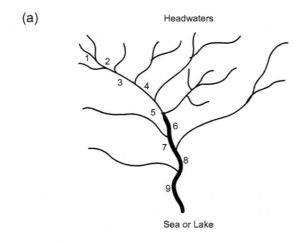

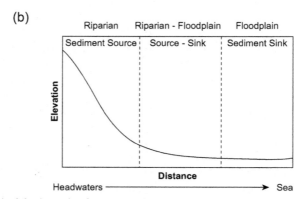

Figure 6.2 (a) Aerial schematic of a watershed showing the arrangement of streams. Lower-order streams are located in the headwaters and upper part of the catchment and increase in number and size downstream. Larger, higher-order streams combine to form rivers in the lower part of the catchment. (b) Generalized change in elevation and forested wetlands (riparian, floodplain) from the headwaters to the receiving water body.

predictable flood events shape the biophysical template of streams and rivers and their associated terrestrial and wetland habitats. On lower-order streams where riparian forests dominate, the flood pulse is brief and unpredictable. On higher-order streams including rivers, the flood pulse is of long duration and predictable. Within a given watershed, the flood pulse generally occurs at the same time each year, corresponding to regional climate patterns such as the monsoonal winds or the "spring" snowmelt (Junk et al., 1989). The flood pulse concept is an outgrowth of the pulsing paradigm of Odum (1969) who first used it to describe the predictable pulsing of the tides on saline coastal marshes and later to describe freshwater marshes such as the Everglades (Florida, USA) where water levels rise and fall in accordance with wet and dry seasons. The seasonal "flood pulse" of floodplain forests is important for maintaining the high productivity and vigor of these and other flood-pulsed wetlands by replenishing soil moisture, creating alternate cycles of drying and wetting that promote coupled processes such as aerobic–anaerobic decomposition and nitrification–denitrification. It is similar to the dry-down phase of freshwater marshes (Weller, 1978) and is critical for germination of seeds of many forested wetland species (Middleton, 1999, 2002).

Vegetation of riparian and floodplain forests is strongly linked to fluvial processes that shape landforms and dictate hydrology. Hupp and Osterkamp (1985) investigated floodplain forest vegetation distribution along Passage Creek in Virginia (USA). They identified four fluvial geomorphic landforms, depositional bar, active-channel shelf, also known as levee, floodplain, and terrace, each of whose vegetation was distinct. Soil properties were less predictive of vegetation composition than fluvial landform, supporting the idea that hydrology and its associated disturbance pulse drive the development of plant communities.

Because of their strong connections to both terrestrial and aquatic ecosystems, riparian and floodplain forests are important for trapping sediment and removing nutrients and this has been an important rationale for restoring them. Riparian and floodplain forests serve as short-term sediment sinks and longer-term sediment storage sites, respectively, storing 14–58% of the total upland sediment production (Phillips, 1989). Kleiss (1996) reported that most sediment deposition in bottomland hardwood floodplain forests occurred not on the natural levee but in the *first bottom*, the lowest portion of the floodplain just inland of it. Distance from the main channel, flood duration, and tree basal area are factors that enhance sediment deposition (Kleiss, 1996; Bannister et al., 2014). Connectivity between the wetland and active channel also determines nutrient (N and P) removal. In floodplains with natural hydrogeomorphology, deposition of N and P was greater than in floodplains where connectivity was restricted by levees (Noe and Hupp, 2005). Substantial P is removed when sediment is deposited in riparian and floodplain forests. Restoration of riparian zones using vegetated filter strips, a type of best management practice, has been widely used to reduce P transport from agricultural lands (Lowrance et al., 1986; Sharpley et al., 2000). About 50–80% of the sediment leaving an agricultural field was trapped in adjacent riparian forests and about 50% of the P exported from the field was deposited with the sediment (Cooper and Gilliam, 1987; Daniels and Gilliam, 1996).

Sediment deposition also buries N that accumulates with soil organic matter (SOM). Studies estimate that $1–8\,g\,N/m^2$ and $0.1–1.5\,g\,P/m^2/year$ are buried in floodplain soils annually (Craft and Casey, 2000; Bannister et al., 2014).

Riparian areas also remove considerable N through denitrification and burial in soil. Riparian areas have been restored to denitrify nitrate in shallow groundwater exiting agricultural land (Lowrance et al., 1984; Jacobs and Gilliam, 1985; Pinay et al., 1993; Hill, 1996; Verchot et al., 1997; Mitsch et al., 2001; Sidle et al., 2000). They can denitrify 3–4 g N/m^2/year (NRC, 2000) with higher rates (13 g N/m^2/year) in areas receiving high nitrate loadings (Brinson et al., 1984; Walbridge and Lockaby, 1994). Denitrification is positively related to wetness with poorly drained and very poorly drained soils exhibiting greater denitrification than somewhat poorly drained and well-drained soils (Groffman et al., 1992; Clement et al., 2002). Denitrification also is positively related to organic matter content of the soil (Clement et al., 2002; Marton et al., 2014). Floodplain forests also are important agents for denitrification. Ullah and Faulkner (2006) compared denitrification potential in a variety of land uses including forested wetland, herbaceous wetland, vegetated and unvegetated ditches, and agricultural land. Forested wetlands exhibited the highest rates of denitrification. Sites that were vegetated, wet, and contained fine-textured (clayey) soils, common features of floodplain forests, had the highest rates of denitrification.

Riparian Forests

Riparian areas exist at the interface between streams and terrestrial land (Figure 6.3). The word *riparian* is defined by Webster as that which is *located on the bank of a natural watercourse (as in a river)* (http://www.merriam-webster.com). The National Research Council (2002) of the US defined riparian areas in the following way.

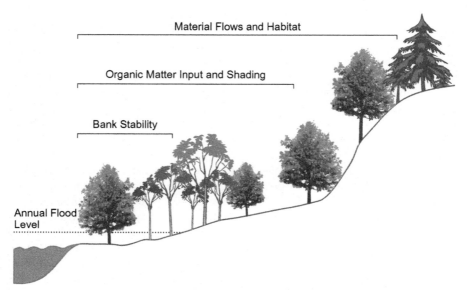

Figure 6.3 Cross-section of a riparian area showing the processes, habitat, organic matter inputs, and bank stability provided by different vegetations.
Adapted from Ilhardt et al. (2000).

*Riparian areas are transitional between terrestrial and aquatic ecosystems and
are distinguished by gradients in biophysical conditions, ecological processes and
biota. They are areas in which surface and subsurface hydrology connect water
bodies with their adjacent uplands. They include those portions of terrestrial
ecosystems that significantly influence exchanges of energy and matter with aquatic
ecosystems. Riparian areas are adjacent to perennial, intermittent, and ephemeral
streams, lakes and estuarine-marine shorelines.*

Riparian areas can be described as transition zones, ecotones, edges, or boundaries, between terrestrial and aquatic ecosystems. Naiman and Decamps (1997) describe riparian areas as being analogous to a semipermeable membrane regulating the flow of energy and materials between adjacent environmental patches. They typically are long and sinuous, paralleling the stream, and have a large edge to area ratio that facilitates energy and material exchange and processing.

Riparian forests shade the stream, dampening oscillations in water temperature (NRC, 2002). They provide organic matter (leaf litter) to the stream community, filter nutrients and sediment, and provide corridors for biotic dispersal and movement (Vannote et al., 1980; Naiman and Decamps, 1997). By serving as corridors, they contribute to the maintenance of regional biodiversity (Naiman et al., 1993). Different portions of the riparian forest provide different functions (Figure 6.3). Trees adjacent to the stream provide bank stability. Those proximal to the channel shade the stream and add organic matter as leaf litter to support in-stream heterotrophic processes. Trees in the entire zone modulate material flows, including intercepting sediment and nutrients and providing habitat and corridors for dispersal and movement. Riparian forests on first-order streams are especially important elements of the landscape. They support high levels of biodiversity and are impacted more by human activities than other types of wetlands (Whigham, 1999).

Riparian forests consist of a number of species though several genera are nearly always represented in different climates and geographic regions (Table 6.1). They include cottonwood (*Populus*), willow (*Salix*), sycamore (*Platanus*), alder (*Alnus*), and birch (*Betula*). Because of their strong connectivity to other ecosystems, invasive species often colonize riparian forests. Invasiveness is facilitated by human disturbance such as grazing and channelization and by alteration, mainly drying, of hydrology caused by surface and groundwater withdrawals. This is especially true in the arid southwestern US and other arid regions where saltcedar (*Tamarix*), Russian olive (*Eleagnus*), and other invasive species have spread (NRC, 2002).

Floodplain Forests

Floodplain forests exist on higher-order streams and on rivers. They are most common in regions where precipitation exceeds evapotranspiration although in drier climates they may be maintained by the flood pulse driven by rain and snowmelt upstream. Webster (http://www.merriam-webster.com) defines floodplain as *an area of low, flat land along a river or stream that may flood*. For this book, we refine the definition to include those forested lands that are submerged by floodwaters during overbank flow from a river channel.

Table 6.1 Common Species of Riparian Forests of the Eastern, Southwestern, Northwestern US and Alaska

Eastern	Southwestern	Northwestern[a]	Alaska
Acer saccharinum	*Alnus oblongofolia*	*Alnus rubra*	*Alnus crispa*
Alnus serrulata	*Platanus wrightii*	*Alnus tenuifolia*	*Betula nana*
Betula nigra	*Populus fremontii*	*Populus angustifolia*	*Salix alaxensis*
Populus deltoides	*Prosopsis juliflora*	*Populus tremuloides*	*Salix arbuculoides*
Populus heterophylla	*Prosopsis pubensis*	*Populus trichocarpa*	*Salix glauca*
Platanus occidentalis	*Salix bonplandiana*	*Salix scouleriana*	*Salix lanata*
Salix nigra	*Salix goodingii*		*Salix pulchra*
	Eleagnus angustifolia[b]		
	Tamarix pentandra[b]		
	Tamarix chiensis[b]		

[a]Includes Rocky Mountains.
[b]Invasive species.
From Brinson et al. (1981), NRC (2002), Stromberg (2001).

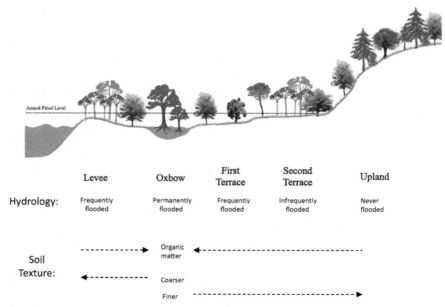

Figure 6.4 Cross-section of a forested floodplain wetland showing the major landforms and height of the annual flood pulse. Different assemblages of trees occupy the different habitats that vary with depth and duration of flooding. See text for detailed explanation of the landforms and habitat types.

Floodplain forests extend across alluvial terraces of varying width, depending on the size of the river and geomorphology (Figure 6.4). They consist of terraces of different ages. The youngest and lowest terrace is closest to the channel and is separated by a naturally formed levee. The lowest lying terraces are flooded several times a year and for several weeks to months at a time. Terraces of increasing age and elevation are found farther away from the channel. They are flooded less frequently and for shorter duration. The depth of inundation also is less. Relic channel features such as oxbows often are found in the floodplain. They are nearly permanently flooded and are dominated by obligate hydrophytes.

The various habitats, levee, terrace, and oxbow, experience different hydrodynamics as well. Overbank flooding of the levee is vigorous whereas flooding of the oxbow is more quiescent. The difference in hydrodynamics results in differences in soil properties. The levee contains more coarse materials such as sand (Faulkner et al., 2011). Finer soil particles, silts and clays, often increase with distance from the channel, especially in low-lying areas (Hodges, 1997; Bannister et al., 2014). Soils of the oxbows and other lower lying areas of the floodplain contain high organic matter content relative to other habitats (Wharton et al., 1982; Bannister et al., 2014).

Hydroperiod also structures the plant communities that exist across the floodplain. In the southeastern and midwestern US, species such as *Salix*, *Platanus*, *Betula*, and *Populus* are common on the levee (Table 6.2; Wharton et al., 1982). Oxbows are dominated by obligate hydrophytes such as *Taxodium* (bald cypress) and *Nyssa* (tupelo gum). Terraces contain the most species-rich communities. Lower terraces are dominated by *Acer*, *Fraxinus*, *Liquidambar*, *Quercus* (oak), and *Carya* (hickory). The higher, drier terraces contain a number of oaks, gums, and other species. Above the floodplain, mesophytic species like *Fagus* (beech), *Liriodendron* (tulip poplar), *Pinus* (pines), and oaks dominate.

Forested Wetlands as Habitat

Riparian forests are important habitat for amphibians, especially salamanders. Crawford and Semlitsch (2007) evaluated the width needed to support four species of stream-breeding salamanders in western North Carolina, USA. The farthest ranging species, traveled 43 m from the stream. Based on this, it was recommended that a buffer width of 93 m was needed, 43 m to accommodate the species range and an additional 50 m to buffer edge effects. Studies of buffer zones and turtle habitat suggested that a buffer of 73 m is needed to support their nesting and hibernation (Burke and Gibbons, 1995).

Floodplain forests are essential habitat for migratory songbirds, especially neotropical migrants. Along the US gulf coast, migratory songbirds stop to regenerate depleted energy stores during the fall and spring migrations. Floodplain forests, especially hardwood forests dominated by trees such as water oak (*Quercus nigra*), black gum (*Nyssa biflora*), and sweetgum (*Liquidambar styraciflua*), are vital habitat for these migrants (Buler et al., 2007; Brittain et al., 2010) and large patches are needed to support healthy populations. The amount of hardwood forest cover, especially the amount within a 5-km radius, was the most important predictor explaining the density of migrants (Buler et al., 2007). Wakeley and Roberts (1996) compared songbird densities across a wetness gradient of floodplain forests of the Cache River, Arkansas. The four habitats, oxbows dominated by tupelo gum and bald cypress, and terraces dominated by overcup oak–water oak, Nuttall oak–willow oak–sweetgum, or sweetgum–water oak–pignut hickory, varied in their use. Overall, tupelo gum–cypress forests

Table 6.2 Common Tree Species of Southeastern and Midwestern US Floodplains and Adjacent Uplands

Levee	*Betula nigra*	River birch
	Populus deltoids	Cottonwood
	Populus heterophylla	Swamp cottonwood
	Platanus occidentalis	Sycamore
	Salix nigra	Black willow
Lower terrace	*Acer rubrum*	Red maple
	Acer saccharinum	Silver maple
	Carya aquatica	Water hickory
	Celtis laevigata	Sugarberry
	Fraxinus pennsylvanicus	Green ash
	Liquidambar styraciflua	Sweetgum
	Nyssa biflora	Black gum
	Quercus lyrata	Overcup oak
	Ulmus americanus	American elm
Oxbow	*Nyssa aquatica*	Tupelo gum
	Taxodium distichum	Bald cypress
Upper terrace	*A. rubrum*	Red maple
	A. saccharinum	Silver maple
	L. styraciflua	Sweetgum
	N. biflora	Black gum
	Quercus michauxii	Swamp chestnut oak
	Quercus nigra	Water oak
	Quercus phellos	Willow oak
Upland	*Cercis Canadensis*	Redbud
	Cornus florida	Dogwood
	Fagus grandifolia	Beech
	Liriodendron tulipifera	Tulip poplar
	Pinus taeda	Loblolly pine
	Prunus serotina	Black cherry
	Quercus alba	White oak
	Quercus virginiana	Live oak

From Wharton et al. (1982), Middleton (2002).

harbored species that were much less abundant elsewhere. Chimney swifts (*Chaetura pelagica*), prothonotary warblers (*Protonotaria citrea*), and great crested flycatchers (*Myiarchus critinus*) were associated with the wetter tupelo gum–cypress forests whereas summer tanagers (*Piranga rubra*) and red-eyed vireos (*Vireo olivaceus*) were associated with the terraces. Overall, a diverse floodplain forest songbird community was positively related to tree species diversity, canopy cover, and mid-story density (Twedt et al., 1999).

Floodplain and riparian forests provide habitat for a number of mammals. Small mammals that use these forests include shrew, mole, bat, squirrel, rabbit, raccoon, mink, otter, and others (DeGraaf and Yamasaki, 2000). Mammals such as deer, moose, and bear require large intact forests such as those found in floodplain forests (Faulkner et al., 2011).

Table 6.3 Physical, Chemical, and Biological Functions of Riparian and Floodplain Forests

Riparian Forests	Floodplain
Physical	
Bank stabilization	Floodwater storage
Chemical	
Sediment trapping	Sediment trapping
P removal	P removal
N removal (burial, denitrification)	N removal (burial, denitrification)
Biological	
Shading	Habitat—migratory songbirds
Detritus (leaf litter, coarse woody debris)	—large mammals
Habitat—herpetofauna	Habitat corridors
Habitat corridors (dispersal)	

Loss of Forested Wetlands

Riparian and floodplain forests provide a number of physical, chemical, and biological functions (Table 6.3) that are lost when these ecosystems are converted to other uses. Loss of the vegetated shoreline leads to stream narrowing and loss of in-stream ecosystem services, including the processing of nutrients (Sweeney et al., 2004). As is the case with wetlands nearly everywhere, there has been dramatic and widespread loss of forested wetlands. In the Mississippi Alluvial Valley (MAV) of the south central US, forest cover constitutes less than 25% of the floodplain today (Twedt and Loesch, 1999) when, historically, the floodplain was completely forested. Most loss is attributed to drainage for agriculture. The remaining forest is fragmented with few areas, less than 100 patches, large enough, greater than 4000 ha, to support self-sustaining densities of breeding birds (Twedt and Loesch, 1999) and large mammals.

The loss of riparian forest is even greater as human activities such as agriculture and urban–suburban development abut streams and waterways. Swift (1984) estimated that there were about 30–40 million hectares of riparian habitat in the 48 conterminous United States prior to European settlement. Today, there are about 10–14 million hectares remaining. Declines are greatest in the Mississippi River Delta, agricultural Midwest, desert Southwest, and California. In the Midwest, estimates of riparian habitat loss ranged from 70% to 95% (Willard et al., 1989). In California, Arizona, and New Mexico, riparian forest has declined by 85–95% with most losses attributed to grazing (NRC, 2002). In Europe, it is estimated that 80% of riparian habitat has disappeared during the past 200 years (Naiman et al., 1993). Worldwide, construction of dams, bank stabilization using hard structures, channelization, levees, ground- and surface-water withdrawal, and vegetation removal are causes for their loss (Brinson et al., 1981; Goodwin et al., 1997; NRC, 2002). Today, riparian and floodplain forests, because of their value to stream water quality and overall stream health, receive high priority for ecosystem restoration (NRC, 2001).

Theory

Succession in riparian and floodplain forests is allogeneic or Gleasonian in nature, structured by their elevation relative to the flood pulse. Assembly rules or the environmental sieve models of succession are most applicable (see Chapter 3, Ecological Theory and Restoration) since flooding, its depth, duration, frequency, and timing, determines the species that colonize and persist on the site. The levee, with its strong hydrodynamics, is maintained in an early successional stage by periodic and vigorous overbank flooding. Oxbows, with their near permanent inundation, may be dominated by only a handful of species whose opportunity for colonization occurs perhaps only once in a decade or more, during drought or extreme low flow. Enrichment of the soil with organic matter is not strongly evident as the cycles of flooding and dry down oxidize organic matter. Episodic flooding also delivers sediment to the soil surface, keeping soils in an early stage of succession.

Keddy (2010) suggested that, for floodplain wetlands, the evidence for autogenic succession is weak as flooding, sedimentation, and disturbance are the primary controls on species colonization. Toner and Keddy (1997) found that, for riparian and floodplain wetlands in Canada, it was the timing of flooding and not flooding depth or number of floods that determined composition of the vegetation. Allogeneic succession is especially important in riparian areas where flooding is intense but brief. In the arid southwestern US, high-energy floods set succession back to a much earlier stage through the effects of a long duration flood recession and deep (>1 m) deposition of sediment (Stromberg and Chew, 2002). Low-energy floods deposit much less sediment (1–10 cm) and the floodwaters recede much more quickly, so succession is set back less during these events. Beavers, by cutting trees and building dams that increase hydroperiod, are important agents of disturbance in forested wetlands and riparian areas (Middleton, 1999).

Restoration of floodplains and riparian areas by natural colonization is stochastic, relying on chance events, rather than deterministic. Trowbridge (2007) compared two restored floodplain wetlands in California, USA. Between 1995 and 1997, the sites were restored by reconnecting the floodplain with the river and reintroducing seasonal flooding. Six years (2000–2005) of monitoring plant communities in permanent plots revealed two distinct trajectories. On one site, the plant community became more similar with time, whereas on the other site vegetation in permanent plots became more dissimilar. Trowbridge (2007) suggested that assembly of plant communities in floodplains is stochastic and that simply restoring physical processes (i.e., hydrology) will not necessarily lead to restoration of the desired endpoint community. Abiotic factors, such as subtle differences in flood frequency, sedimentation, soils, and disturbance history, interact to produce floodplain forests with dissimilar composition and structure (Figure 6.5).

Hodges (1997) describes three patterns of succession for floodplain forests of the southern and southeastern US that are driven by flooding and sedimentation. On permanently flooded sites with little sediment deposition, cypress–tupelo forests dominate. Succession is arrested on these sites and the forest may exist unchanged for 200–300 years. On poorly drained sites at low elevation, sediment deposition

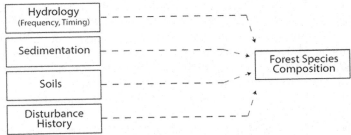

Figure 6.5 Abiotic factors that drive allogeneic succession in floodplain and riparian forests.

slows or even arrests succession, whereas on higher elevation, well-drained sites, succession proceeds along the lines of a terrestrial forest.

Floodplain and riparian forests, in particular, are susceptible to invasion by exotic species. The combination of frequent natural disturbance, high connectivity to other ecosystems, and human alteration of the stream bank makes these areas ripe for colonization. A number of plant species are problems. They include saltcedar in arid regions (NRC, 2002), grasses such as Japanese stilt grass (*Microstegium vimineum*) and woody shrubs such as *Elaeagnus* spp., and *Lagustrum* spp. in temperate regions (Vidra et al., 2006; Osland et al., 2009; Ho and Richardson, 2013), and horticultural introductions such as English ivy (*Hedera helix*), periwinkle (*Vinca minor*), and Japanese honeysuckle (*Lonicera japonica*) in urban areas.

Practice

Early forested wetland restoration projects focused on timber production and afforestation. In the US, wetland trees were planted along the margins of fluctuating-level reservoirs for timber in the 1940s (Silker, 1948). Species such as bald cypress (*Taxodium distichum*), tupelo gum (*Nyssa aquatica*), green ash (*Fraxinus pennsylvanica*), sweetgum (*L. styraciflua*), sycamore (*Platanus occidentalis*), Atlantic white cedar (*Chamacyparis thyoides*), and oaks (*Quercus nigra, Quercus phellos*) were planted and recommendations were made regarding species-specific elevation and flooding regimes (Silker, 1948). Today restoration of forested wetlands is widespread and usually involves restoring hydrology by reconnecting floodplains with the river or stream, then seeding or planting woody species. On agricultural lands, hydrology is reintroduced using water control structures such as flashboard risers and adding microtopographic features (see *Importance of Microtopography* below) (Tweedy and Evans, 2001). Following disturbances such as logging and harvesting, replanting often is done. Restoration of riparian areas involves reestablishing streamside woody vegetation for bank stabilization and water quality improvement (NRC, 1992, 2002; Vellidis et al., 1994, 2003).

In the US, government programs such as the Conservation Reserve Program and Wetland Reserve Program restore floodplain and riparian forest by focusing heavily on reforestation. Much work has been undertaken in the MAV where extensive drainage and logging of bottomland hardwood forests occurred (Faulkner et al., 2011).

The floodplain forest historically extended from southern Illinois to the Louisiana coast and comprised the largest contiguous floodplain forest in North America (Haynes, 2004). Today, its area is reduced by 90% (Faulkner et al., 2011). In the MAV, more than 41,000 ha of wetlands have been restored by restoring hydrology and seeding/planting native species. More than 15,000 ha of riparian forest buffer in the region have been restored by natural and artificial regeneration of native trees and shrubs (Faulkner et al., 2011), much of it on marginal or abandoned agricultural land (Toliver, 1986; Twedt, 2004).

In Europe, forested wetlands are restored by reconnecting rivers with their floodplains and riparian areas. In Denmark, re-meandering of streams is done to connect riparian wetlands with their channel to improve water quality (Pederson et al., 2007; Hoffman et al., 2011). In large European rivers such as the Rhone (Henry et al., 2002), Rhine (Cals et al., 1998), and Danube Rivers (Tockner et al., 1998), efforts have been made to reconnect the river with the floodplain to provide habitat for endangered plants and animals. For example, on the Danube River, 16% of the 653 vascular plant species and 50% of the 19 amphibian species require floodplain habitat (Tockner et al., 1998). Restoration efforts include reconnecting oxbows or old channels with the main channel, dredging fine sediment to reduce eutrophication, and rehabilitating river banks and littoral zones (Cals et al., 1998; Acreman et al., 2007). River and floodplain restoration also has been carried out on smaller rivers in the United Kingdom, the Netherlands, Italy, and elsewhere (Mant and Janes, 2006). A number of river restoration projects were undertaken beginning in the 1980s by WWF (World Wide Fund for Nature) under the project *Wise Use of Floodplains* and supported by the EU (European Union) Life-Environment Programme (Zockler, 2000). River and floodplain wetland restoration efforts are well underway in Japan as well (Yoshimura et al., 2005). The challenges are greater there than in other countries due to the high population density and degree of urbanization, the rice culture that converts the floodplain into arable land, and the mountainous topography with its steep floodplains and flashy flows (Nakamura et al., 2006).

In the US, the best examples of large-scale reconnection of river and floodplain are the Kissimmee River (Florida) and the Colorado River. The Kissimmee River restoration involves restoring 70 km of river channel and connecting it to 11,000 ha of floodplain wetlands (Toth et al., 2002). Restoration of the river, which was channelized between 1962 and 1971, is underway with the goal of restoring floodplain vegetation and biota, including aquatic invertebrates, amphibians and reptiles, fish, and avian communities. In the Colorado River, controlled releases of water from the Glen Canyon dam upstream of the Grand Canyon have been used to restore sandbars and riparian habitat for endangered species, including the Kanab ambersnail, humpback chub, and southwestern willow flycatcher (Wuethrich, 1998; Cohn, 2001; Pennisi, 2004). Reduced duration and depth of flooding following construction of the dam actually enhanced riparian habitat (Webb et al., 1999), but native riparian vegetation dominated by cottonwood (*Populus*) and willows (*Salix*) disappeared and was replaced by invasive saltcedar (*Tamarix ramosissima*) (Cohn, 2001). Releases of water and sediment from Lake Powell were undertaken in 1996 and again in 2004. The controlled floods buried herbaceous riparian vegetation under a meter or more of sand but had little adverse effect on woody riparian vegetation (Stevens et al., 2001). Without annual or near-annual releases of water and adequate sediment, though, permanent restoration of riparian habitat along the river will not be successful.

Restoration of forested wetlands is more difficult as compared to herbaceous wetlands such as shallow marsh. Much of this is due to the long time needed to produce a mature forest but also to the difficulty in restoring the proper hydrology at some sites. Brown and Veneman (2001) reported that, in Massachusetts (USA), 71% of wetland impacts were to forested wetlands and that restoration efforts to mitigate for the loss produced no wetlands (39%), wet meadows (37%), or some other wetland type (24%). Robb (2002) reported similar results for forested wetland mitigation projects in Indiana (USA) where the failure rate for forested wetlands was 71%. These studies highlight the difficulty in restoring forested wetlands and point to the need to preserve existing ones.

Floodplain Forest

Once flooding is reintroduced, restoration usually consists of reforestation through direct seeding and planting (Allen, 1990; King and Keeland, 1999). Bare root seedlings typically are planted (Figure 6.6(a); King and Keeland, 1999). Mostly hard

Figure 6.6 (a) Bare root cypress (*Taxodium distichum*) seedlings being planted at McMillan Creek, Penholloway Swamp, Georgia. (b) The same plantings 8 years later.
Photo (a) courtesy of Mark Jicha.

mast, large-seeded species such as oaks (*Quercus*) are planted to enhance wildlife habitat and to accelerate succession (Clewell and Lea, 1989; Schoenholtz et al., 2001; Haynes, 2004). However, such plantings typically result in even aged forests with low species diversity (Faulkner et al., 2011). More recent efforts involve planting mixtures of large-seeded species and small, light-seeded species such as ash (*Fraxinus pennsylvanica*), bald cypress (*T. distichum*) (Figure 6.6(b)), and bitter pecan (*Carya aquatica*) to mimic a more natural regeneration process (Schoenholtz et al., 2001).

A review of different planting techniques (direct seeding vs bare root seedling) and season of planting revealed no clear preference for one technique or season over the other (Schoenholtz et al., 2001). Other studies, however, show generally superior establishment and growth of planted stands (Allen, 1990; Haynes, 2004) though it is more expensive than direct seeding. Planting also produces stands of greater height and diameter growth. Allen (1990) compared direct seeded and planted stands of oaks on the Yazoo River (MS) that were 4–8 years old. The planted stands were significantly taller and diameter growth was greater than in seeded stands. Poor growth of seeded stands was attributed to dense ground cover that impeded seedling development. Planted seedlings, in contrast, are little affected by competition from herbaceous vegetation (McLeod et al., 2000). Rather, environmental factors during seedling establishment, including drought during the growing season, late freeze following planting, high temperatures, standing water, and herbivory by rodents, rabbits, or deer, affect the success of seedling and sapling establishment (Haynes and Moore, 1988).

Natural colonization of restored sites often is limited to species such as ash (*Fraxinus*), sweetgum (*Liquidambar*), and others that produce light seeds (Allen, 1990; Shear et al., 1996; Twedt, 2004) and that dramatically slows the development of a species-rich forest community. Studies suggest that even after 25–50 years, natural colonization cannot be counted on to produce a mixed species forest (Allen, 1997; Haynes, 2004). This is especially true on sites that are distant, more than 60 m, from existing floodplain forest. Twedt (2004) suggested that to increase forest species diversity, greater diversity of planted species and thinning of planted trees during canopy closure is needed. Allen (1997) suggested that practices which slow the rate of canopy closure and create more opportunities for natural colonization should be implemented to enhance woody diversity. These practices include direct seeding rather than planting, leaving gaps, and planting heavy-seeded species such as hickory (*Carya*) that are not typically planted.

Natural colonization also depends on conditions favorable for germination of seeds. Flooding and sedimentation are the two disturbances that most affect germination and establishment of vegetation on floodplains (Middleton, 1999; Pierce and King, 2007; Keddy, 2010) and riparian areas (Brooks et al., 2005). Various species respond differently to these disturbances. In a greenhouse experiment, Walls et al. (2005) evaluated the effects of flooding and sedimentation on germination and sapling growth of three floodplain trees, red maple (*Acer rubrum*), green ash (*Fraxinus pennsylvanica*), and pin oak (*Quercus palustris*). Both flooding and sedimentation delayed germination of green ash and pin oak, but overall germination was reduced only in the oak (Figure 6.7). Growth of 2-year-old saplings was unaffected by periodic inundation, but the combination of continuous flooding and sedimentation decreased sapling growth. Sedimentation also led to the production of adventitious roots in the three species. Pierce and King

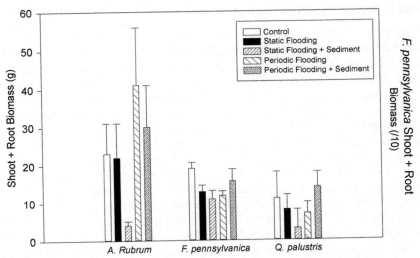

Figure 6.7 Effects of flooding and sedimentation on biomass of three floodplain trees. Adapted from Walls et al. (2005).

(2007) evaluated the effects of flooding (0, 15, 30 days duration), sedimentation (2, 8 cm), and sediment type (topsoil, sand) on germination of two oak species. Germination of overcup oak (*Quercus lyrata*) seedlings increased in response to flooding duration. In contrast, germination of swamp chestnut oak (*Quercus michauxii*) seedlings decreased in the 30-day flood treatment. Sedimentation was of secondary importance, mostly reducing the height of seedlings. Overall, these studies highlight the allogeneic nature of succession in floodplain forests.

Riparian Forest

Riparian areas are restored by stabilizing stream banks and planting vegetation. Bank plantings using willow (*Salix*) stakes are employed for stream bank stabilization (NRC, 1992), whereas single and multispecies plantings are used to restore functions described in Table 6.3. In the eastern and midwest US, species that are planted include red maple (*A. rubrum*), silver maple (*Acer saccharinum*), river birch (*Betula nigra*), sweetgum (*L. styraciflua*), green ash (*Fraxinus pennsylanica*), hawthorn (*Crataegus* sp.), and others (Willard et al., 1989; Niswander and Mitsch, 1995). In arid and semi-arid regions of the US, *Salix* spp., *Populus fremontii*, *Platanus wrightii*, *Prosopsis* spp., *Fraxinus velutina*, and *Juglans major* (walnut) are planted (Carothers et al., 1990).

Typically, one-gallon potted plants are used for planting. Saplings are planted along the stream bank, between the stream or river and agricultural fields or adjacent to existing riparian forest. The width of the restored buffer varies depending on the land available but a buffer of at least 15 m (50 ft) is considered necessary to protect streams (Castelle et al., 1994; NRC, 2002) though a much wider buffer, 100 m or more, is necessary to provide dispersal corridors and support migratory songbirds.

Importance of Microtopography

A characteristic feature of floodplain forests and other forested wetlands is their pronounced microtopography that consists of hummocks and hollows (Barry et al., 1996; Day et al., 2007). These features exert a strong influence on elevation and, hence, flooding. Bledsoe and Shear (2000) found that, in floodplain forests of North Carolina, an elevation change of as little as 10 cm resulted in a 20% difference in flooding frequency during the growing season. Microtopography lends spatial variability in plant communities (Titus, 1990), soils (Stolt et al., 2001), plant productivity, and biogeochemical processes (Schilling and Lockaby, 2005) that enhance plant species diversity and nutrient cycling (Courtwright and Findley, 2011; Bannister et al., 2014). Hummocks and other higher elevation microsites are important sites for establishment of woody species (Titus, 1990; Vivian-Smith, 1997) (Figure 6.8).

Created and restored forested wetlands often lack microtopography, leading to reduced spatial variability of vegetation, soil, and microbial processes. This is the result of drainage for agriculture and, sometimes, activities such as grading that are used to prepare the site (Unghire et al., 2011). Bruland and Richardson (2005b) compared spatial variability of soil properties of created, restored, and natural forested wetlands in North Carolina. Natural forested wetlands contained greater fine-scale variability in bulk density, SOM, and pH than created and restored forests. Microbial processes such as denitrification potential also were spatially more variable in natural forests and that was related to spatial distribution of nitrate in soil (Bruland et al., 2006).

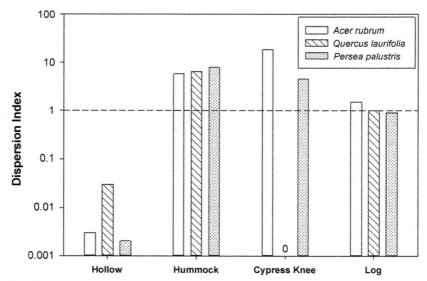

Figure 6.8 Dispersion index of three floodplain forest seedlings in low- (hollow) and high-elevation microsites. A dispersion index greater than one indicates that the species is overrepresented on a given microsite. Values less than one indicate underrepresentation. Adapted from Titus (1990).

Introducing microtopography leads to greater spatial variation in hydrology, soil temperature, and nutrient cycling. In a forested wetland restored on agricultural land in North Carolina, roughing the soil surface to create microtopography led to a 30% reduction in outflow of water from the site (Tweedy and Evans, 2001). Bruland and Richardson (2005a) compared hydrology, vegetation, and nutrient cycling in a 3-year-old restored forested wetland where microtopography, hummocks, hollows, and flats were introduced. Hollows were characterized by wetter and cooler soil conditions relative to restored sites where microtopography was not introduced. Aboveground biomass also was greater in hollows and was attributed to greater growth of herbaceous vegetation. Colonization and species richness of woody vegetation was greater on hummocks that were inundated less, producing soils that were more aerobic relative to hollows. Available nitrate and ammonium also were greater in soils of hummocks. The findings illustrate the importance of microtopography when restoring community structure and ecosystem function of forested wetlands.

Amendments

Amendments sometimes are used to accelerate ecosystem development of forested wetlands. Most additions involve adding topsoil to increase SOM and improve fertility. Bruland and Richardson (2004) evaluated the effects of topsoil additions on created forested wetlands in Virginia. Topsoil additions increased SOM, moisture content, and phosphorus sorption capacity. Bailey et al. (2007) compared organic matter loading rates on forested wetland creation in the same area. Composted wood and yard waste was added at rates from 0 to 336 Mg/ha. Soil organic C, total N, and available P increased with increasing rate of addition. The optimum rate of addition for wetland woody species was at the intermediate rate of 112 Mg/ha. Higher rates increased soil surface elevation and promoted growth of terrestrial trees. There was no effect of amendments on diversity of woody species. Bailey et al. (2007) suggested that intermediate rates provided additional soil nutrients without increasing surface elevation to the point where inundation is reduced and soils dry out. As part of the same study, Bruland et al. (2009) evaluated the effects of increased organic matter loadings on water quality improvement functions, microbial biomass, denitrification, and phosphorus sorption. Microbial biomass increased linearly with organic matter additions whereas denitrification peaked at intermediate loadings (Figure 6.9). Phosphorus sorption, in contrast, decreased with organic matter additions.

Studies where organic matter (compost) amendments were added to restored riparian wetlands yielded similar results. Soil nutrient and microbial processes, including denitrification, were enhanced by compost additions but there was a limited response of the vegetation (Sutton-Grier et al., 2009). The cumulative outcome of these studies is that topsoil and organic matter additions accelerate ecosystem development but that different functions or services respond differently. Ideally, moderate loadings are best for optimizing some services without compromising others.

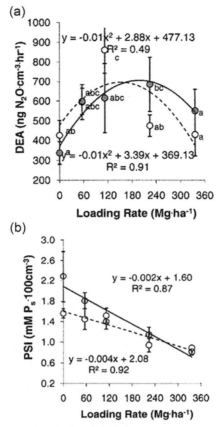

Figure 6.9 Regressions of (a) denitrification enzyme activity, and (b) phosphorus sorption index versus organic amendment loading rate in restored floodplain forests.
Reprinted with permission from Bruland et al. (2009). Reproduced with permission of the Society of Wetlands Scientists.

Ecosystem Development

Establishment of woody vegetation, as one would expect, is slow. Planted seedlings establish best on moist but not flooded soils. Niswander and Mitsch (1995) evaluated woody sapling establishment on a 2-year-old afforested wetland in Ohio. A variety of species (*A. rubrum*, *A. saccharinum*, *B. nigra*, *Fraxinus pennsylvanica*, *L. styraciflua*, *Nyssa sylvatica*, *Q. nigra*) were planted. After 2 years, most individuals were alive except for those planted in standing water.

Studies of young created and restored forests indicate rapid accrual of biomass, similar to young aggrading natural forests. Clewell (1999) compared vegetation and soils of an 11-year-old forested wetland created on reclaimed phosphate-mined lands in Florida. After 11 years, planted trees, including *Acer*, *Fraxinus*, *Liquidambar*, *Quercus*, and *Taxodium*, dominated the site. Some trees were as tall as 12.5 m and some produced seeds and seedlings. A healthy ground cover had developed as well as an A (topsoil) horizon and an incipient B (subsoil) horizon. Seventy-three percent of the woody and herbaceous species at the restored site were found in a nearby reference forest. Battaglia et al. (2002) periodically evaluated forested wetland development on farmland that was abandoned 16 years earlier in northeastern Louisiana. A number of species of oaks and hickories as well as bald cypress were planted. After 16 years, the forest was dominated by green ash (*Fraxinus pennsylvanica*) whose seeds dispersed from trees on the nearby levee. Planted species were present but not common. Height of trees ranged from about 3 to 5 m. Shear et al. (1996) compared forest development in a 50-year-old planted floodplain forest, sites of similar age that colonized naturally, and a mature reference forest in western Kentucky. Both planted and naturally colonized forests contained less aboveground biomass (181–194 t/ha) than the mature forest (239 t/ha). Planted and naturally colonized forests had a much higher rate of biomass accumulation suggesting that the C sequestration function of the forests would soon achieve equivalence with the mature stands. However, forest composition was different as the restored forests lacked the heavy-seeded oaks and hickories needed to support wildlife habitat. These studies illustrate the difficulty of heavy-seeded species such as oaks and hickories to recruit to the site and the need to plant them.

Oelbermann and Gordon (2000) compared litterfall input to streams of restored and natural riparian forests in southern Ontario, Canada. Between 1985 and 1990, the riparian area was planted with alder (*Alnus* spp.), poplar (*Populus* spp.), and maple (*A. saccharinum*). Litterfall was measured in 1996 and 1997 in three planting treatments that varied in tree density and buffer width (2–5 m, 50–100 m). Not surprisingly, the wider, densely planted buffer produced the most litterfall (161 g/m^2/year) but it only was about 50% of the litterfall measured in a mature riparian forest (314 g/m^2/year). Development of a mature forest with a closed canopy also benefits in stream organisms. Parkyn et al. (2003) compared macroinvertebrates using the quantitative macroinvertebrate community index (QMCI) in 2- to 24-year-old planted riparian forests in New Zealand. The presence of maturing riparian forest improved the QMCI score, but even in the 24-year-old forest, it was not sufficient to buffer warm stream temperatures enough to support a high quality "clean water" macroinvertebrate assemblage.

Many forested wetlands are restored on former agricultural land that retains legacies of past agricultural practices. Bruland et al. (2003) evaluated vegetation and soils of a 2-year-old forested wetland restored by cessation of agriculture, filling 3300 m of ditches, and planting 192,000 woody seedlings. The site, a 250-ha area, historically was an isolated forested wetland, a Carolina Bay, with portions of it drained and cultivated beginning in the 1960s. The site was planted with a number of wetland species, including bald cypress, tupelo gum, water ash (*Fraxinus caroliniana*), and a number of oak species. After 2 years, the restored site met the definition of a jurisdictional wetland (see Chapter 2, Definitions). However, soils of the restored site contained high bulk density, pH, phosphorus, and base cations (Ca, Mg) relative to a reference Carolina Bay, reflecting the chemical legacies of past agricultural activities.

Restoring wetlands on former agricultural land often leads to P export when soils are reflooded. This phenomenon has been observed in a number of forest and marsh restorations including projects in Florida (D'Angelo and Reddy, 1994; Pant and Reddy, 2003),

Oregon (Aldous et al., 2005), and North Carolina (Ardon et al., 2010b). In North Carolina, a large (440 ha) forested wetland restored in 2007 by filling ditches, planting saplings, and installing a water control structure exported large amounts of PO_4, especially during summer dry down (Ardon et al., 2010a). It was suggested that more than a decade may elapse before all legacy P is released (Ardon et al., 2010a).

Soils of restored floodplain forests typically contain less SOM than natural forests. This is because many sites were restored on agricultural lands that were drained and cultivated many years ago. Bruland and Richardson (2006) compared SOM in 11 created and restored forested wetlands with paired natural forested wetlands in North Carolina. Natural forested wetlands contained nearly three times more SOM (30%) than created and restored forested wetlands (12%). Natural forests also contained more inorganic N, NH_4^+, and NO_3^- than restored forests (Bruland et al., 2006). Stolt et al. (2000) compared soils of three 4- to 7-year-old constructed forested wetlands with natural reference wetlands in Virginia. Constructed forested wetlands contained less soil organic C and N than the reference wetlands. Cation exchange capacity and percent silt and clay also were lower in constructed wetland soils. A restored 10-year-old riparian forest in Indiana also contained less soil organic than natural forest even though the site exhibited similar hydrology (Vidon et al., 2014). In contrast to these studies, Hunter et al. (2008) reported no differences in soil organic C and N in young (4–7 years old) restored floodplain forests and natural forests in the Mississippi alluvial plain, Louisiana.

Based on chronosequence studies, SOM and N increase with time since restoration. Wigginton et al. (2000) compared soil carbon in 7-, 9-, and 11-year-old restored bottomland hardwood forests with a 75-year-old natural forest in South Carolina. Soil C (0–0.7 m deep) increased from 15.6 kg/m² in the 7-year-old forest to 23.3 kg/m² in the 11-year-old forest but it still was considerably less than in the mature forest (55.9 kg/m²). Stanturf et al. (2001) reported that soil organic C increased from 2% to 3.5% along a chronosequence of planted 5- to 8-year-old and 18- to 25-year-old Nuttall oak (*Quercus nuttalli*) and 60-year-old natural forests of the MAV.

Increased organic matter leads to greater microbial activity and nutrient (P) retention. D'Angelo et al. (2005) compared soil carbon, moisture, and microbial activity in early successional restored forests with eight late successional bottomland floodplain forests in western Kentucky. Restored forests contained less litter and soil C, gravimetric water content, microbial respiration, and microbial biomass than mature forests. Reduced microbial activity in restored forest soils was attributed to lower moisture content, shorter duration of wetness, and less anaerobic microbial activity. D'Angelo (2005) compared P sorption in the same young (<10-years old) restored floodplain forests and mature floodplain forests. Phosphorus sorption was much greater in natural forests and it was attributed to the high organic matter content and organically bound aluminum in the soil. Using ^{31}P nuclear magnetic resonance spectroscopy, Sundareshwar et al. (2009) found that organic forms of P (monoesters, polyesters, polyphosphates) increased with age of restored forests.

Hunter and Faulkner (2001) and Hunter et al. (2008) compared heterotrophic microbial activity and denitrification potentials in 5- to 10-year-old forested wetlands restored by planting and in natural floodplain forests. Hydrology was restored to some sites but not to others. The restored forest where hydrology was not reintroduced had significantly lower denitrification potentials and less sediment deposition than hydrologically restored floodplain forests and natural forests (Hunter and Faulkner, 2001; Hunter et al., 2008). In contrast, denitrification did not differ between hydrologically restored planted floodplain forests and natural forests that had similar hydrology of 97–104 days of inundation/saturation (Hunter et al., 2008). There was no difference in overall heterotrophic microbial activity among all sites. Studies in the southeastern and midwestern US suggest that denitrification in soils of recently restored riparian forests also are comparable to natural riparian forests (Lowrance et al., 1995; Marton et al., 2014). These studies collectively suggest that, with reintroduction of hydrology and adequate SOM, anaerobic microbial processes such as denitrification are quickly restored.

Limited information is available regarding invertebrate use of restored floodplain and riparian forests. Ettema et al. (1998) measured nematode density and diversity in a 3-year-old restored riparian forest in Georgia. A diverse and abundant nematode population was found with 68 taxa identified. Individual taxa varied in number from 10,000 to 1,000,000 individuals per square meter.

Herpetofauna quickly use ponds associated with restored forested wetlands. Petranka et al. (2003a,b) compared 22 constructed pond wetlands with reference ponds in western North Carolina. Breeding populations, based on the presence of egg masses of two species (wood frog, *Rana sylvatica* and spotted salamander, *Ambystoma maculatum*), were observed within a few months of filling (Petranak et al., 2003b). There was no difference in juvenile production among restored and reference ponds. During the first year of filling, abundance and species richness were positively correlated with pond size, depth, and hydroperiod (Petranaka et al., 2003a). Even though the sites were geographically isolated from other populations, restoration of a number of ponds on-site led to long-term persistence of populations in spite of multiyear drought that affected the region (Petranka et al., 2007).

Restoration of forested wetlands, given time, can lead to remarkable use by herpetofauna. Gibbons et al. (2006) measured amphibian—frogs, toads, salamanders—of a 10-ha forested Carolina Bay, Ellenton Bay, in South Carolina, USA. Prior to 1951, the site was an agricultural field, growing cotton, peanuts, and corn. Since then, agriculture was abandoned and the wetland and its surrounding terrestrial land were allowed to recolonize naturally. A total of 24 amphibian species were identified. The number of individuals (>360,000) and biomass (>1400 kg) also was remarkable. This study illustrates two key points: (1) isolated forested wetlands and their associated riparian buffers are critical habitat for herpetofauna, and (2) restoration of the forested wetland–upland complex brings back high levels of both herpetofauna productivity and biodiversity.

Creation and restoration of forested wetlands results in prompt use by avifauna, but, not surprisingly, assemblages differ between the young aggrading restored forests and mature forested wetlands, and between plantings of different species. Twedt et al. (2002) compared bird use in 2- to 10-year-old restored wetlands planted with either fast-growing cottonwood (*Populus deltoides*) or slower growing oaks (*Quercus* spp.) in the MAV. Rapid vertical growth of cottonwood quickly produced a forest with considerable structure that produced greater species richness and diversity as compared to oak plantings. In a study in Virginia, Snell-Rood and Cristol (2003) compared avian communities of 8-year-old planted floodplain forests with mature reference floodplain forests. The six planted forests had significantly lower species richness, diversity, and community composition than reference forests. Neotropical migrants also were less abundant in the planted forests. Reference forests contained species with greater habitat specificity, wetland dependency, and trophic level than created forests. Differences between restored and reference forests were attributed to reduced forest structure, to be expected in an 8-year-old forest, and differences in hydrology. Snell-Rood and Cristol (2003) observed that the created forests contained areas of deep open water not found in reference forests but that are common in many wetland restoration projects. Based on these differences, they suggested that created forests were developmentally behind naturally regenerating forests of the same age or were following a different developmental trajectory.

Berkowitz (2013) employed a rapid assessment method to evaluate ecosystem development of a number of forested wetlands in the MAV restored by replanting. The planted forests ranged in age from 1 to 20 years old. Seventeen variables were measured and four variables, shrub-sapling density, herbaceous vegetation cover, and O (organic) and A horizon development, yielded positive trajectories across the 20-year chronosequence. Surprisingly, there was no relationship between site age and measures of forest composition, including tree density, basal area, and canopy composition.

In summary, because of the large size and longevity of woody species, ecosystem development of restored forest wetlands is necessarily slow (Table 6.4), depending on recruitment or introduction of heavy-seeded species important for wildlife and

Table 6.4 Development of Forested Wetland Ecosystem Services Following Creation or Restoration

	Time (years) to Equivalence to Mature Natural Marshes
Productivity and habitat functions	
Primary production	10–15
Litterfall	40–50
Woody biomass	>50
Tree diversity	100s (if not seeded or planted with heavy-seeded species)
In-stream habitat	>50
Microbial activity	>10
Benthic invertebrates	30 to >50
Herpetofauna	30 to >50
Migratory songbirds	50 to 100 (if not seeded or planted)
Regulation functions	
Sedimentation	High at first, declining to sustainable levels after 10–20 years
Nutrient (N, P) retention	10–20
C sequestration[a]	>50
N cycling/denitrification	1–3 with proper hydrology
Soil formation	10s to 100s

See text for references.
[a]Most C is sequestered in wood rather than in soil.

a closed canopy to provide in-stream benefits and habitat for migratory songbirds (Table 6.4). Functions such as primary production and microbial activity develop within 10–15 years. Little data are available regarding sedimentation and N and P retention in restored forested floodplains, but it is likely that these processes develop along with forest productivity and biomass so long as hydrology is restored. Sedimentation may be high initially as herbaceous ground cover becomes established, but rates decline as the tree canopy develops and shades the forest floor. Perhaps 10–20 years are needed to store N and P in accruing biomass and bury it in the soil. Woody biomass and C sequestration and litter inputs to the stream require a maturing forest that may take more than 50 years to develop. It is thought that 40–60 years is needed for a floodplain forest to develop a closed canopy (Stanturf et al., 2001) and support healthy wildlife populations (Haynes and Moore, 1988). Benthic invertebrate and herpetofauna communities that also depend on a closed canopy to shade the soil take about the same amount of time, 30–60 years. The development of a diverse forest with high woody species richness and the migratory songbirds that use them may take 100 years or more unless heavy-seeded species are introduced. Soil formation in forested wetlands is slowed by the annual cycles of flooding and drying. Inputs of sediment from overbank flooding also keep floodplain soils in a relatively young stage of succession as does high intensity, short duration flooding of riparian forests.

Keys to Ensure Success

Successful restoration of forested riparian and floodplain wetlands involves reintroducing the flood pulse by reconnecting the floodplain with the river. This is done by removing levees, reconnecting channels, and restoring the seasonal timing of flooding (Table 6.5). Select sites that are contiguous with other forested stands to provide large patches for migratory songbirds and large mammals. For riparian areas, ensure that the width of the forest buffer is adequate, 50–100 m, to protect streams and provide corridors for dispersal of herpetofauna and other animals. If a seed source is nearby, light-seeded species such as ash (*Fraxinus*), maple (*Acer*), cottonwood (*Populus*), and others will quickly reestablish. Heavy-seeded species, such as oaks (*Quercus*), hickories (*Carya*), and walnut (*Juglans*) that are important for wildlife but do not disperse readily must be seeded or planted. Take care to protect seedlings and saplings from herbivory. Control of invasive plants, both herbaceous and woody, is necessary, especially during seedling and sapling establishment. On former agricultural land, introducing microtopography (hummocks and hollows) is necessary to re-create the spatial heterogeneity of nutrient cycling and plant species diversity and to enhance woody seedling germination and survival. Moderate additions of topsoil or organic compost can enhance microtopography, soil fertility, and nutrient removal (denitrification), but larger amounts increase elevation to the point where recruitment of wetland woody species may be inhibited. Forest diversity is enhanced by direct seeding, creating unplanted gaps, and introducing more heavy-seeded species. The time needed for a restored forested wetland to achieve equivalence to natural forested wetlands ranges from 40 to 60 years to a century or more in order to achieve the closed canopy and species-rich forest needed to support migratory songbirds, herpetofauna, and mammals.

Table 6.5 **Key Characteristics to Ensure Success When Restoring Forested Wetlands**

Site selection

- Select sites with appropriate hydrology, connectivity between the river and the floodplain, and appropriate flood pulse
- Select sites that are contiguous to and serve as corridors to other forested wetlands. For riparian wetlands, ensure that there is adequate, at least 50 m, buffer width
- Ensure there is a seed source available for recruitment of light- and, ideally, heavy-seeded species

Construction considerations

- Restore hydrology and the flood pulse by removing levees and channels to reconnect the floodplain/riparian area with stream/river
- Reintroduce the flood pulse on dammed rivers
- Incorporate microtopography of hummocks and hollows, especially for floodplain forests
- Modest additions of topsoil or organic compost can enhance microtopography and soil fertility

Planting considerations

- Seeds and plants selected for appropriate inundation regime and geographic region
- Seed or plant heavy-seeded species (e.g., oaks, *Quercus*; hickories, *Carya*; and others) important to wildlife
- Protect young transplants from herbivory by rodents, rabbits, deer, others

Maintenance

- Monitoring and control of herbivores during the seedling and sapling stages
- Monitoring and removal of invasive species, especially in riparian wetlands

References

Acreman, M.C., Fisher, J., Stratford, C.J., Mould, D.J., Mountford, J.O., 2007. Hydrological science and wetland restoration: some case studies from Europe. Hydrology and Earth System Sciences 11, 158–169.

Aldous, A., McCormick, P., Ferguson, C., Graham, S., Craft, C., 2005. Hydrologic regime controls soil phosphorus fluxes in restoration and undisturbed wetlands. Restoration Ecology 13, 341–347.

Allen, J.A., 1990. Establishment of bottomland oak plantations on the Yazoo National Wildlife Refuge Complex. Southern Journal of Applied Forestry 14, 206–210.

Allen, J.A., 1997. Reforestation of bottomland hardwoods and the issue of woody species diversity. Restoration Ecology 5, 125–134.

Ardon, M., Montanari, S., Morse, J.L., Doyle, M.W., Bernhardt, E.S., 2010a. Phosphorus export from a restored wetland ecosystem in response to natural and experimental hydrologic fluctuations. Journal of Geophysical Research: Biogeosciences 115, G4.

Ardon, M., Morse, J.L., Doyle, M.W., Bernhardt, E.S., 2010b. The water quality consequences of restoring wetland hydrology in a large agricultural watershed in the southeastern Coastal plain. Ecosystems 13, 1060–1078.

Bailey, D.E., Perry, J.E., Daniels, W.L., 2007. Vegetation dynamics in response to organic matter loading rates in a created freshwater wetland in southeastern Virginia. Wetlands 27, 936–950.

Bannister, J.M., E.R. Herbert, C.B., Craft. 2014. Spatial variability in sedimentation, carbon sequestration, and nutrient accumulation in an alluvial floodplain forest. In: Vymazal, J., (Ed.), The Role of Natural and Constructed Wetlands in Nutrient Cycling and Retention on the Landscape. Springer. New York, pp.41–55.

Barry, W.J., Garlo, A.S., Wood, C.A., 1996. Duplicating the mound and pool microtopography of forested wetlands. Restoration and Management Notes 14, 15–21.

Battaglia, L.L., Minchin, P.R., Prichett, D.W., 2002. Sixteen years of old-field succession and reestablishment of a bottomland hardwood forest in the Lower Mississippi Alluvial Valley. Wetlands 22, 1–17.

Berkowitz, J.F., 2013. Development of restoration trajectory metrics in reforested bottomland hardwood forests applying a rapid assessment approach. Ecological Indicators 34, 600–606.

Bledsoe, B.P., Shear, T.H., 2000. Vegetation along hydrologic and edaphic gradients in a North Carolina coastal plain creek bottom: Implications for restoration. Wetlands 20, 126–147.

Brinson, M.M., 1993. Changes in the functioning of wetlands along environmental gradients. Wetlands 13, 65–74.

Brinson, M.M., Bradshaw, H.D., Kane, E.S., 1984. Nutrient assimilative capacity of an alluvial floodplain swamp. Journal of Applied Ecology 21, 1041–1058.

Brinson, M.M., Swift, B.L., Plantico, R.C., Barclay, J.S., 1981. Riparian Ecosystems: Their Ecology and Status. FWS/OBS-81/17. U.S. Fish and Wildlife Service, Kearneysville, West Virginia.

Brittain, R.A., Meretsky, V., Craft, C.B., 2010. Avian communities of the Altamaha River estuary in Georgia USA. The Wilson Journal of Ornithology 122, 532–544.

Brooks, R.P., Wardrop, D.H., Cole, C.A., Campbell, D.A., 2005. Are we purveyors of wetland homogeneity? A model of degradation and restoration to improve wetland mitigation performance. Ecological Engineering 24, 331–340.

Brown, S.C., Veneman, P.L.M., 2001. Effectiveness of compensatory wetland mitigation in Massachusetts, USA. Wetlands 21, 508–518.

Bruland, G.L., Richardson, C.J., 2004. Hydrologic gradients and topsoil additions affect soil properties of Virginia created wetlands. Soil Science Society of America 68, 2069–2077.

Bruland, G.L., Richardson, C.J., 2005a. Hydrologic, edaphic, and vegetative responses to microtopographic reestablishment in a restored wetland. Restoration Ecology 13, 515–523.

Bruland, G.L., Richardson, C.J., 2005b. Spatial variability of soil properties in created, restored, and paired natural wetlands. Soil Science Society of America Journal 69, 273–284.

Bruland, G.L., Richardson, C.J., 2006. Comparison of soil organic matter in created, restored, and paired natural wetlands in North Carolina. Wetlands Ecology and Management 14, 245–251.

Bruland, G.L., Hanchey, M.F., Richardson, C.J., 2003. Effects of agriculture and wetland restoration on hydrology, soils, and water quality of a Carolina Bay complex. Wetlands Ecology and Management 11, 141–156.

Bruland, G.L., Richardson, C.J., Daniels, W.L., 2009. Microbial and geochemical responses to organic matter amendments in a created wetland. Wetlands 29, 1153–1165.

Bruland, G.L., Richardson, C.J., Whalen, S.C., 2006. Spatial variability of denitrification potential and related soil properties in created, restored, and paired natural wetlands. Wetlands 26, 1042–1056.

Buler, J.J., Moore, F.R., Woltmann, S., 2007. A multi-scale examination of stopver habitat use by birds. Ecology 88, 1789–1802.

Burke, V.J., Gibbons, J.W., 1995. Terrestrial buffer zone sand wetland conservation: a case study of freshwater turtles in a Carolina Bay. Conservation Biology 9, 1365–1369.

Cals, M.J.R., Postma, R., Buijse, A.D., Marteijn, E.C.L., 1998. Habitat restoration along the Rhine River in the Netherlands: putting ideas into practice. Aquatic Conservation: Marine and Freshwater Ecosystems 8, 61–70.

Carothers, S.W., Mills, G.S., Johnson, R.R., 1990. The creation and restoration of riparian habitat in southwestern arid and semi-arid regions. In: Kusler, J.A., Kentula, M.E. (Eds.), Wetland Creation and Restoration: The Status of the Science. EPA/600/3089/038. U.S. Environmental Protection Agency, Corvallis, Oregon, pp. 359–376.

Castelle, A.J., Johnson, A.W., Conolly, C., 1994. Wetland and stream buffer size requirements: a review. Journal of Environmental Quality 23, 878–882.

Clement, J.C., Pinay, G., Marmonier, P., 2002. Seasonal patterns of denitrification along topohydrosequences in three different riparian wetlands. Journal of Environmental Quality 31, 1025–1037.

Clewell, A.F., 1999. Restoration of riverine forest at Hall Branch on phosphate-mined land, Florida. Restoration Ecology 7, 1–14.

Clewell, A.F., Lea, R., 1989. Creation and restoration of forested wetland vegetation in the southeastern United States. In: Kusler, J.A., Kentula, M.E. (Eds.), Wetland Creation and Restoration: The Status of the Science. EPA/600/3089/038. U.S. Environmental Protection Agency, Corvallis, Oregon, pp. 199–237.

Cohn, J.P., 2001. Resurrecting the dammed: a look at the Colorado River restoration. BioScience 51, 998–1003.

Cooper, J.R., Gilliam, J.W., 1987. Phosphorus redistribution from cultivated fields into riparian areas. Soil Science Society of America Journal 51, 1600–1604.

Courtwright, J., Findlay, S.E.G., 2011. Effects of microtopography on hydrology, physiochemistry, and vegetation in a tidal swamp of the Hudson River. Wetlands 31, 239–249.

Craft, C.B., Casey, W.P., 2000. Sediment and nutrient accumulation in floodplain and depressional freshwater wetlands of Georgia, USA. Wetlands 20, 323–332.

Crawford, J.A., Semlitsch, R.D., 2007. Estimation of core terrestrial habitat for stream-breeding salamanders and delineation of riparian buffers for protection of biodiversity. Conservation Biology 21, 152–158.

D'Angelo, E.M., 2005. Phosphorus sorption capacity and exchange by soils from mitigated and late successional bottomland forest wetlands. Wetlands 25, 297–305.

D'Angelo, E.M., Reddy, K.R., 1994. Diagenesis of organic matter in a wetland receiving hypereutrophic lake water: I. Distribution of dissolved nutrients in the soil and water column. Journal of Environmental Quality 23, 928–936.

D'Angelo, E.M., Karathanasis, A.D., Sparks, E.J., Ritchey, S.A., Wehr-McChesney, S.A., 2005. Soil carbon and microbial communities at mitigated and late successional bottomland forest wetlands. Wetlands 25, 162–175.

Daniels, R.B., Gilliam, J.W., 1996. Sediment and chemical load reduction by grass and riparian filters. Soil Science Society of America Journal 60, 246–251.

Day, R.H., Williams, T.M., Swarzenski, C.M., 2007. Hydrology of tidal freshwater forested wetlands of the southeastern United States. In: Conner, W.H., Doyle, T.W., Krauss, K.W. (Eds.), Ecology of Tidal Freshwater Forested Wetlands of the Southeastern United States. Springer, New York, pp. 29–63.

DeGraaf, R.M., Yamasaki, M., 2000. Bird and mammal habitat in riparian areas. In: Verry, E.S., Horneck, J.W., Dollof, C.A. (Eds.), Riparian Management in Forests of the Continental Eastern United States. Lewis Publishers, Boca Raton, Florida. pp. 139–156.

Ettema, C.H., Coleman, D.C., Vellidis, G., Lowrance, R., Rathbun, S.L., 1998. Spatiotemporal distributions of bacterivorous nematodes and soil resources in a restored riparian wetland. Ecology 79, 2721–2734.

Faulkner, S., Barrow Jr., W., Keeland, B., Walls, S., Telesco, D., 2011. Effects of conservation practices on wetland ecosystem services in the Mississippi Alluvial Valley. Ecological Applications 21, S31–S48.

Gibbons, J.W., Winne, C.T., Scott, D.E., Willson, J.D., Glaudas, X., Andrews, K.M., Todd, B.D., Fedewa, L.A., Wilkinson, L., Tsaliasgos, R.N., Harper, S.J., Greene, J.L., Tuberville, T.D., Metts, B.S., Dorcas, M.E., Nestor, J.P., Young, C.A., Akre, T., Reed, R.N., Buhlmann, K.A., Norman, J., Croshaw, D.A., Hagen, C., Rothermel, B.T., 2006. Remarkable amphibian biomass and abundance in an isolated wetland: Implications for wetland conservation. Conservation Biology 20, 1457–1465.

Goodwin, C.N., Hawkins, C.P., Kershner, J.L., 1997. Riparian restoration in the western United States: overview and perspective. Restoration Ecology 5, 4–14.

Groffman, P.M., Gold, A.J., Simmons, R.C., 1992. Nitrate dynamics in riparian forests: microbial studies. Journal of Environmental Quality 21, 666–671.

Haynes, R.J., Moore, L., 1988. Reestablishment of bottomland hardwoods within National Wildlife Refuges in the southeast. In: Zelazny, J., Fierabend, J.S. (Eds.), Increasing Our Wetland Resources. National Wildlife Federation, Washington, D.C, pp. 95–103.

Haynes, R.J., 2004. The development of bottomland forest restoration in the lower Mississippi alluvial valley. Ecological Restoration 22, 170–182.

Henry, C.P., Amoros, C., Roset, N., 2002. Restoration ecology of riverine wetlands: a 5-year post-operation survey on the Rhone River, France. Ecological Engineering 18, 543–554.

Hill, A.C., 1996. Nitrate removal in stream riparian zones. Journal of Environmental Quality 23, 743–755.

Ho, M., Richardson, C.J., 2013. A five year study of the floristic succession in a restored urban wetland. Ecological Engineering 61B, 511–518.

Hodges, J.D., 1997. Development and ecology of bottomland hardwood sites. Forest Ecology and Management 90, 117–125.

Hoffman, C.C., Kronvang, B., Audet, J., 2011. Evaluation of nutrient retention in four Danish riparian wetlands. Hydrobiologia 674, 5–24.

Hunter, R.G., Faulkner, S.P., 2001. Denitrification potentials in restored and natural bottomland hardwood wetlands. Soil Science Society of America Journal 65, 1865–1872.

Hunter, R.G., Faulkner, S.P., Gibson, K.A., 2008. The importance of hydrology in restoration of bottomland hardwood wetland functions. Wetlands 28, 605–615.

Hupp, C.R., Osterkamp, W.R., 1985. Bottomland vegetation distribution along Passage Creek, Virginia, in relation to fluvial landforms. Ecology 66, 670–681.

Ilhardt, B.L., Verry, E.S., Palik, B.J., 2000. Defining riparian areas. In: Verry, E.S., Horneck, J.W., Dollof, C.A. (Eds.), Riparian Management in Forests of the Continental Eastern United States. Lewis Publishers, Boca Raton, Florida, pp. 23–42.

Jacobs, T.C., Gilliam, J.W., 1985. Riparian losses of nitrate from agricultural drainage waters. Journal of Environmental Quality 14, 472–478.

Junk, W.J., Bayley, P.B., Sparks, R.E., 1989. The flood pulse concept in river-floodplain systems. Canadian Special Publication of Fisheries and Aquatic Sciences 106, 110–127.

Keddy, P.A., 2010. Wetland Ecology: Principles and Conservation. Oxford University Press, Cambridge, UK.

King, S.L., Keeland, B.D., 1999. Evaluation of reforestation in the lower Mississippi River Alluvial Valley. Restoration Ecology 7, 348–359.

Kleiss, B.A., 1996. Sediment retention in a bottomland hardwood wetland in eastern Arkansas. Wetlands 16, 321–333.

Lowrance, R., Vellidis, G., Hubbard, R.K., 1995. Denitrification in a restored riparian forest wetland. Journal of Environmental Quality 24, 808–815.

Lowrance, R., Sharpe, J.K., Sheridan, J.M., 1986. Long-term sediment deposition in the riparian zone of a coastal plain watershed. Journal of Soil and Water Conservation 41, 266–271.

Lowrance, R., Todd, R., Fail Jr., J., Hendrickson Jr., O., Leonard, R., Asmussen, L., 1984. Riparian forests as nutrient filters in agricultural watersheds. BioScience 34, 374–377.

Mant, J., Janes, M., 2006. Restoration of rivers and floodplains. In: Van Andel, J., Aronson, J. (Eds.), Restoration Ecology: The New Frontier. Blackwell Publishing, Malden, Massachusetts, pp. 141–157.

Marton, J.M., Fennessy, S., Craft, C.B., 2014. Functional differences between natural and restored wetlands in the Glaciated Interior Plains. Journal of Environmental Quality 43, 409–417.

McLeod, K.W., Reed, M.R., Wike, L.D., 2000. Elevation, competition control, and species affect bottomland forest restoration. Wetlands 20, 162–168.

Middleton, B., 1999. Wetland restoration: flood pulsing and disturbance dynamics. John Wiley and Sons, New York.

Middleton, B., 2002. Flood pulsing in the regeneration and maintenance of species in riverine forested wetlands of the southeastern United States. In: Middleton, B. (Ed.), Flood Pulsing in Wetlands: Restoring the Natural Hydrological Balance. John Wiley and Sons, New York, pp. 223–294.

Mitsch, W.J., Day Jr., J.W., Gilliam, J.W., Groffman, P.M., Hey, D.L., Randall, G.W., Wang, N., 2001. Reducing nitrogen loading to the Gulf of Mexico from the Mississippi River basin: strategies to counter a persistent ecological problem. BioScience 51, 373–388.

Naiman, R.J., Decamps, H., 1997. The ecology of interfaces: riparian zones. Annual Review of Ecology and Systematics 28, 621–658.

Naiman, R.J., Decamps, H., Pollock, M., 1993. The role of riparian corridors in maintaining regional biodiversity. Ecological Applications 3, 209–212.

Nakamura, K., Tockner, K., Amano, K., 2006. River and wetland restoration: lessons from Japan. BioScience 56, 419–429.

National Research Council, 1992. Restoration of Aquatic Ecosystems. National Academy Press, Washington, D.C.

National Research Council, 2000. Clean Coastal Waters: Understanding and Reducing the Effects of Nutrient Pollution. National Academy Press, Washington, D.C.

National Research Council, 2001. Compensating for Wetland Losses under the Clean Water Act. National Academy of Sciences, Washington, DC.

National Research Council, 2002. Riparian Areas: Functions and Strategies for Management. National Academy Press, Washington, D.C.

Niswander, S.F., Mitsch, W.J., 1995. Functional analysis of a two-year-old created in-stream wetland: hydrology, phosphorus retention, and vegetation growth and survival. Wetlands 15, 212–225.

Noe, G.B., Hupp, C.R., 2005. Carbon, nitrogen and phosphorus accumulation in floodplains of Atlantic Coastal Plain Rivers. Ecological Applications 15, 1178–1190.

Odum, E.P., 1969. The strategy of ecosystem development. Science 164, 262–270.

Oelbermann, M., Gordon, A.M., 2000. Quantity and quality of litterfall into a rehabilitated agricultural stream. Journal of Environmental Quality 29, 603–611.

Osland, M.J., Pahl, J.W., Richardson, C.J., 2009. Native bamboo [*Arundinaria gigantea* (Walter) Muhl., Poaceae] establishment and growth after the removal of an invasive non-native shrub (*Ligustrum sinense* Lour., Oleaceae): implications for restoration. Castanea 74, 247–258.

Pant, H.K., Reddy, K.R., 2003. Potential internal loading of phosphorus in a wetland constructed in agricultural land. Water Research 37, 965–972.

Parkyn, S.M., Davies-Colley, R.J., Halliday, N.J., Costley, K.J., Crocker, G.F., 2003. Planted riparian buffer zones in New Zealand: do they live up to expectation? Restoration Ecology 11, 436–447.

Pedersen, M.L., Andersen, J.M., Nielsen, K., Linnemann, M., 2007. Restoration of Skjern River and its valley: project description and general ecological changes in the area. Ecological Engineering 30, 131–144.

Pennisi, E., 2004. The grand (Canyon) experiment. Science 306, 1884–1886.

Petranka, J.W., Kennedy, C.A., Murray, S.S., 2003a. Response of amphibians to restoration of a southern Appalachian wetland: a long-term analysis of community dynamics. Wetlands 23, 1030–1042.

Petranka, J.W., Murray, S.S., Kennedy, C.A., 2003b. Response of amphibians to restoration of a southern Appalachian wetland: perturbations confound post-restoration assessment. Wetlands 23, 278–290.

Petranka, J.W., Harp, E.M., Holbrook, C.T., Hamel, J.A., 2007. Long-term persistence of amphibian populations in a restored wetland complex. Biological Conservation 138, 371–380.

Phillips, J.D., 1989. Fluvial sediment storage in wetlands. Water Resources Bulletin 25, 867–873.

Pierce, A.R., King, S.L., 2007. The effects of flooding and sedimentation on seed germination of two bottomland hardwood tree species. Wetlands 27, 588–594.

Pinay, G., Roques, L., Fabre, A., 1993. Spatial and temporal patterns of denitrification in a riparian forest. Journal of Applied Ecology 30, 581–591.

Robb, J.T., 2002. Assessing wetland compensatory mitigation sites to aid in establishing mitigation ratios. Wetlands 22, 435–440.

Schilling, E.B., Lockaby, B.G., 2005. Microsite influences on productivity and nutrient circulation within two southeastern floodplain forests. Soil Sciences Society of America Journal 69, 1185–1195.

Schoenholtz, S.H., James, J.P., Kaminski, R.M., Leopold, B.D., Ezell, A.W., 2001. Afforestation of bottomland hardwoods in the lower Mississippi alluvial valley: status and trends. Wetlands 21, 602–613.

Sharpley, A., Foy, B., Withers, P., 2000. Practical and innovative measures for the control of agricultural phosphorus losses to water: an overview. Journal of Environmental Quality 29, 1–9.

Shear, T.H., Lent, T.J., Fraver, S., 1996. Comparison of restored and mature bottomland hardwood forests of southwestern Kentucky. Restoration Ecology 4, 111–123.

Sidle, W.C., Roose, D.L., Yzerman, V.T., 2000. Isotope evaluation of nitrate attenuation in restored and native riparian zones in the Kankakee watershed, Indiana. Wetlands 20, 333–345.

Silker, T.H., 1948. Planting of water-tolerant trees along margins of fluctuating-level reservoirs. Iowa State College Journal of Science 22, 431–448.

Snell-Rood, E.C., Cristol, D.A., 2003. Avian communities of created and natural wetlands: bottomland forests in Virginia. The Condor 105, 303–315.

Stanturf, J.A., Schoenholtz, S.H., Schweitzer, C.J., Shephard, J.P., 2001. Achieving restoration success: myths in bottomland hardwood forests. Restoration Ecology 9, 189–200.

Stevens, L.E., Ayers, T.J., Bennett, J.B., Christensen, K., Kearsley, M.J.C., Meretsky, V.J., Phillips I.I.I., A.M., Parnell, R.A., Spence, J., Sogge, M.K., Springer, A.E., Wegner, D.L., 2001. Planned flooding and Colorado River riparian trade-offs downstream from Glen Canyon Dam, Arizona. Ecological Applications 11, 701–710.

Stolt, M.H., Genthner, M.H., Daniels, W.L., Groover, V.A., 2001. Spatial variability in palustrine wetlands. Soil Science Society of America Journal 65, 527–535.

Stolt, M.H., Genthner, M.H., Daniels, W.L., Groover, V.A., Nagle, S., Haering, K.C., 2000. Comparison of soil and other environmental conditions in constructed and adjacent palustrine reference wetlands. Wetlands 20, 671–683.

Stromberg, J.C., 2001. Biotic integrity of *Platanus wrightii* riparian forests in Arizona: first approximation. Forest Ecology and Management 142, 251–266.

Stromberg, J.C., Chew, M.K., 2002. Flood pulses and restoration of riparian vegetation in the American southwest. In: Middleton, B. (Ed.), Flood Pulsing in Wetlands: Restoring the Natural Hydrological Balance. John Wiley and Sons, New York, pp. 11–50.

Sundareshwar, P.V., Richardson, C.J., Gleason, R.A., Pellechia, P.J., Honomichl, S., 2009. Nature versus nurture: functional assessment of restoration efforts on wetland services using Nuclear Magnetic Resonance Spectroscopy. Geophysical Research Letters 36, L03402.

Sutton-Grier, A.E., Ho, M., Richardson, C.J., 2009. Organic amendments improve soil conditions and denitrification in a restored riparian wetland. Wetlands 29, 343–352.

Sweeney, B.W., Bott, T.L., Jackson, J.K., Kaplan, L.A., Newbold, J.D., Standley, L.J., Hession, W.C., Horwitz, R.J., 2004. Riparian deforestation, stream narrowing, and loss of stream ecosystem services. Proceedings of the National Academy of Sciences 101, 14132–14137.

Swift, B.L., 1984. Status of riparian ecosystems in the United States. Water Resources Bulletin 20, 223–228.

Titus, J.H., 1990. Microtopography and woody plant regeneration in a hardwood floodplain swamp in Florida. Bulletin of the Torrey Botanical Club 117, 429–437.

Tockner, K., Schiemer, F., Ward, J.V., 1998. Conservation by restoration: the management concept for a river-floodplain system on the Danube River in Austria. Aquatic Conservation: Marine and Freshwater Ecosystems 8, 71–86.

Toliver, J.R., 1986. Survival and growth of hardwoods planted on abandoned fields. Louisiana Agriculture 30, 10–11.

Toner, M., Keddy, P., 1997. River hydrology and riparian wetlands: a predictive model for ecological assembly. Ecological Applications 7, 236–246.

Toth, L.A., Koebel Jr., J.W., Warne, A.G., Chamberlain, J., 2002. Implications of reestablishing prolonged flood pulse characteristics of the Kissimmee River and floodplain ecosystem. In: Middleton, B. (Ed.), Flood Pulsing in Wetlands: Restoring the Natural Hydrological Balance. John Wiley and Sons, New York, pp. 191–221.

Trowbridge, W.B., 2007. The role of stochasticity and priority effects in floodplain restoration. Ecological Applications 17, 1312–1324.

Tweedy, K.L., Evans, R.O., 2001. Hydrologic characterization of two prior converted wetland restoration sites in eastern North Carolina. Transactions of the American Society of Agricultural Engineers 44, 1135–1142.

Twedt, D.J., 2004. Stand development on reforested bottomlands in the Mississippi alluvial valley. Plant Ecology 172, 251–263.

Twedt, D.J., Loesch, C.R., 1999. Forest area and distribution in the Mississippi alluvial valley: Implications for breeding bird conservation. Journal of Biogeography 26, 1215–1224.

Twedt, D.J., Wilson, R.R., Henne-Kerr, J.L., Grosshuesch, D.A., 2002. Avian response to bottomland hardwood restoration: the first 10 years. Restoration Ecology 10, 645–655.

Twedt, D.J., Wilson, R.R., Henne-Kerr, J.L., Hamilton, R.B., 1999. Impact of forest type and management strategy on avian densities in the Mississippi Alluvial Valley. USA. Forest Ecology and Management 123, 261–274.

Ullah, S., Faulkner, S.P., 2006. Denitrification potential of different land-use types in an agricultural watershed, lower Mississippi valley. Ecological Engineering 28, 131–140.

Unghire, J.M., Sutton-Grier, A.E., Flanagan, N.E., Richardson, C.J., 2011. Spatial impacts of stream and wetland restoration on riparian soil properties in the North Carolina Piedmont. Restoration Ecology 19, 738–746.

Vannote, R.L., Minshall, G.W., Cummins, K.W., Sedell, J.R., Cushing, C.E., 1980. The river continuum concept. Canadian Journal of Fisheries and Aquatic Sciences 37, 130–137.

Vellidis, G., Lowrance, R., Smith, M.C., 1994. A quantitative approach for measuring N and P concentration changes in surface runoff from a restored riparian forest wetland. Wetlands 14, 73–81.

Vellidis, G., Lowrance, R., Gay, P., Hubbard, R.K., 2003. Nutrient transport in a restored riparian wetland. Journal of Environmental Quality 32, 711–726.

Verchot, L.V., Franklin, E.C., Gilliam, J.W., 1997. Nitrogen cycling in vegetated filter zones: II. Subsurface nitrate removal. Journal of Environmental Quality 26, 337–347.

Vidon, P., Jacinthe, P.A., Liu, X., Fisher, K., Baker, M., 2014. Hydrobiogeochemical controls on riparian nutrient and greenhouse gas dynamics: 10 years post-restoration. Journal of the American Water Resources Association 50, 639–652.

Vidra, R.L., Shear, T.H., Wentworth, T.R., 2006. Testing the paradigms of exotic species invasion in urban riparian forests. Natural Areas Journal 26, 339–350.

Vivian-Smith, G., 1997. Microtopographic heterogeneity and floristic diversity in experimental wetland communities. Journal of Ecology 85, 71–82.

Wakeley, J.S., Roberts, T.H., 1996. Bird distributions and forest zonation in a bottomland hardwood forest. Wetlands 16, 296–308.

Walbridge, M.R., Lockaby, B.G., 1994. Effects of forest management on biogeochemical functions in southern forested wetlands. Wetlands 14, 10–17.

Walls, R.L., Wardrop, D.H., Brooks, R.P., 2005. The impact of experimental sedimentation and flooding on the growth and germination of floodplain trees. Plant Ecology 176, 203–213.

Webb, R.H., Wegner, D.L., Andrews, E.D., Valdez, R.A., Patten, D.T., 1999. Downstream effects of glen Canyon dam on the Colorado river in the grand Canyon: a review. In: Webb, R.H., Schmidt, J.C., Marzolf, G.R., Valdez, R.A. (Eds.), The Controlled Flood in the Grand Canyon. Geophysical Monograph 110. American Geophysical Union, Washington, D.C, pp. 1–21.

Weller, M.W., 1978. Management of freshwater marshes for wildlife. In: Good, R.E., Whigham, D.F., Simpson, R.L. (Eds.), Freshwater Wetlands: Ecological Processes and Management Potential. Academic Press, New York, pp. 267–284.

Wharton, C.H., Kitchens, W.M., Sipe, T.W., 1982. The Ecology of Bottomland Hardwood Swamps of the Southeast: A Community Profile. FWS/OBS-81/37. U.S. Fish and Wildlife Service, Biological Services Program, Washington, D.C.

Whigham, D.F., 1999. Ecological issues related to wetland preservation, restoration, creation and assessment. The Science of the Total Environment 240, 31–40.

Wigginton, J.D., Lockaby, B.G., Trettin, C.C., 2000. Soil organic matter formation and sequestration across a forested floodplain chronosequence. Ecological Engineering 15, S141–S155.

Willard, D.E., Finn, V.M., Levine, D.A., Klarquist, J.E., 1989. Creation and restoration of riparian wetlands in the agricultural Midwest. In: Kusler, J.A., Kentula, M.E. (Eds.), Wetland Creation and Restoration: The Status of the Science. EPA/600/3089/038. U.S. Environmental Protection Agency, Corvallis, Oregon, pp. 333–358.

Wuethrich, B., 1998. Deliberate flood renews habitat. Science 272, 344–345.

Yoshimura, C., Omura, T., Furumai, H., Tockner, K., 2005. Present state of rivers and streams in Japan. River Research and Applications 21, 93–112.

Zockler, C., 2000. Wise Use of Floodplains: Review of River Restoration Projects in a Number of European Countries: A Representative Sample of WWF Projects. WWF European Freshwater Programme, Cambridge, UK.

Peatlands

7

Chapter Outline

Introduction

Peatlands form from the accumulation and burial of organic matter derived from plant detritus. They develop under conditions of near continuous soil saturation that leads to anaerobic conditions which dramatically slows decomposition. Peatlands usually are associated with freshwater and with cool or wet climates including the high latitudes of the northern hemisphere, oceanic environments, and moist tropical areas. Canada and Russia have the most extensive areas of peatlands although extensive areas are found in northern Europe, especially Fennoscandia, and in the tropics of Indonesia. Peatland soils encompass a wide range of organic matter content and thickness though it frequently contains 80–100% organic matter that is several meters or more thick. The International Peat Society (IPS) defines peat as "sedentarily accumulated material consisting of at least 30% (dry mass) of dead organic material" (Joosten and Clark, 2002). While there is no absolute minimum thickness, 30 cm is used when assessing global inventories of peat (Joosten and Clark, 2002). Peat may be derived from herbaceous or woody vegetation or from the moss *Sphagnum* from which horticultural peat moss is produced. In cool climates, *Sphagnum* is the primary builder of peat, whereas in warmer climates graminoids and woody vegetation provide most of the organic matter. Many tropical peatlands may form under mangrove forests (see Chapter 9, Mangroves).

Peatlands are referred to by various names such as bogs, fens, and mires. According to the IPS, a *mire* refers to a peatland where peat is actively being formed (Table 7.1). A *bog*, also known as an *ombrogenous mire*, is raised above the surrounding landscape and receives water only from precipitation. A *fen*, or *geogenous mire*, is situated in depressions and receives water that has been in contact with mineral bedrock or soil.

Creating and Restoring Wetlands. http://dx.doi.org/10.1016/B978-0-12-407232-9.00007-5

Table 7.1 **Common Terms Used to Describe Peatlands**

Mire	A peatland where peat is actively being formed
Bog or ombrogenous mire	A peatland that is raised above the surrounding landscape and that receives water only from precipitation
Fen or geogenous mire	A peatland that is situated in a depression and receives water that has been in contact with mineral bedrock or soil
Soligenous mire	A fen that receives water from precipitation and surface water
Lithogenous fen	A fen that receives water from precipitation and deep groundwater
Histosol	A peatland whose soil contains at least 12–18% organic C and whose thickness is at least 40 cm

From Joosten and Clark (2002) and USDA (1999).

Geogenous mires may be further subdivided into those that are *soligenous*, receiving water from precipitation and surface runoff, and *lithogenous*, receiving mostly deep groundwater (Joosten and Clark, 2002). In the US, peat sometimes is referred to as organic soil or histosol (Table 7.1). The U.S. Department of Agriculture (USDA) definition of a histosol is more precise than that of the IPS, requiring a minimum organic C content of at least 12% and thickness of 40 cm or more but less if directly atop bedrock (USDA, 1999).

Classification systems for peatlands are based on water source (ombrogenous, geogenous) that changes with time during succession and that governs water and nutrient chemistry. Minerotrophic or "nutrient-rich" peatlands, also known as *fens*, have strong connections to groundwater. These peatlands often have circumneutral pH, an abundance of base cations, Ca and Mg, and are dominated by grasses, sedges, and rushes. As the peat grows thicker with time, it becomes increasingly isolated from groundwater, relying more on precipitation for water and nutrients. Grasses, sedges, and rushes are replaced by *Sphagnum* and woody vegetation. In these older mesotrophic peatlands, the peat is thicker, contains higher organic matter content, is more acidic, and contains less Ca and Mg. Bog vegetation, with *Sphagnum* as the keystone species, receives nearly all water from precipitation. It is an acidic, nutrient-poor peatland and represents the successional endpoint along the minerotrophic to oligotrophic continuum. The combination of anaerobic soils, acidic conditions, and low nutrient availability serve to preserve the annual production that *Sphagnum* and other species add to the soil. Because bogs and fens represent a gradient or continuum of environmental conditions, Bridgham et al. (1995) suggest describing them based on water source (ombrogenous, geogenous) and pH (circumneutral, moderately acid, strongly acid) rather than on nutrient availability. The global distribution and variation in peat properties and vegetation types defies a precise way of classifying peatlands. For this reason, many authorities suggest using the generic term *peatland* rather than bog or fen when describing them (Bridgham et al., 1995; Joosten and Clark, 2002).

Succession in peatlands is autogenic, relying on accumulating organic matter to increase elevation that alters water source, nutrient availability, and species composition. Peat accretion is a slow process with the peat growing vertically at perhaps

1–2 mm/year over the long term. Thus, restoration of peatlands or, at least, building a substantive layer of peat, requires centuries or more. Reestablishing *Sphagnum* and other peatland species, however, may take considerably less time and may be accelerated using nurse or companion plants.

Many opportunities exist to restore peatlands, including sites drained for agriculture and where peat has been mined. Restoration of hydrology is critical to successful restoration of peatlands but it is more difficult than restoring hydrology in wetlands underlain by mineral soils. When peats are drained, the organic matter that comprises peat is oxidized to CO_2 and is permanently lost from the system. Over time, the peat subsides, with resultant soil compaction and subsidence, making it difficult to restore proper hydrology without water table management.

Peat—Origin, Composition, and Geographic Distribution

Peatlands are found throughout the world in warm, cold, wet, and dry climates and in wet and dry environments. They mostly are found in areas where precipitation exceeds evapotranspiration (van Breeman, 1995) which occurs in cooler climates such as the far north latitudes, moist oceanic environments (Taylor, 1983), and in moist to humid mountainous regions (Darlington, 1943). Peat accumulates under a variety of plant communities. The peat of many northern peatlands such as bogs form from *Sphagnum*. The harsh conditions including cold temperature, saturated soil, limited nutrients, and low pH exclude most species except for acid-tolerant woody plants such as heath, ericaceous shrubs, and conifers (Heinselman, 1970). Fens that receive groundwater in addition to precipitation tend to be dominated by grasslike or graminoid species such as sedges (Heinselman, 1970). Here, the peat is formed mostly from the accumulation of belowground roots and rhizomes (Lindeman, 1941). Under the right conditions, woody species build peat. This is true of mangroves and freshwater coastal forests. Here, the dense woody material produced by the trees is resistant to decomposition and is the primary contributor to peat formation (Chimner and Ewel, 2005) (see Chapter 9, Mangroves). Peat produced by woody vegetation contains more aboveground material, leaves and twigs, than peat produced by herbaceous vegetation (Cohen, 1973). Tidal marshes also build peat but this is confined to freshwater and brackish environments, cooler climates, and areas with low tidal energy and sediment input (Craft, 2007).

Peat is like a sponge. It consists mostly of empty spore space, compact and lightweight when dry, but, when wet, it expands as it fills with water. Peat is classified based on its fiber content and degree of decomposition. There are a number of methods that involve either rubbing, centrifugation, sieving, or squeezing (Malterer et al., 1992). Two commonly used methods are the Von Post and the USDA soil taxonomy methods. The Von Post method is used to assess the degree of peat decomposition and humification. It involves squeezing an egg-shaped lump of peat with one hand and observing the composition of the free water that is wrung out (Verry et al., 2011). The scale ranges from 1 to 10, with 1 being essentially undecomposed material and 10 being highly decomposed material (Table 7.2). Generally, peat of fen and minerotrophic

Table 7.2 Von Post Method for Characterizing the Degree of Decomposition of Peat

Von Post Value	Volume of Peat Passing through Fingers	Composition of Free Water
1	0	Clear
2	0	Almost clear
3	0	Muddy brown
4	0	Turbid, muddy
5	2–10	Very turbid and muddy
6	26–35	
7	46–55	Thick, soupy, very dark
8	66–75	Almost no free water, all amorphous material
9	76–95	No free water, all amorphous material
10	95–100	No free water, all amorphous material

From Verry et al. (2011).

peatlands is more decomposed, or more humified, than ombrotrophic peat from bogs. Peat formed from woody vegetation may be difficult to classify using the Von Post method (Malterer et al., 1992).

In the US, the USDA (1999) taxonomic classification of soils characterizes peat into fibric, hemic, and sapric materials based on the degree of decomposition. This is done by rubbing the peat between the forefinger and thumb for a short period of time, like 30 s. Fibrous or fibric peat is undecomposed or only slightly decomposed. It contains more than 2/3–3/4 fibrous materials after rubbing. *Sphagnum* often produces fibric peat. Hemic material is intermediate in terms of its degree of decomposition, 1/6–2/3 fibers after rubbing, and is often associated with herbaceous emergent vegetation. Sapric material is highly decomposed. It contains less than 1/6 fibers by volume after rubbing. Sapric materials often are found deeper in the peat profile, having been exposed to decomposition, albeit slowly, over a much longer period of time than freshly deposited surface peat. The USDA method is simpler than the Von Post method and is useful in regions where peatlands are not common. For regions with extensive peatlands such as Canada, Russia, and northern Europe, the Von Post method is more appropriate, reflecting the abundance and diversity of peatlands in these regions. The degree of decomposition is important from a restoration standpoint. Typically, many restoration sites were drained in the past and so were exposed to long-term oxidation of the peat that becomes more highly decomposed with time. As a result, on many restoration sites the initial peat material will be highly decomposed or more sapric than it was originally.

The peat profile often is separated into two layers, a surface layer known as the *acrotelm* and a subsurface layer, the *catotelm*, as proposed by the Russian mire hydrologist K.E. Ivanov (Ingram, 1978). The acrotelm is the uppermost layer that consists of relatively undecomposed peat. It is characterized by relatively porous peat with high hydraulic conductivity or water transmission, and a fluctuating water table and is subject to periodic air entry (Ingram, 1978; Glaser, 1987). It is rich in aerobic bacteria and contains growing plants. The acrotelm is considered the layer of maximum biological activity. The catotelm is the underlying subsurface layer. It contains highly decomposed peat that is permanently saturated. In the catotelm, hydraulic conductivity is low and biological activity is minimal (Glaser, 1987). The catotelm is not subject to air entry and lacks aerobic microorganisms (Ingram, 1978).

Canada and Russia contain most of the world's peatlands (Kivinen and Pakarinen, 1981; Taylor, 1983). The IPS estimates that Canada (1,235,000 km^2) and Russia (1,390,000 km^2) contain 67% of the world's 3,935,000 km^2 of peatlands (Joosten and Clark, 2002). These areas include the Hudson Bay lowlands of Canada and the west Siberian plain of Russia. In the tropics, Indonesia contains 270,000 km^2 or about 7% of the world's peatlands. Tropical peatlands include mangroves as well as palm swamps that are common in Indonesia and Central America (Anderson, 1964; Morley, 1981; Ellison, 2004). In the tropics, *Sphagnum* is not an important component of lowland peatlands but it may be found in montane and alpine habitats (Whinam et al., 2003). Palm swamps exhibit characteristics of ombrotrophic bogs with thin peat around the periphery of a nutrient-poor interior of raised peat (Troxler, 2007). Though small in area, oceanic environments such as the British Isles, Fennoscandia, eastern Canada, and western Alaska contain extensive peatlands (Heinselman, 1975; Taylor, 1983).

Peatland Succession

Peatlands form by two processes, terrestrialization and paludification (Gorham, 1957; Glaser, 1987). Terrestrialization, or hydrarch succession, involves the filling of a pond or lake with sediment and peat over an extended period of time (Lindeman, 1941; Darlington, 1943) (Figure 7.1). Initially, the pond becomes shallower as sediment and calcareous aquatic organisms, algae, and diatoms, are deposited on the bottom. In the early stages of hydrarch succession, wetland vegetation grows only around the periphery of the pond, depositing peat there (Figure 7.1(a)). Over time, peat begins to grow outward toward the center and, in the middle stages of infilling, wetland vegetation forms a floating mat around the increasingly small pond center (Figure 7.1(b)). Eventually, the pond fills with peat with herbaceous emergent vegetation characteristic of a fen (Figure 7.1(c)). With continued peat deposition, the wetland surface increasingly becomes isolated from groundwater and *Sphagnum* begins to dominate. The endpoint or "climax" community is a precipitation-fed, nutrient-poor, acidic *Sphagnum* bog that may also contain acid-loving woody shrubs and trees (Figure 7.1(d)). Paludification involves the accumulation of peat directly atop flat to gently sloping mineral soils, either synchronous with or following terrestrialization (Klinger, 1996) (Figure 7.1(e)). These peats tend to be shallower than those produced during terrestrialization and often are referred to as blanket bogs as they form a continuous but not necessarily a thick layer of peat over large areas (Moore and Chater, 1969).

Peat chemistry changes with increasing isolation from groundwater, greater reliance on precipitation, and succession of the plant community. Organic content and acidity increase, and bulk density and nutrient (N, P) content decrease with time (Table 7.3). The degree of decomposition of the surface peat also decreases.

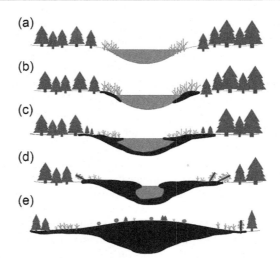

Figure 7.1 Peatland succession: terrestrialization or hydrarch succession where a pond fills in with time to form a peatland. (a) Growth of emergent vegetation around the periphery; (b) Colonization by *Sphagnum* and incipient accumulation of peat; (c) Floating mat forms toward the center; (d) Near-complete infilling and development of a fen with paludification into the adjacent forest; and (e) Complete infilling and formation of a raised bog.

Table 7.3 Peat Properties (pH, Von Post Scale, Bulk Density, C, N, P) as a Function of Succession from (Minerotrophic) Fen to (Ombrotrophic) Bog. Measured Values are from the 0–25 cm Depth

	pH	Von Post Scale	Bulk Density (g/cm³)	C (%)	N (%)	P (µg/g)
Fen	6.0	Nm	0.09	44	2.76	765
Cedar swamp	5.6	6.7	0.12	43	1.92	770
Intermediate fen	5.2	4.4	0.09	41	2.52	790
Poor fen	4.1	3.7	0.02	42	1.51	630
Bog	3.8	2.9	0.05	43	1.14	510

Nm, not measured.
Fen from the author (unpublished data). All other sites from Bridgham et al. (1998).

Once established, the unique characteristics of *Sphagnum* promote its persistence that hinders other species from colonizing (van Breeman, 1995) (Table 7.4). The recalcitrant nature of *Sphagnum* inhibits herbivory in the short term and builds peat in the long term. Its hyaline cells and pendant branches enhance water use and storage and produce a finely porous, impermeable peat. *Sphagnum*, since it draws nutrients (N) from precipitation, outcompete vascular plants that draw more N from mineralization (Aldous, 2002; Malmer et al., 2003). The collective result is a low nutrient, acidic, anaerobic, and cold soil environment that is unfavorable for establishment of vascular plants.

Table 7.4 **Characteristics of *Sphagnum* that Promote Its Persistence and Inhibit Its Competitors**

Characteristic	Short-Term Benefit	Long-Term Benefit
Organochemical composition	Inhibits herbivory	Builds peat
Hyaline cells	Conserves water	Finely porous, impermeable, and insulating peat that creates cold soil conditions
Pendant branches	Capillary water	Collapses easily to produce dense peat
High nutrient retention	Efficient nutrient use	Low nutrient supply to environment
Relies on N from precipitation	Outcompetes vascular plants for N	High nutrient retention that maintains *Sphagnum*'s dominance

From van Breeman (1995) and Malmer et al. (2003).

Theory

Peatland development is guided by autogenic succession, the result of accumulating organic matter from incomplete decomposition of plant production. Peat accretion, the rate at which peat builds vertically and is expressed as millimeter/year or centimeter/100 years, is a slow process. ^{14}C dating of basal peat yields accretion rates of less than 1 mm/year (Botch et al., 1995) amounting to less than 10 cm of peat buildup every 100 years. The long-term rate of carbon accumulation based on ^{14}C ranges from <10 g C/m^2/year to 80 g/m^2/year (Schlesinger, 1997; Francez and Vasander, 1995; Belyea and Warner, 1996; Turunen and Tolonen, 1996; Roulet et al., 2007; Yu, 2012). Studies of net ecosystem exchange (NEE) are similar, yielding 20–105 g C/m^2/year (Roulet et al., 2007; Nilsson et al., 2008; Koehler et al., 2011; Yu, 2012). The rate of peat buildup is generally lower in subarctic climates than in boreal climates (Ovenden, 1990; Botch et al., 1995). There is some evidence that peat accumulation varies with vegetation type with greater accumulation in fen and hardwood forests than in bogs (Botch et al., 1995).

Nurse plants such as cotton grass and the moss, *Polytrichum*, may accelerate colonization and growth of *Sphagnum* and woody seedlings (Robert et al., 1999; Groeneveld and Rochefort, 2002). This is especially true in cold climates where frost heaving damages or kills young plants. Reintroducing fragments of *Polytrichum* to disturbed peatlands has been shown to reduce frost heaving (Groeneveld and Rochefort, 2005). Other benefits of nurse plants include serving as a seed trap, facilitating seedling establishment, binding loose soil, preventing the formation of a surface crust, and creating a favorable microclimate (Groeneveld and Rochefort, 2002; Groenveld et al., 2007). The insulating effects of *Polytrichum* lead to increased water content and reduced temperature fluctuations. In field experiments, the presence of a *Polytrichum* carpet kept *Sphagnum* fragments more humid than bare peat alone (Groeneveld et al., 2007). Over time, *Polytrichum*, which prefers drier sites, is replaced by *Sphagnum*.

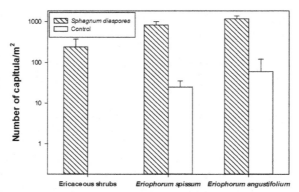

Figure 7.2 Role of nurse plants (ericaceous shrubs and *Eriophorum*) in establishment of
Sphagnum diaspores. Note the log scale on the y-axis.
From Boudreau, S., Rochefort, L., 1998. Restoration of post-mined peatlands: effect of vascular
pioneer species on *Sphagnum* establishment. In: Proceedings of the International Peat Symposium:
Peatland Restoration and Reclamation: Techniques and Regulatory Considerations, 1998,
Jyvaskyla, Finland, pp. 39–43. Reproduced with permission of the International Peat Society.

Sphagnum colonization also occurs under cotton grass, *Eriophorum* (Grosvernier
et al., 1995) and it facilitates *Sphagnum* colonization in experimental studies (Ferland
and Rochefort, 1997). Diaspores of *Sphagnum* introduced to post-mined peatlands
reestablished in greater numbers beneath the cover of two species of cotton grass,
Eriophorum spissum and *Eriophorum angustifolium*, than beneath ericaceous shrubs
(Figure 7.2) (Boudreau and Rochefort, 1998). Twenty years following abandonment
of a cutaway peatland, tussocks of cotton grass were focal points for colonization by a
variety of woody, ericaceous, and moss species (Tuittila et al., 2000b). Not all studies
support the role of cotton grass as a nurse plant. Lavoie et al. (2005b) reported that
establishment of mosses on vacuum-mined peatlands in Quebec was related more to
hydrological characteristics of the peat than the presence of *Eriophorum*, as *Sphagnum*
was more likely to colonize peat that contained greater than 85% volumetric water
content. *Eriophorum* has a number of characteristics that make it a capable nurse plant
(Figure 7.3). It readily germinates on bare peat and mineral soil surfaces (Wein and
MacLean, 1973; Gartner et al., 1986) and it creates a more humid microclimate than
bare peat alone (Grosvernier et al., 1995). Shading by *Eriophorum* reduces the drying
effect caused by solar insolation, wind, and the low albedo of peat. *Eriophorum* is a
deep-rooted plant (Bliss, 1956; Hunter and Knight, 1958) and is able to assimilate
nutrients and concentrate them in surface peat (Chapin et al., 1979), making them available to other species. The role of mycorrhizae in facilitating establishment of vascular
plants in peatlands is unclear. Thormann et al. (1999) observed mycorrhizal infection in
a number of plants including cotton grass, *Eriophorum vaginatum*, ericaceous species,
and black spruce, *Picea mariana*.

Sphagnum is a *keystone* species in many bogs, especially in northern (cooler) climates (Rochefort, 2000). It is the key builder of peat in these wetlands yet there are
several factors that constrain its ability to do so. It is a late successional plant during
the evolution from fen to bog and so is a latecomer to the site. It also is a stress tolerator

Figure 7.3 Sumarsky Most, a mined peatland in the Czech Republic that was abandoned in 2000. The site was restored between 2000 and 2004 by blocking drainage ditches, introducing *Sphagnum* diaspores and mulch (Jongepierová et al., 2012). (a) Naturalization colonization of the site in 2011; and (b) Clumps of *Eriophorum*, a magnet for colonization by other species. Photos by the author.

(Grime, 1977) that grows slowly. In spite of its importance as a keystone species, the dispersal requirements of *Sphagnum* are poorly understood (Rochefort, 2000). Reintroduction of *Sphagnum* is essential to establishing the species on restoration sites (Rochefort et al., 1995). On sites where peat was harvested, *Sphagnum* failed to recolonize even after 15–20 years (Lavoie and Rochefort, 1996; Desrochers et al., 1998). Instead, many sites become dominated by woody species such as birch and poplar or ericaceous shrubs such as *Kalmia*, *Ledum*, and *Molinia* (Rochefort, 2000).

There are many species of *Sphagnum*. Some are acclimated to lower elevation microsites, in hollows, whereas others are adapted to the higher elevations, hummocks and lawns (Smolders et al., 2003). Hummock and lawns species are slow colonizers compared to hollow species (Smolders et al., 2003). *Sphagnum* may disperse to restoration sites either as spores or as diaspores, fragments or pieces. Diaspores that naturally disperse are either too few or too small to regenerate the *Sphagnum* lawn or carpet characteristic of many bogs (Rochefort, 2000). Dispersal of spores is more likely but their ability to colonize is poorly documented. From the constraints described above, it is necessary that, on sites targeted for restoration, *Sphagnum* will need to be deliberately reintroduced.

Practice

Restoration of peatlands is implemented on lands that were drained for agriculture and forestry and where peat was harvested, either by hand cutting in the past or, more recently, on sites where vacuum mining was employed. Goals of restoration include restoring mires (i.e., peat-accumulating systems) to support biodiversity and peatland functions (Smart et al., 1989; Graf and Rochefort, 2008) but also to create buffer zones between forestry operations and sensitive aquatic ecosystems (Sallantaus et al., 1998; Vasander et al., 2003). Restoration of harvested peatlands has become a common practice, beginning with block-cut peatlands in Europe in the 1980s (Blankenburg and Kunze, 1986; Meade, 1992; Salonen and Laaksonen, 1994), expanding to vacuum-mined peatlands in Canada in the 1990s (Rochefort and Price, 2003). Eutrophied peatlands, usually fens, are targeted for restoration, especially in western Europe.

Restoring peatlands first involves reintroducing the proper water source to support the appropriate physicochemical conditions. For bogs, water derived from precipitation that is slightly acidic and low in nutrients needs to be introduced. For fens, groundwater is needed to produce the unique chemistry of these systems, slightly acidic to circumneutral pH with abundant base (Ca) cations and low nutrient (P) availability that supports high plant species diversity.

Mined Peatlands

Restoration of peatlands on sites where peat was harvested is especially problematic because of the extractive nature of the disturbance. Worldwide, peat harvesting is concentrated in Europe, Russia, and North America (Lappalainen, 1996). About half is used for fuel, mostly in northern Europe and Russia, and half for horticultural purposes or as an absorbent (Nyronen, 1996). Peat is used to generate electricity in Ireland, Finland, and Russia (Gottlich et al., 1993). Nearly all peat harvested in North America is used in horticulture and horticultural products (Lappalainen, 1996).

Prior to harvesting, the water table first is lowered by ditching. This increases the effective thickness of the acrotelm, drying the peat and making it easier to harvest. Drainage associated with both agriculture and harvesting leads to oxidation of the peat, compaction, and subsidence that lowers the elevation of the peat surface and dramatically reduces hydraulic conductivity of the soil. The lack of viable seeds and propagules in the seedbank (Bakker et al., 1996; Jansen et al., 1996) and isolation from intact peatlands, which inhibits dispersal, further compound the problem. Abandoned mined peatlands typically remain barren or contain only sparse vegetation for several years and are not usually recolonized by *Sphagnum* (Ferland and Rochefort, 1997). Furthermore, harvesting often removes the ombrotrophic peat, leaving behind minerotrophic peat or mineral soil that is characteristic of an earlier stage of succession. Because of these antecedent conditions, it may be wiser to restore a bog back to an earlier stage of succession such as a fen (Meade, 1992) then let succession take over.

Harvesting of peat in the past was done by hand. Block-cutting was widely employed before 1970 (Famous et al., 1991). It creates a distinctive topography with parallel trenches 1–3 m deep alternating with drier raised ridges. Only a small portion of the peat surface is disturbed at any one time. Abandonment of these sites may eventually lead to colonization by *Sphagnum* and woody species once the ditches are plugged and, after a century or more, vegetation of abandoned peatlands may come to resemble uncut peatlands in the region.

Vacuum mining became the predominant method of harvesting after 1970 (Famous et al., 1991). Here, the peat is harvested over a much larger area, leaving a landscape devoid of vegetation. On vacuum-mined peatlands, the peat first is drained by ditching and removal of vegetation (Crum, 1992; Frilander et al., 1996). The peat surface then is milled to a depth of 5–10 cm to break up and dry the peat and harvested using mechanical means of harrowing, ridging, and loading, or using a vacuum collector. In North American mined peatlands, up to 5 cm of peat is harvested each year (Crum, 1992). In Finland, a similar technique is used. After drainage and vegetation removal, the peat is milled to a depth of 2 cm and then removed (Frilander et al., 1996). After abandonment, without hydrologic restoration, the site is slowly colonized by woody species characteristic of drier habitats but with a near-absence of *Sphagnum* (Kaine et al., 1995).

Harvesting results in greater contact between floodwaters and the underlying older peat or mineral soil, altering both water and peat chemistry. The result is that harvested peat bogs are more similar to fens, with higher pH, dissolved base cations (Ca, Mg, K, Na), sulfate, and chloride relative to unharvested raised bogs (Wind-Mulder et al., 1996). Harvested peatlands also contain more water-soluble and plant-available NH_4^+ and NO_3^- (Wind-Mulder and Vitt, 2000). Lally et al. (2012) reported that the partial and complete peat removal in mined Irish peatlands reconnected the water with alkaline subsoils that increased pH, effectively turning the chemistry back toward a fen. By turning succession back to an earlier stage, these measures slow the establishment of *Sphagnum* and add considerable cost to the restoration (Lally et al., 2012).

The surface environment of cutover and vacuum-mined peats is harsh for the establishment of *Sphagnum* (Table 7.5). Because of drainage, the acrotelm is thicker and subject to large variation in the depth to the water table. The absence of vegetation and low albedo of the surface peat leads to wide swings in day- versus night-time temperature, high rates of evaporation, frost heaving, and burial of diaspores by wind (Salonen, 1987; Campbell et al., 2002). Faubert and Rochefort (2002) reported that burial at depths of as little as 10 mm was enough to inhibit establishment of most moss species.

Table 7.5 Substrate and Dispersal Constraints on Natural Colonization of *Sphagnum* on Mined Peatlands

Substrate	Dispersal
Low moisture High temperature Frost heaving Wind-driven burial Low phosphorus	No/few viable propagules in seedbank Absence of nearby donor sites

Drained Peatlands

Hydrologic restoration of drained peatlands typically involves blocking or filling ditches to rewet the peat (Heikkila and Lindholm, 1995; Mawby, 1995; Price et al., 2003; Acreman et al., 2007) and building dikes to retain water (Francez et al., 2000). Other measures include using bunds, polders, or terraces to retain water; establishing buffer zones between the restoration site and sites of active peat mining; and reconfiguring the surface by creating shallow basins (Price et al., 2002, 2003). Abandoned peatlands often show wide fluctuations in the water table (Schouwenaars, 1993; Price, 1996) that must be remediated.

Peatlands drained for forestry are easier to restore than mined peatlands. Hydrology is restored by filling ditches after harvesting. In Finland, restoration of a minerotrophic fen in this way led to rapid colonization by cotton grass (*E. vaginatum*) and, within 3 years, coverage reached 50% (Jauhiainen et al., 2002). Colonization by peatmosses (*Sphagnum balticum*) and woody vegetation (*Calluna vulgaris*) was slower to develop following harvesting and ditch filling of a restored bog. Jauhiainen et al. suggested that spontaneous revegetation is more likely on harvested than mined peatlands because of the presence of residual propagules.

Many *Sphagnum* species grow well in flooded conditions but do poorly or do not grow at all on higher, drier sites (Mawby, 1995). Price and Whitehead (2001) evaluated hydrologic conditions on an abandoned cutover bog in Quebec where, after 30 years after abandonment, *Sphagnum* covered only 10% of the site. Areas where *Sphagnum* colonized were characterized by low slope, water table within 25 cm of the surface, and soil moisture greater than 50%. Areas where these thresholds did not occur generally contained no *Sphagnum*.

Other techniques such as adding straw mulch have been tested to enhance soil moisture. Application of straw mulch maintained moister surface soil and a higher water table as compared to bare peat (Price et al., 1998). The presence of straw also enhanced the establishment and spread of introduced diaspores of *Sphagnum* (Boudreau and Rochefort, 1998) (Figure 7.4), especially during the first year (Rochefort et al., 1997). Creation of microtopography, hummocks and hollows, fostered moister conditions, especially in the hollows, produced a more favorable situation for establishment of *Sphagnum* (Figure 7.5) (Campeau et al., 2004). Creating shallow, 20 cm deep, basins of varying size also enhanced establishment of *Sphagnum* (Campeau et al., 2004). After three to four growing seasons, *Sphagnum* cover in shallow depressions ranged from 20% to 25% with mechanical reintroduction of diaspores and 40–60% with manual reintroduction. Price et al. (2002) suggested that the additional water retained by the basins protects *Sphagnum* against dessication during dry periods.

Drained and mined peatlands often contain low soluble P and K relative to unmined peatlands (van Duren et al., 1997; Wind-Mulder and Vitt, 2000; Andersen et al., 2006; Basiliko et al., 2007), so fertilizer additions may be needed to accelerate the establishment of *Sphagnum*. Studies indicate a negative effect of liming on reestablishing *Sphagnum* cover but positive effect of P additions (Money, 1995). Additions of P to mined peatlands containing *Sphagnum* fragments produced greater cover of *Sphagnum* (Ferland and Rochefort, 1997). Rochefort et al. (2003) and Sottocornola et al. (2007) reported that P additions also favored establishment of nurse species like *Polytrichum strictum*.

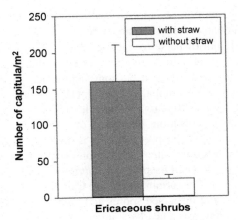

Figure 7.4 The effect of straw on establishment of *Sphagnum* diaspores under ericaceous shrubs.
Reprinted with permission from Boudreau, S., Rochefort, L., 1998. Restoration of post-mined peatlands: effect of vascular pioneer species on *Sphagnum* establishment. In: Proceedings of the International Peat Symposium: Peatland Restoration and Reclamation: Techniques and Regulatory Considerations, 1998, Jyvaskyla, Finland, pp. 39–43. Reproduced with permission of the International Peat Society.

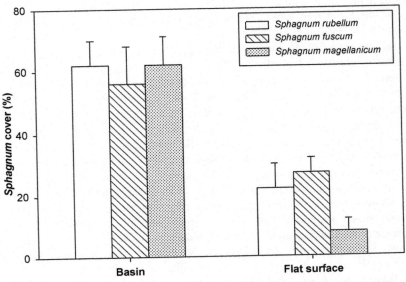

Figure 7.5 The effects of microtopography (creating 20-cm deep basins) on establishment of three species of *Sphagnum*.
Reprinted with permission from Campeau, S., Rochefort, L., Price, J.S., December 2004. On the use of shallow basins to restore cutover peatlands: plant establishment. Restoration Ecology 12 (4), 471–482. Reproduced with permission of John Wiley and Sons.

Spontaneous revegetation of abandoned peatlands with characteristic bog species rarely occurs. The seedbank often is depauperate on mined and abandoned peatlands and diaspores often do not disperse to the site (Salonen, 1987; Salonen and Setala, 1992). In the foothills of the Alps, southern Germany, diaspores of all the important species of *Sphagnum* were present but long abandoned peatlands contained few mire species (Poschlod, 1995). Huopalainen et al. (1998) evaluated the seedbank of a block-cut peatland abandoned 20 years ago in Finland. The seedbank consisted mostly of woody and ericaceous vegetation. Diaspores of a few species of brown mosses, *Polytrichum*, were present but diaspores of only a single species of *Sphagnum* were found. When diaspores reach the site, germination usually fails because of the harsh soil conditions (Salonen, 1987). Diaspores of *Sphagnum* will disperse to abandoned vacuum-mined peatlands under certain situations. Campbell et al. (2003) reported that mosses, owing to their wind-dispersed spores and high fecundity, had relatively high potential to disperse if there was a nearby source.

Because of the smaller size of its disturbance, spontaneous revegetation is more commonly observed in block-cut than in vacuum-mined peatlands (Robert et al., 1999; Farrell and Doyle, 2003; Lavoie et al., 2003). Vascular plants readily colonize abandoned block-cut peatlands but colonization by *Sphagnum* occurs much more slowly (Girard et al., 2002). Farrell and Doyle (2003), however, reported rapid colonization of *Sphagnum* on a block-cut blanket bog in Ireland as, within 2 years, *Sphagnum* cover on the site increased from 1% to 31%. Not surprisingly, spontaneous revegetation is more likely to occur on fens, where *Sphagnum* is less common, than on bogs. Graf et al. (2008) observed 50–70% coverage of vascular plants in block-cut fens which is much greater than the 25% cover reported for block-cut bogs (Poulin et al., 2005).

In the absence of deliberate reintroduction of vegetation, given time, succession will occur in abandoned peatlands, albeit slowly. Konvalinkova and Prach (2010) evaluated plant colonization on 17 peatlands in the Czech Republic mined by block-cutting and/or vacuum mining. They found a general tendency for succession to occur on block-cut sites that were greater than 50 years old. In Estonia, milled peatlands abandoned 26–31 years ago contained a large number of plant species in the seedbank but germination was inhibited by low soil moisture (Triisberg et al., 2013). These studies highlight the need to restore hydrology and adequate soil moisture before reintroducing plant propagules and diaspores.

In a survey of abandoned peatlands, Poulin et al. (2005) found that block-cut peatlands contained 50% or more cover of ericaceous shrubs as compared to only 16% in vacuum-mined peatlands. Whereas herbaceous cover of abandoned peatlands was comparable to natural peatlands, cover of *Sphagnum* was very low. In vacuum-mined peatlands, *Sphagnum* cover was only 2%, whereas, in block-cut peatlands, cover ranged from 2% in less deeply mined peat (baulks) to 30% in trenches (Poulin et al., 2005). Where spontaneous recolonization on vacuum-mined peatlands does occur, cotton grass, *E. vaginatum*, often is a pioneer species. The slow rate of recovery of vegetation in vacuum-mined peatlands is the result of the large scale of extraction (Money, 1995; Lavoie and Rochefort, 1996) and the intense ditching and drainage needed to support mechanized tractors (Lavoie et al., 2003).

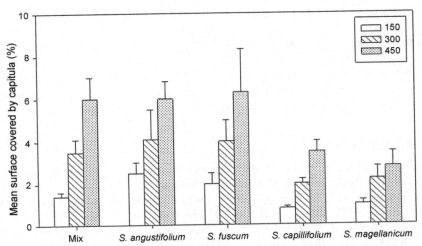

Figure 7.6 The effect of initial introduction of *Sphagnum* diaspores (number per square meter) percent surface cover after one growing year. Four species and a mixture of them were tested.
Reprinted with permission from Campeau, S., Rochefort, L., June 1996. Sphagnum regeneration on bare peat surfaces: field and greenhouse experiments. Journal of Applied Ecology 33 (3), 599–608. Reproduced with permission of Blackwell Publishing Ltd.

Because unassisted establishment of *Sphagnum* on abandoned peatlands is difficult to impossible, deliberate reintroduction of diaspores is needed. Experimental studies show that introduction of fragments or diaspores is superior to introducing entire plants (Money, 1995) and that establishment of *Sphagnum* increases with increasing density of introduced diaspores (Figure 7.6). The effect of diaspore size, however, does not affect recolonization success. Campeau and Rochefort (1996) reported that diaspore fragments of 0.5, 1, and 2 cm in size had similar recolonization success. Moisture was the primary determinant of recolonization, with moist peat having a much higher rate of colonization than dry peat (Campeau and Rochefort, 1996).

Studies suggest that successful establishment of characteristic peatland species depends more on the species introduced rather than the number of diaspores. Chirino et al. (2006) introduced diaspores of four species of *Sphagnum* on mined peat surfaces and tracked their development for 4 years. Of the four species introduced, *Sphagnum fuscum*, a common species of peat moss, was the most successful in reestablishing. Other researchers reported that under humid conditions *Sphagnum fallax* was more successful than other species of *Sphagnum*. *Sphagnum fallax* also serves as a springboard for colonization by other species that are characteristic of bogs (Grosvernier et al., 1997; Buttler et al., 1998). Under drought conditions, *S. fuscum* is superior (Chirino et al., 2006). It also is more resistant to burial (Faubert and Rochefort, 2002).

One means to accelerate colonization of *Sphagnum* is by encouraging the formation of floating rafts of peat. This technique has been tested in Europe where block-cut bogs are rewetted. The combination of poorly humified peat low in lignin, phenolics, and C:N and inundation with alkaline waters creates buoyant mats suitable for establishment of *Sphagnum* (Smolders et al., 2002; Tomassen et al., 2003). Anaerobic conditions to support high rates of methane production were needed to ensure buoyancy of the mat (Lamers et al., 1999). Methane production was enhanced by the poorly humified, labile peat and the high pH of floodwaters (Tomassen et al., 2004). This technique may not be applicable to many abandoned peatlands where the proper hydrologic conditions (i.e., flooding to depths to support floating rafts), water chemistry, and peat quality do not exist.

Rewetting of Fens

A common practice for restoring peatlands in northern and central Europe is rewetting of degraded fens and heathlands. These wetlands were drained many years ago for agriculture (Pfadenhauer and Klötzli, 1996; Pfadenhauer and Grootjans, 1999; Richert et al., 2000; Lamers et al., 2002a; Timmerman et al., 2006; Malson et al., 2008; Schultz et al., 2011) or the water level was lowered by pumping groundwater for public water supplies (Kemmers et al., 2003). In addition to the usual consequences of drainage, oxidation, compaction, and subsidence of peat, hydraulic conductivity is reduced (Okrusko, 1995) and soil acidification occurs as base-rich groundwater is replaced by precipitation (van Diggelen et al., 1991; Beltman et al., 1995). Drainage transforms organic bound phosphorus (P) to Fe-bound P that is sensitive to changing redox potential. Upon rewetting and the establishment of anaerobic conditions, the peat soil releases P (Zak et al., 2004) in a process known as *internal eutrophication* (Venterink et al., 2002). The process is compounded by increased acidity and sulfate inputs from polluted groundwater and atmospheric deposition where S preferentially binds to Fe (Bobbink et al., 1998; Lamers et al., 2002b) and by the highly decomposed surface peat with its low C:P ratio (Zak et al., 2010). The cumulative effect of these stressors is a dramatic decline in plant species diversity in these historically species-rich wetlands. Without interventions such as liming, high water column, P concentrations may persist for decades (Zak et al., 2008).

Restoration of eutrophied fens consists of blocking ditches to restore hydrology to increase contact with groundwater and to enhance moisture content of the surface peat (Pfadenhauer and Klötzli, 1996; Cooper et al., 1998; Cobbaert et al., 2004) and introduction of diaspore-rich hay (Patzelt et al., 2001; Graf and Rochefort, 2008). If groundwater contains sufficient calcium, the site may be successfully restored. Other studies suggest that reintroduction of groundwater may not be sufficient to restore these systems. van Dijk et al. (2004) found that restoring seepage conditions to an acidified peat soil in the Netherlands led to a doubling of P availability in the soil. Decomposition also was stimulated and was linked to increased soil pH. Lamers et al. (2014) reviewed the stressors affecting fens and the challenges associated with restoring them. Stressors included drainage, acidification caused by drainage, P eutrophication from external sources, and internal eutrophication caused by oxidation of the peat. Restoration measures consist of rewetting to reintroduce hydrology and buffer soil acidity. Eutrophication is ameliorated by halting external inputs of P. To reduce internal loading, sod removal (discussed below) may be required. Species reintroduction by propagules or by planting also may be needed.

Seeding rewetted fens using hay cut from high-quality reference sites often is used to increase species richness (Pfadenhauer and Grootjans, 1999; Rasran et al., 2007). Cobbaert et al. (2004) tested the introduction of diaspores and mulch on establishment of vegetation following restoration of minerotrophic fens in Quebec. Both measures accelerated establishment of fen species. The combination of topsoil removal and diaspore-rich hay from a donor site increased vegetation development on abandoned bare peat fens in Germany (Patzelt et al., 2001). In Sweden, reintroducing diaspores to restore characteristic fen brown mosses enhanced their establishment (Malson and Rydin, 2007). In field experiments, additions of lime, litter (to provide cover), and diaspores increased survival and cover of four brown moss species on the site (Malson and Rydin, 2007). In the Rocky Mountains of the US, Cooper and MacDonald (2000) investigated restoring high elevation fens through rewetting and planting. They found that planting was necessary in these stressful cold, semiarid environments. They suggested that, even after planting, full restoration of these systems would require hundreds of years.

Eutrophied Peatlands: Sod Cutting, Topsoil Removal, and Mowing

Where rewetting with unpolluted water fails, additional measures such as sod cutting or topsoil removal to remove excess nutrients, especially P, are needed (Beltman et al., 1996; Jansen et al., 1996; Snow and Marrs, 1997; van Duren et al., 1998; Pfadenhauer and Grootjans, 1999; Mitchell et al., 2000; Tallowin and Smith, 2001; Rasran et al., 2007; Malson et al., 2010). Sod cutting also exposes the bare soil to seed germination and favors establishment of heathland vegetation (De Graaf et al., 1998; Britton et al., 2000). In the UK, Tallowin and Smith (2001) investigated topsoil removal to restore fen (*Molinia caerulea–Circium dissectum*) vegetation on agriculturally improved pasture. Removal of 15–20 cm of topsoil led to an 85% reduction in total P, reduced available P, and promoted development of a *Molinia–Cirsium* meadow. Working in the Netherlands, Verhagen et al. (2001) evaluated plant community development on former agricultural soil where topsoil was removed to facilitate restoration. Over a 9-year period, vegetation in the plots increased in similarity toward reference sites. However, many target species did not colonize after 9 years even though they were present in the local species pool, suggesting that dispersal was the limiting factor. Beltman et al. (1996) investigated the effects of sod cutting and drainage to remove acidic surface water on restoration of characteristic fen species in the Netherlands. Sod cutting removed excess nutrients but only temporarily removed characteristic bog mosses from the site. Both drainage and sod cutting resulted in removal of bryophytes and facilitated establishment of fen species. Sod cutting also has been shown to increase species richness of acid-tolerant species but both sod cutting and liming are needed to reestablish acid-sensitive species (Dorland et al., 2005a). Klimkowska et al. (2007) conducted a meta-analysis of the effects of rewetting, topsoil removal, and introduction of diaspores on restoration of fens and floodplain meadows in Western Europe. Topsoil removal and diaspore transfer had the greatest positive effects on plant species richness whereas rewetting alone was of no benefit.

In severe cases of eutrophication, liming may be necessary (De Graaf et al., 1998; Dorland et al., 2004; van Diggelen et al., 2015). In the Netherlands, a catchment liming experiment was undertaken to increase plant species diversity in heathland and moorland pools (Dorland et al., 2005b). Liming with 2–6 tons/ha resulted in increased pH and base cation concentrations. Endangered *Red List* species increased in abundance in the moorlands but the response was much more muted in the heathlands. The positive effects of liming were detected for at least 6 years after liming.

Mowing and prescribed fire also are employed to increase species richness of fens and other graminoid wetlands. Billeter et al. (2007) compared mowed and unmowed plots in 15 montane fen meadows in Switzerland that had been abandoned 4–35 years ago. After 2 years, mowed plots contained an average of 11% more plant species and a 15% increase in fen "indicator" species compared to plots that were not mowed. van Diggelen et al. (2015) investigated the effects of mowing, fire, and liming on acidified and eutrophied fen vegetation in the Netherlands. Summer mowing led to an increase in poor fen vegetation, including bryophytes such as *Sphagnum*. Winter mowing followed by burning counteracted acidification but increased nutrient availability and led to dominance by highly productive graminoids uncharacteristic of rich fens. Liming produced the desired conditions of reduced acidity but maintenance of these conditions required repeated applications. Bakker and Berendse (1999) concluded that restoration of ecological diversity of graminoid wetlands, including fens, heathlands, and wet grasslands, is hampered by the abiotic constraints of eutrophication and acidification and the biotic constraints of a depauperate seedbank and inability to disperse in the highly fragmented landscape in which these wetlands are found.

Ecosystem Development

Many efforts have been undertaken to restore peatlands in Europe and Canada to reproduce a functioning, peat-accumulating wetland. Extensive research has been conducted in Canada to evaluate limitations and controls on restoration. Finland and other European countries, including Germany, Estonia, England, and Ireland, also have focused on restoration strategies and on measurements to evaluate ecosystem development following restoration.

Famous et al. (1991) evaluated revegetation patterns of 35 peatlands in Canada and the US that were abandoned between 4 and 92 years ago. Sites where the peat was block-cut revegetated much faster than vacuum-mined peatlands. Tree species, black spruce (*P. mariana*), white birch (*Betula papyrifera*), and larch (*Larix laricinia*), colonized the ridges while *Sphagnum* and ericaceous shrubs colonized the trenches. Vacuum-mined sites where peat was harvested beginning in 1970 were larger and colonized by cotton grass (*Eriophorum*) and white birch. *Sphagnum* was rare on these sites (Famous et al., 1991). Soro et al. (1999) compared plant species richness in 50-year-old abandoned block-cut peatlands with natural peatlands in Sweden. Even after 50 years, species richness was less in the abandoned peatlands. Lavoie et al. (2005a) assessed vegetation dynamics of a vacuum-mined peatland in southern Quebec 14 years following abandonment. During the 5-year monitoring period, cotton grass decreased in cover while ericaceous shrubs increased on the site. Two decades after abandonment, vegetation cover was still sparse and there was no evidence to indicate that the plant community would eventually resemble that of an undisturbed peatland. These findings highlight the slow rate of recovery following abandonment unless measures to accelerate succession are implemented.

Salonen (1990) compared succession on two block-cut peatlands in Finland that were abandoned 8–9 years ago. The two sites differed with respect to the thickness of the remaining peat. One site had more than 1 m of peat remaining whereas peat thickness at the other site was only 17 cm. Sixty-two species colonized the site with shallow peat including 16 moss species and two species of *Sphagnum*. At the site with deeper peat, only 16 species colonized including three mosses but not *Sphagnum*. Greater colonization of the shallow peat was attributed to differences in seed dispersal as well as greater access of roots to nutrient-rich mineral soil.

Peat mining inevitably alters water chemistry and is caused by increased contact of floodwaters with mineral substrate. The result is water and peat chemistry that is more characteristic of a fen than a bog (Lally et al., 2012). A bog whose water chemistry was monitored for 7 years following restoration had greater pH, base cations, and electrical conductivity than unmined bogs (Andersen et al., 2010a). After 7 years, there was no evidence that water chemistry was converging toward that of an unmined bog. Tuittila et al. (2000a) followed changes in vegetation 1 year before and 4 years following rewetting of a vacuum-mined ombrotrophic bog. *Eriophorum*, *Carex*, and other fen species quickly colonized the site. After 4 years, *Sphagnum* established in waterlogged areas, but in higher areas, mosses of disturbed peat surfaces colonized. Tuittila et al. (2000a) pointed out that restoration of minerotrophic vegetation occurs faster than ombrotrophic vegetation so using surface runoff and fertilizer additions should be considered to initiate succession to create a minerotrophic fen that, given time, will become a bog. Pfadenhauer and Klötzli (1996) went further, stating that a minerotrophic fen represents an earlier and more attainable state of succession so restoration efforts should target nutrient demanding fens and transitional bogs to quickly restore efficient peat production.

Harvested peatlands, especially vacuum-mined peatlands, have a pronounced effect on microbial communities. Post vacuum-mined peatlands in Canada contained fewer total bacteria, cellulose-degrading bacteria, and microbial biomass carbon (MBC) as compared to natural ombrotrophic bogs (Croft et al., 2001). Nitrogen cycling on the restored sites was enhanced with greater N mineralization and NH_4 content. Andersen et al. (2006) compared microbial biomass and CO_2 production on a 20-year-old abandoned vacuum-mined peatland that was restored by blocking drainage ditches and introducing *Sphagnum* propagules. Three years after restoration, microbial biomass and soil respiration was lower in the restored and unrestored sites relative to a natural peatland. There was no difference in microbial community structure among the three sites based on phospholipid fatty analysis (Andersen et al., 2010b). On a site that was monitored for 6 years following restoration, there was a trend of increasing MBC with time, but even after 6 years MBC was more similar to unrestored vacuum-mined peatlands than to intact ombrotrophic bogs (Croft et al., 2001). Studies of restored fens also suggest that nutrient cycles do not equilibrate to levels found in intact fens. Richert et al. (2000) reported that mineral nitrogen (N) levels in a fen restored by rewetting declined, but, after 3 years, inorganic N still was much higher than in natural fens.

Unsurprisingly, rewetting increases CH_4 emissions. In Finland, Tuittila et al. (2000c) compared CH_4 fluxes on peatlands before and after rewetting. CH_4 fluxes increased following restoration but not to the high levels found in intact mires. Tuittila et al. (1999) also measured CO_2 fluxes following rewetting of the same peatland. Submerged areas achieved a positive C balance (i.e., the area sequestered C) within 2 years and it increased more during the third year following rewetting. Higher drier areas, however, continued to be a net source of CO_2 to the atmosphere. In Quebec, restored block-cut and vacuum-mined peatlands produced less CO_2 and more CH_4 than natural peatlands. Differences were attributed to wetter conditions of the restored site, in contrast to natural peatlands that were surrounded by extensive drained and mined areas (Basiliko et al., 2007).

Waddington et al. (2002) measured NEE in young (less than 2 years old) and old (7–8 years old) abandoned vacuum-mined peatlands in Quebec. Even after 8 years, the abandoned peatlands were a source of CO_2. Waddington and Warner (2001) compared NEE in bare (no restoration), restored, and unmined peatlands. On the restored site, diaspores of *Sphagnum* were spread over the site and, at the time of sampling, produced about 50% cover. Gross photosynthesis was two times greater in the restored peatland than in the unmined site and almost three times greater than in the bare peatland. Even so, the restored site was a net source of CO_2 to the atmosphere. These studies collectively suggest that, even with efforts to reintroduce vegetation, abandoned peatlands will continue to be a net source of CO_2 to the atmosphere for years following restoration.

Similar to vegetation, reestablishment of fauna communities is slow following peatland restoration. van Duinen et al. (2003) measured aquatic macroinvertebrate diversity in 27 block-cut peatlands restored by rewetting in the Netherlands. The restored sites ranged in age from 1 to 28 years old and were compared with 20 remnant peatlands that were cut more than 50 years ago and were in various stages of secondary succession. Compared to restored sites, remnant sites contained characteristic and rare bog species not found in restored sites. Species richness also was greater in the remnant sites. On the restored sites, there was no clear association between most measures of macroinvertebrate diversity and characteristic bog vegetation. However, there were positive correlations between the number of rare species and fauna species quality with site age (van Duinen et al., 2003). Active restoration efforts, however, have been shown to accelerate development of invertebrate communities. In New Zealand, restoration of mined peatlands using direct seeding and sod transfer resulted in a rapid development of beetle (Coleoptera) communities (Watts et al., 2008). Within 7–9 years, beetle communities of restored peatlands converged with unmined peat bogs. Sod transfer resulted in faster convergence than direct seeding. In eastern New Brunswick, Canada, creation of shallow pools during bog restoration enhanced arthropod and amphibian use though, 4 years after restoration, animal and plant assemblages differed between pools of restored and natural bogs (Mazerolle et al., 2006).

Desrochers et al. (1998) compared avian communities of abandoned block-cut and vacuum-mined peatlands with natural sites in Quebec. After 20 years, vegetation cover was less in harvested peatlands, especially in vacuum-mined sites, than in natural peatlands. Avian abundance and richness was less on abandoned vacuum-mined peatlands but did not differ between block-cut and natural peatlands. Since block-cutting is no longer practiced in this region, Desrochers et al. (1998) suggested that, for vacuum-mined peatlands, restoration efforts should include preserving areas of unmined peatlands to compensate for the enduring effects of vacuum mining on avian communities.

Efforts to accelerate restoration by introducing diaspores, mulch, and fertilizer accelerate development of peatland structure and function beyond rewetting alone. Waddington et al. (2003) evaluated *Sphagnum* production and decomposition on a peatland restored by rewetting and introduction of diaspores and mulch in experimental plots. Eight years following restoration, a thin veneer of *Sphagnum* had developed and production rates on fully covered areas were comparable to a nearby natural peatland. The surprisingly high level of productivity on the restored site, whose hydrology was drier than the natural peatland, was attributed to the higher density of capitula in the restored plots. Although decomposition rates were measured only in the restored plots, the rate of decomposition (13–17% over the 2-year study period) was comparable to published studies from unmined peatlands (Waddington et al., 2003).

In summary, block-cut peatlands, because of their small size and waterlogged trenches, are capable of spontaneous revegetation whereas industrially mined vacuum and milled peatlands are not (Table 7.6). Colonization occurs first by nurse plants such as *Eriophorum* and the moss *Polytrichum*. Plant productivity and diversity, especially *Sphagnum*, require decades to become similar to unmined peatlands. Microbial

Table 7.6 Ecosystem Development of Peatlands Following Rewetting of Block-Cut and Vacuum-Mined Peatlands

	Time (years) to Equivalence to Natural Peatlands	
	Block-Cut	**Vacuum-Mined**
Productivity and Habitat Functions		
Primary production	3–10	>20 (faster in waterlogged trenches)
Plant cover	<20	>20–50
Plant diversity		>50
Aquatic invertebrates	>30	>50[a]
Avian communities	<20	>20 (contingent on vegetation)
Regulation Functions		
Microbial activity	Sooner	>10–20
CH$_4$ flux	Sooner	>20
Peat/carbon accumulation	Sooner	>20 (contingent on vegetation)
Nitrogen cycling	Sooner	>20

[a]Faster with active restoration efforts such as seeding and sod transfer.

processes, C and N cycling and C accumulation, also require considerable time to develop. Overall, the key measure of success is the restoration of a peat-accumulating system on the site.

Keys to Ensure Success

Since peat accumulation is a slow process, successful restoration of these peatlands will require considerably more time than restoration of other types of wetlands. Reintroducing hydrology, modifying water chemistry including base status and phosphorus, improving microclimate, and introducing diaspores may be needed to restore the site (Gorham and Rochefort, 2003). On some sites, especially in Europe, alteration of the chemical environment by rewetting with sulfate-rich water and atmospheric S deposition complicates the restoration process. Here, remedial measures such as liming to bind P (Zak et al., 2008) may be required to increase species diversity (Dorland et al., 2005a,b). In the case of bogs, it may be necessary to step back to an earlier stage of succession (Lally et al., 2012) to create a fen when the effects of base- and sulfate-rich waters cannot be mitigated. Studies suggest that restoration of vacuum-mined peatlands is more difficult than block-cut peatlands because of the larger spatial footprint and greater intensity of drainage (Lavoie et al., 2003).

Much effort has been devoted to developing techniques to accelerate peatland restoration, especially in Europe and Canada. Sliva and Pfadenhauer (1999) tested a variety of techniques to restore cutover milled bogs in Germany, including hydrologic restoration, seeding, transplanting, reintroduction of *Sphagnum* sods, seed covering, and fertilization. Based on their experiments, Sliva and Pfadenhauer (1999) recommended the following steps for restoring cutover bogs: (1) rewetting; (2) introducing diaspores; (3) mulching; and (4) P fertilization. Poschlod (1995) further suggested removing the diaspore-rich "top-spit" of the peat prior to harvesting, storing it wet, and then spreading it over the site once mining ceases.

In Canada, Rochefort and coworkers developed techniques to restore mined peatlands that are detailed in a special issue of *Wetlands Ecology and Management* (Rochefort et al., 2003, Rochefort and Price, 2003). The Canadian methods consist of the following steps: restoring hydrology, collecting and introducing diaspores, diaspore protection, and fertilization. A detailed guide to peatland restoration was developed that includes estimates of man-hours and cost (Quinty and Rochefort, 2003). The step-by-step recommendations consist of restoring hydrology by blocking ditches, creating shallow basins, polders, or contouring surfaces as appropriate to retain water (Price et al., 2003), collecting and spreading diaspores, applying straw mulch, and fertilizing with P (Table 7.7). An additional key recommendation is to avoid excavating to mineral soil because it enriches the peat with mineral nutrients and promotes colonization by non-peatland species (Quinty and Rochefort, 2003). In summary, successful restoration of peatlands will depend on introducing appropriate hydroperiod and water sources (ombrotrophic for bogs, minerotrophic for fens), the proper substrate (highly humified peat for bogs and poorly humified peat for fens), reintroducing diaspores and providing them proper protection from the elements (including nurse and companion plants to facilitate *Sphagum* establishment), and much time to allow peat to accumulate.

Table 7.7 **The North American Approach to Restoring Cutover** *Sphagnum* **Peatlands**

Goal	Activity
Site preparation	Block ditches using wet peat, construct dikes, contour to create convex surfaces and shallow basins 75–150 m^2 in size
Diaspore collection	Identify donor site with high (>50%) *Sphagnum* cover, remove the upper 10 cm, collect fragments <2 cm in size manually or mechanically with a rototiller
Diaspore introduction	Spread diaspores at a ratio of 1:10 *surface collected:surface restored* using a manure spreader
Diaspore protection	Spread straw uniformly at the rate of 3000 kg/ha
Fertilization	Apply rock phosphate at the rate of 150 kg/ha (19.5 kg P/ha), avoid fertilizers rich in calcium as it is detrimental to *Sphagnum*

From Rochefort et al. (2003), Price et al. (2003), and Quinty and Rochefort (2003). A similar strategy for European cutover bogs was recommended by Sliva and Pfadenhauer (1999).

References

Acreman, M.C., Fisher, J., Stratford, C.J., Mould, D.J., Mountford, J.O., 2007. Hydrological science and wetland restoration: some case studies from Europe. Hydrology and Earth System Sciences 11, 158–169.

Aldous, A.R., 2002. Nitrogen translocation in *Sphagnum* mosses: effects of atmospheric N deposition. New Phytologist 156, 241–253.

Andersen, R., Francez, A.J., Rochefort, L., 2006. The physicochemical and microbiological status of a restored bog in Quebec: identification of relevant criteria to monitor success. Soil Biology and Biochemistry 38, 1375–1387.

Andersen, R., Rochefort, L., Poulin, M., 2010a. Peat, water and plant tissue chemistry monitoring: a seven-year case study in a restored peatland. Wetlands 30, 159–170.

Andersen, R., Grasset, L., Thormann, M.N., Rochefort, L., Francez, A.J., 2010b. Changes in microbial community structure and function following *Sphagnum* peatland restoration. Soil Biology and Biochemistry 42, 291–301.

Anderson, J.A.R., 1964. The structure and development of peat swamps of Sarawak and Brunei. Journal of Tropical Geography 18, 7–16.

Bakker, J.P., Berendse, F., 1999. Constraints in the restoration of ecological diversity in grassland and heathland communities. Trends in Ecology and Evolution 14, 63–68.

Bakker, J.P., Poschlod, P., Strystra, R.J., Bekker, R.M., Thompson, K., 1996. Seed banks and seed dispersal: implications for restoration. Acta Botanica Neerlandica 45, 461–490.

Basiliko, N., Blodau, C., Roehm, C., Bengtson, P., Moore, T.R., 2007. Regulation of decomposition and methane dynamics across natural, commercially mined and restored peatlands. Ecosystems 10, 1148–1165.

Beltman, B., van den Broek, T., Bloemen, S., 1995. Restoration of acidified rich fen ecosystems in the Vechtplassen area: successes and failures. In: Wheeler, B.D., Shaw, S.C., Fojt, W.J., Robertson, R.A. (Eds.), Restoration of Temperate Wetlands. John Wiley and Sons, Chichester, UK, pp. 273–286.

Beltman, B., Van Den Broek, T., Bloemen, S., Witsel, C., 1996. Effects of restoration measurement on nutrient availability in a formerly nutrient-poor floating fen after acidification and eutrophication. Biological Conservation 78, 271–297.

Belyea, L.R., Warner, B.G., 1996. Temporal scale and the accumulation of peat in a *Sphagnum* bog. Canadian Journal of Botany 74, 366–377.

Billeter, R., Peintinger, M., Diemer, M., 2007. Restoration of montane fen meadows by mowing remains possible after 4–35 years of abandonment. Botanica Helvatica 117, 1–13.

Blankenburg, J., Kunze, H., 1986. Aspects of future utilization of cut over bogs in western Germany. Socio-economic impacts of the utilization of peatlands in industry and forestry. In: Proceedings of the International Peat Society Symposium, Oulu, Finland, pp. 219–226.

Bliss, L.C., 1956. A comparison of plant development in micro-environments of arctic and alpine tundra. Ecological Monographs 26, 303–337.

Bobbink, R., Hornung, M., Roelofs, J.G.M., 1998. The effects of air borne nitrogen pollutants on species diversity in natural and semi-natural European vegetation. Journal of Ecology 86, 717–738.

Botch, M.S., Kobak, K.I., Vinson, T.S., Kolchugina, T.P., 1995. Carbon pools and accumulation in peatlands of the former Soviet Union. Global Biogeochemical Cycles 9, 37–46.

Boudreau, S., Rochefort, L., 1998. Restoration of post-mined peatlands: effect of vascular pioneer species on *Sphagnum* establishment. In: Proceedings of the International Peat Symposium, Peatland Restoration and Reclamation: Techniques and Regulatory Considerations, Jyvaskyla, Finland, pp. 39–43.

van Breeman, N., 1995. How *Sphagnum* bogs down other plants. Trends in Ecology and Evolution 10, 270–275.

Bridgham, S.D., Pastor, J., Janssens, J.A., Chapin, C., Malterer, T.J., 1995. Multiple limiting gradients in peatlands: a call for a new paradigm. Wetlands 16, 45–65.

Bridgham, S.D., Updegraff, K., Pastor, J., 1998. Carbon, nitrogen and phosphorus mineralization in northern wetlands. Ecology 79, 1545–1561.

Britton, A.J., Marrs, R.H., Cary, P.D., Pakeman, R.J., 2000. Comparison of techniques to increase *Calluna vulgaris* cover on heathland invaded by grasses in Breckland, south east England. Biological Conservation 95, 227–232.

Buttler, A., Grosvernier, P., Matthey, Y., 1998. Development of *Sphagnum fallax* diaspores on bare peat with implications for the restoration of cut-over bogs. Journal of Applied Ecology 35, 800–810.

Campbell, D.R., Lavoie, C., Rochefort, L., 2002. Wind erosion and surface stability in abandoned milled peatlands. Canadian Journal of Soil Science 82, 85–95.

Campbell, D.R., Rochefort, L., Lavoie, C., 2003. Determining the immigration potential of plants colonizing disturbed environments: the case of milled peatlands in Quebec. Journal of Applied Ecology 40, 78–91.

Campeau, S., Rochefort, L., 1996. *Sphagnum* regeneration on bare peat surfaces: field and greenhouse experiments. Journal of Applied Ecology 33, 599–608.

Campeau, S., Rochefort, L., Price, J.S., 2004. On the use of shallow basins to restore cutover peatlands: plant establishment. Restoration Ecology 12, 471–482.

Chapin III, F.S., Van Kleve, K., Chapin, M.C., 1979. Soil temperature and nutrient cycling in the tussock growth form of *Eriophorum vaginatum*. Journal of Ecology 67, 169–189.

Chimner, R.A., Ewel, K.C., 2005. A tropical freshwater wetland: II. Production, decomposition, and peat formation. Wetlands Ecology and Management 13, 671–684.

Chirino, C., Campeau, S., Rochefort, L., 2006. *Sphagnum* establishment on bare peat: the importance of climatic variability and *Sphagnum* species richness. Applied Vegetation Science 9, 285–294.

Cobbaert, D., Rochefort, L., Price, J.S., 2004. Experimental restoration of a fen plant community after peat mining. Applied Vegetation Science 7, 209–220.

Cohen, A.D., 1973. Petrology of some Holocene peat sediments from the Okefenokee swamp-marsh complex of Southern Georgia. Geological Society of America Bulletin 84, 3867–3878.

Cooper, D.J., MacDonald, L.H., 2000. Restoring the vegetation of mined peatlands in the southern Rocky Mountains of Colorado, USA. Restoration Ecology 8, 103–111.

Cooper, D.J., MacDonald, L.H., Wenger, S.K., Woods, S.W., 1998. Hydrologic restoration of a fen in Rocky Mountain National Park, Colorado, USA. Wetlands 18, 335–345.

Craft, C.B., 2007. Freshwater input structures soil properties, vertical accretion and nutrient accumulation of Georgia and United States (U.S.) tidal marshes. Limnology and Oceanography 52, 1220–1230.

Croft, M., Rochefort, L., Beauchamp, C.J., 2001. Vacuum-extraction of peatlands disturbs bacterial population and microbial biomass carbon. Applied Soil Ecology 18, 1–12.

Crum, H., 1992. A Focus on Peatlands and Peatmosses. The University of Michigan Press, Ann Arbor, Michigan.

Darlington, H.C., 1943. Vegetation and substrate of Cranberry Glades, West Virginia. Botanical Gazette 163, 371–393.

De Graaf, M.C.C., Verbeek, P.J.M., Bobbink, R., Roelofs, J.G.M., 1998. Restoration of species-rich dry heaths: the importance of appropriate soil conditions. Acta Botanica Neerlandica 47, 89–111.

Desrochers, A., Savard, J.P.L., Rochefort, L., 1998. Avian recolonization in eastern Canadian bogs after peat mining. Canadian Journal of Zoology 76, 989–997.

van Diggelen, R., Grootjans, A.P., Kemmers, R.H., Kooyman, A.M., Soccow, M., De Vries, N.P.J., Van Wirdum, G., 1991. Hydroecological analysis of the fen system Lieper Posse (eastern Germany). Journal of Vegetation Science 2, 465–476.

van Diggelen, J.M.H., Bense, I.H.M., Brouwer, E., Limpens, J., van Schie, J.M.M., Smolders, A.J.P., Lamers, L.P.M., 2015. Restoration of acidized and eutrophied fens: long-term effects of traditional management and experimental liming. Ecological Engineering 75, 208–216.

van Dijk, J., Stroetgenga, M., Bos, L., Van Bodegom, P.M., Verhoef, H.A., Aerts, R., 2004. Restoring natural seepage conditions on former agricultural grasslands does not lead to reduction of soil organic matter decomposition and soil nutrient dynamics. Biogeochemistry 71, 317–337.

Dorland, E., Kerkof, A.C., Rulli, J.M., Bobbink, R., 2004. Mesocosm seepage experiment to restore the buffering capacity of acidified wet heath soils. Ecological Engineering 23, 221–229.

Dorland, E., Hart, M.A.C., Vermeer, M.L., Bobblink, R., 2005a. Assessing the success of wet heath restoration by combined sod cutting and liming. Applied Vegetation Science 8, 209–218.

Dorland, E., van den Berg, L.J.L., Brouwer, E., Roelofs, J.G.M., Bobbink, R., 2005b. Catchment liming to restore degraded, acidified heathlands and moorland pools. Restoration Ecology 13, 302–311.

van Duinen, G.J.A., Brock, A.M.T., Kuper, J.T., Leuven, R.S.E.W., Peeters, T.M.J., Roelofs, J.G.M., van der Velde, G., Verberk, W.C.E.P., Esselink, H., 2003. Do restoration measures rehabilitate faunal diversity in raised bogs? A comparative study on aquatic macroinvertebrates. Wetlands Ecology and Management 11, 447–459.

van Duren, I.C., Boeye, D., Grootjans, A.P., 1997. Nutrient limitations in an extant and drained poor fen: implications for restoration. Plant Ecology 133, 91–100.

van Duren, I.C., Strystra, R.J., Grootjans, A.P., ter Heerdt, G.N.J., Pegtel, D.M., 1998. A multidisciplinary evaluation of restoration measures in a degraded *Cirsio-Molinietum* fen meadow. Applied Vegetation Science 1, 115–130.

Ellison, A.M., 2004. Wetlands of Central America. Wetlands Ecology and Management 12, 3–55.

Famous, N.C., Spencer, M., Nilsson, H., 1991. Revegetation patterns in harvested peatlands in central and eastern North America. In: Proceedings of the International Peat Symposium, Duluth, MN, pp. 48–66.

Farrell, C.A., Doyle, G.J., 2003. Rehabilitation of industrial cutaway blanket bog in County Mayo, north-west Ireland. Wetlands Ecology and Management 11, 21–35.

Faubert, P., Rochefort, L., 2002. Response of peatland mosses to burial by wind-dispersed peat. The Bryologist 105, 96–103.

Ferland, C., Rochefort, L., 1997. Restoration techniques for *Sphagnum*-dominated peatlands. Canadian Journal of Botany 75, 1110–1118.

Francez, A.J., Vasander, H., 1995. Peat accumulation and peat decomposition after human disturbance in French and Finnish mires. Acta Oecologia 16, 599–608.

Francez, A.J., Gogo, S., Josselin, N., 2000. Distribution of potential CO_2 and CH_4 productions, denitrification and microbial biomass C and N in the profile of a restored peatland in Brittany (France). European Journal of Soil Science 36, 161–168.

Frilander, P., Leinonen, A., Lakangas, E., 1996. Peat production technology. In: Vasander, H. (Ed.), Peatlands in Finland. Finnish Peat Society, Helsinki, pp. 99–106.

Gartner, B.L., Chapin III, F.S., Shaver, G.R., 1986. Reproduction of *Eriophorum vaginatum* by seed in Alaskan tussock tundra. Journal of Ecology 74, 1–18.

Girard, M., Lavoie, C., Thériault, M., 2002. The regeneration of a highly disturbed ecosystem: a mined peatland in southern Quebec. Ecosystems 5, 274–288.

Glaser, P.H., 1987. The Ecology of Patterned Boreal Peatlands of Northern Minnesota: A Community Profile. U.S. Fish and Wildlife Service Report. 85 (7.14).

Gorham, E., Rochefort, L., 2003. Peatland restoration: a brief assessment with special reference to *Sphagnum* bogs. Wetlands Ecology and Management 11, 109–119.

Gorham, E., 1957. The development of peat lands. The Quarterly Review of Biology 32, 145–166.

Gottlich, K.H., Richard, K.-H., Kuntze, H., Eggelsmann, R., Gunther, J., Eichelsdorfer, D., Briemle, G., 1993. Mire utilization. In: Heathwaite, A.L. (Ed.), Mires: Process, Exploitation, and Conservation. Wiley, Chichester, UK, pp. 325–415.

Graf, M.D., Rochefort, L., 2008. Techniques for restoring fen vegetation on cut-away peatlands in North America. Applied Vegetation Science 11, 521–528.

Graf, M.D., Rochefort, L., Poulin, M., 2008. Spontaneous revegetation of cutaway peatlands of North America. Wetlands 28, 28–39.

Grime, J.P., 1977. Evidence for the existence of three primary strategies in plants and its relevance to ecological and evolutionary theory. The American Naturalist 111, 1169–1194.

Groeneveld, E.V.G., Rochefort, L., 2002. Nursing plants in peatland restoration: on their potential use to alleviate frost heaving problems. Suo 53, 73–85.

Groeneveld, E.V.G., Rochefort, L., 2005. *Polytrichum strictum* as a solution to frost heaving in disturbed ecosystems: a case study with milled peatlands. Restoration Ecology 13, 74–82.

Groeneveld, E.V.G., Masse, A., Rochefort, L., 2007. *Polytrichum strictum* as a nurse-plant in peatland restoration. Restoration Ecology 15, 709–719.

Grosvernier, P., Matthey, Y., Butler, A., 1995. Microclimate and physical properties of peat: new clues to the understanding of bog restoration processes. In: Wheeler, B.D., Shaw, S.C., Fojt, W.J., Robertson, R.A. (Eds.), Restoration of Temperate Wetlands. John Wiley and Sons, Chichester, UK, pp. 435–450.

Grosvernier, P., Matthey, Y., Buttler, A., 1997. Growth potential of three *Sphagnum* species in relation to water table level and peat properties with implications for their restoration in cut-over bogs. Journal of Applied Ecology 34, 471–483.

Heikkila, H., Lindholm, T., 1995. The basis of mire restoration in Finland. In: Wheeler, B.D., Shaw, S.C., Fojt, W.J., Robertson, R.A. (Eds.), Restoration of Temperate Wetlands. John Wiley and Sons, Chichester, UK, pp. 549–559.

Heinselman, M.L., 1970. Landscape evolution, peatland types, and the environment in the Lake Agassiz peatlands natural area, Minnesota. Ecological Monographs 40, 235–261.

Heinselman, M.L., 1975. Boreal peatlands in relation to environment. In: Hasler, A.D. (Ed.), Coupling of Land and Water Systems. Springer-Verlag, New York, pp. 93–103.

Hunter, R.F., Knight, A.H., 1958. Studies of the root development in the field using radioactive tracers. II. Communities growing in deep peat. Journal of Ecology 44, 629–639.

Huopalainen, M., Tuittila, E.-S., Laine, J., Vasander, H., 1998. Seed and spore bank in a cut-away peatland twenty years after abandonment. International Peat Journal 8, 42–51.

Ingram, H.A.P., 1978. Soil layers in mires: functions and terminology. Journal of Soil Science 29, 224–227.

Jansen, A.J.M., De Graaf, M.C.C., Roelofs, J.G.M., 1996. The restoration of species-rich heathland communities in the Netherlands. Vegetatio 126, 73–88.

Jauhiainen, S., Laiho, R., Vasander, H., 2002. Ecohydrological and vegetational change in a restored bog and fen. Annales Botanici Fennici 39, 185–199.

Jongepierová, I., Pešout, P., Jongepier, J.W., Prach, K. (Eds.), 2012. Ecological Restoration in the Czech Republic. Nature Conservation Agency of the Czech Republic, Prague.

Joosten, H., Clark, D., 2002. Wise Use of Mires and Peatlands – Background and Principles Including a Framework for Decision-Making. International Mire Conservation Group and International Peat Society. Saarijarven Offset Oy, Saarijarvi, Finland.

Kaine, J., Vasander, H., Laiho, R., 1995. Long-term effects of water level drawdown on the vegetation of drained pine mires in southern Finland. Journal of Applied Ecology 32, 785–802.

Kemmers, R.H., van Delft, S.P.J., Jansen, P.C., 2003. Iron and sulfate as possible key factors in the restoration ecology of rich fens in discharge areas. Wetlands Ecology and Management 11, 367–381.

Kivinen, E., Pakarinen, P., 1981. Peatland areas and the proportion of virgin peatlands in different countries. In: Proceedings of the 6th International Peat Congress, Duluth, MN, pp. 52–54.

Klimkowska, A., Van Diggelen, R., Bakker, J.P., Grootjans, A.B., 2007. Wet meadow restoration in western Europe: a quantitative assessment of the effectiveness of several techniques. Biological Conservation 140, 318–328.

Klinger, L.F., 1996. The myth of the classic hydrosere model of bog succession. Arctic and Alpine Research 28, 1–9.

Koehler, A.-K., Sottocornola, M., Kiely, G., 2011. How strong is the current carbon sequestration of an Atlantic blanket bog? Global Change Biology 17, 309–319.

Konvalinkova, P., Prach, K., 2010. Spontaneous succession in mined peatlands: a multi-site study. Preslia 82, 423–435.

Lally, H., Gormally, M., Higgens, T., Colleran, E., 2012. Evaluating different wetland creation approaches for Irish cutaway peatlands using water chemical analysis. Wetlands 32, 129–136.

Lamers, L.P.M., Farhoush, C., van Groenendael, J.M., Roelofs, J.G.M., 1999. Calcareous groundwater raises bogs: the concept of ombrotrophy revisited. Journal of Ecology 87, 639–648.

Lamers, L.P.M., Smolders, A.J.P., Roelofs, J.G.M., 2002a. The restoration of fens in the Netherlands. Hydrobiologia 478, 107–130.

Lamers, L.P.M., Falla, S.J., Samborska, E.M., van Dulken, I.A.R., van Hengstrum, G., Roelofs, J.G.M., 2002b. Factors controlling the extent of eutrophication and toxicity in sulfate-polluted freshwater wetlands. Limnology and Oceanography 47, 585–593.

Lamers, L.P.M., Vile, M.A., Grootjans, A.P., Acreman, M.C., van Diggelen, R., Evans, M.G., Richardson, C.J., Rochefort, L., Kooijman, A.M., Roelofs, J.G.M., Smolders, A.J.P., 2014. Ecological restoration of rich fens in Europe and North America: from trial and error to an evidence-based approach. Biological Reviews. http://dx.doi.org/10.1111/brv.12102.

Lappalainen, E., 1996. General review on world peat and peatland resources. In: Lappalainen, E. (Ed.), Global Peat Resources. International Peat Society, Jyska, Finland, pp. 53–56.

Lavoie, C., Rochefort, L., 1996. The natural revegetation of a harvested peatland in southern Quebec: a spatial and dendroecological analysis. Ecoscience 3, 101–111.

Lavoie, C., Grosvernier, P., Girard, M., Marcoux, K., 2003. Spontaneous revegetation of mined peatlands: a useful restoration tool. Wetlands Ecology and Management 11, 97–107.

Lavoie, C., Saint-Louis, A., Lachance, D., 2005a. Vegetation dynamics on an abandoned vacuum-mined peatland: 5 years of monitoring. Wetlands Ecology and Management 13, 621–633.

Lavoie, C., Marcoux, K., Saint-Louis, A., Price, J.S., 2005b. The dynamics of a cotton-grass (*Eriophorum vaginatum* L.) cover expansion in a vacuum-mined peatland, southern Quebec, Canada. Wetlands 25, 64–75.

Lindeman, R.L., 1941. The developmental history of Cedar Creek bog. American Midland Naturalist 25, 101–112.

Malmer, N., Albinsson, C., Svennson, B.M., Wallen, B., 2003. Interferences between *Sphagnum* and vascular plants: effects on plant community structure and peat formation. Oikos 100, 469–482.

Malson, K., Rydin, H., 2007. The regeneration capabilities of bryophytes for rich fen restoration. Biological Conservation 135, 435–442.

Malson, K., Backeus, I., Rydin, H., 2008. Long-term effects of drainage and initial effects of hydrological restoration on rich fen vegetation. Applied Vegetation Science 11, 99–106.

Malson, K., Sundberg, S., Rydin, H., 2010. Peat disturbance, mowing and ditch blocking as tools in rich fen restoration. Restoration Ecology 18, 469–478.

Malterer, T.J., Verry, E.S., Erjavec, J., 1992. Fiber content and degree of decomposition in peats: a review of national methods. Soil Science Society of America Journal 56, 1200–1211.

Mawby, F.J., 1995. Effects of damming peat cuttings on Glassom Moss and Wedholme Flow, two lowland raised bogs in North-West England. In: Wheeler, B.D., Shaw, S.C., Fojt, W.J., Robertson, R.A. (Eds.), Restoration of Temperate Wetlands. John Wiley and Sons, Chichester, UK, pp. 349–357.

Mazerolle, M.J., Poulin, M., Lavoie, C., Rochefort, L., Desrochers, A., Drolet, B., 2006. Animal and vegetation patterns in natural and man-made bog pools: implications for restoration. Freshwater Biology 51, 333–350.

Meade, R., 1992. Some early changes following the rewetting of a vegetated cutover peatland surface at Danes Moss, Cheshire, UK, and their relevance to conservation management. Biological Conservation 61, 31–40.

Mitchell, R.J., Auld, M.H.D., Hughes, J.M., Marrs, R.H., 2000. Estimates of nutrient removal during heathland restoration in successional sites in Dorset, southern England. Biological Conservation 95, 233–246.

Money, R.P., 1995. Re-establishment of a *Sphagnum*-dominated flora on cut-over lowland raised bogs. In: Wheeler, B.D., Shaw, S.C., Fojt, W.J., Robertson, R.A. (Eds.), Restoration of Temperate Wetlands. John Wiley and Sons, Chichester, UK, pp. 405–422.

Moore, P.D., Chater, E.H., 1969. Studies in the vegetational history of mid-Wales. I. The post-glacial period in Cardiganshire. New Phytologist 68, 183–196.

Morley, R.J., 1981. Development and vegetation dynamics of a lowland ombrogenous peat swamp in Kalimantan Tengah, Indonesia. Journal of Biogeography 8, 383–404.

Nilsson, M., Sagerfors, J., Buffam, I., Laudon, H., Eriksson, T., Grelle, A., Klemedtsson, L., Weslien, P., Lindroth, A., 2008. Contemporary carbon accumulation in a boreal oligotrophic minerogenic mire – a significant sink after accounting for all C fluxes. Global Change Biology 14, 2317–2332.

Nyronen, T., 1996. Peat production. In: Lappalainen, E. (Ed.), Global Peat Resources. International Peat Society, Jyska, Finland, pp. 315–319.

Okrusko, H., 1995. Influence of hydrological differentiation of fens on their transformation after dehydration and on possibilities for restoration. In: Wheeler, B.D., Shaw, S.C., Fojt, W.J., Robertson, R.A. (Eds.), Restoration of Temperate Wetlands. John Wiley and Sons, Chichester, UK, pp. 113–119.

Ovenden, L., 1990. Peat accumulation in northern wetlands. Quaternary Research 33, 377–386.

Patzelt, A., Wild, U., Pfadenhauer, J., 2001. Restoration of wet fen meadows by topsoil removal: vegetation development and germination biology of fen species. Restoration Ecology 9, 127–136.

Pfadenhauer, J., Grootjans, A., 1999. Wetland restoration in Central Europe: aims and methods. Applied Vegetation Science 2, 95–106.

Pfadenhauer, J., Klötzli, F., 1996. Restoration experiments in European wet terrestrial ecosystems: an overview. Vegetatio 126, 101–115.

Poschlod, P., 1995. Diaspore rain and diaspore bank in raised bogs and implications for the restoration of peat-mined sites. In: Wheeler, B.D., Shaw, S.C., Fojt, W.J., Robertson, R.A. (Eds.), Restoration of Temperate Wetlands. John Wiley and Sons, Chichester, UK, pp. 471–494.

Poulin, M., Rochefort, L., Quinty, F., Lavoie, C., 2005. Spontaneous revegetation of mined peatlands in eastern Canada. Canadian Journal of Botany 83, 539–557.

Price, J.S., Whitehead, G.S., 2001. Developing hydrologic thresholds for *Sphagnum* recolonization on an abandoned cutover bog. Wetlands 21, 32–40.

Price, J., Rochefort, L., Quinty, F., 1998. Energy and moisture considerations on cutover peatlands: surface microtopography, mulch cover and *Sphagnum* revegetation. Ecological Engineering 10, 293–312.

Price, J.S., Rochefort, L., Campeau, S., 2002. Use of shallow basins to restore cutover peatlands: hydrology. Restoration Ecology 10, 259–266.

Price, J.S., Heathwaite, A.L., Baird, A.J., 2003. Hydrological processes in abandoned and restored peatlands: an overview of management approaches. Wetlands Ecology and Management 11, 65–83.

Price, J.S., 1996. Hydrology and microclimate of a partly restored cutover bog, Quebec. Hydrological Processes 10, 1263–1272.

Quinty, F., Rochefort, L., 2003. Peatland Restoration Guide, second ed. Canadian Sphagnum Peat Moss Association and New Brunswick Department of Natural Resources and Energy, Quebec, Quebec.

Rasran, L., Vogt, K., Jensen, K., 2007. Effects of topsoil removal, seed transfer with plant material and moderate grazing on restoration of riparian fen grasslands. Applied Vegetation Science 10, 451–460.

Richert, M., Dietrich, O., Koppisch, D., Roth, S., 2000. The influence of rewetting on vegetation development and decomposition in a degraded fen. Restoration Ecology 8, 186–195.

Robert, E.C., Rochefort, L., Garneau, M., 1999. Natural revegetation of two block-cut mined peatlands in eastern Canada. Canadian Journal of Botany 77, 447–459.

Rochefort, L., Price, J., 2003. Restoration of *Sphagnum* dominated peatlands. Wetlands Ecology and Management 11, 1–2.

Rochefort, L., Gauthier, R., Lequere, D., 1995. *Sphagnum* regeneration – towards an optimization of bog restoration. In: Wheeler, B.D., Shaw, S.C., Fojt, W.J., Robertson, R.A. (Eds.), Restoration of Temperate Wetlands. John Wiley and Sons, Chichester, UK, pp. 423–434.

Rochefort, L., Quinty, F., Campeau, S., 1997. Restoration of peatland vegetation: the case of damaged or completely removed acrotelm. International Peat Journal 7, 20–28.

Rochefort, L., Quinty, F., Campeau, S., Johnson, K., Malterer, T., 2003. North American approach to the restoration of *Sphagnum* dominated peatlands. Wetlands Ecology and Management 11, 3–20.

Rochefort, L., 2000. *Sphagnum* – a keystone genus in habitat restoration. The Bryologist 103, 503–508.

Roulet, N.T., Lafleurs, P.M., Richard, P.J.H., Moor, T.R., Humphreys, E.R., Bubier, J., 2007. Contemporary carbon balance and late Holocene carbon accumulation in a northern peatland. Global Change Biology 13, 397–411.

Sallantaus, T., Vasander, H., Laine, J., 1998. Prevention of detrimental impacts of forestry operations on water bodies using buffer zones created from drained peatlands. Suo 49, 125–133.

Salonen, V., Laaksonen, M., 1994. Effects of fertilization, liming, watering and tillage on plant colonization of bare peat surfaces. Annales Botanici Fennici (Finland) 31, 29–36.

Salonen, V., Setala, H., 1992. Plant colonization of bare peat surface – relative importance of seed availability and soil. Ecography 15, 199–204.

Salonen, V., 1987. Relationship between the seed rain and the establishment of vegetation in two areas abandoned after peat harvesting. Holarctic Ecology 10, 171–174.

Salonen, V., 1990. Early plant succession in two abandoned cut-over peatland areas. Holarctic Ecology 13, 217–223.

Schlesinger, W.H., 1997. Biogeochemistry: An Analysis of Global Change. Academic Press, New York.

Schouwenaars, J.M., 1993. Hydrological differences between bog and bog-relicts and consequences for bog restoration. Hydrobiologia 265, 217–224.

Schultz, K., Timmerman, T., Steffenhagen, P., Zerbe, S., Succow, M., 2011. The effect of loading on carbon and nutrient stocks of helophyte biomass in rewetted fens. Hydrobiologia 674, 25–40.

Sliva, J., Pfadenhauer, J., 1999. Restoration of cut-over raised bogs in southern Germany – a comparison of methods. Applied Vegetation Science 2, 137–148.

Smart, P.J., Wheeler, B.D., Willis, A.J., 1989. Revegetation of peat excavations in a derelict raised bog. New Phytologist 111, 733–748.

Smolders, A.J.P., Tomassen, H.B.M., Lamers, L.P.M., Lomans, B.P., Roelofs, J.G.M., 2002. Peat bog restoration by floating mat formation: the effects of groundwater and peat quality. Journal of Applied Ecology 39, 391–401.

Smolders, A.J.P., Tomassen, H.B.M., van Mullekom, M., Lamers, L.P.M., Roelofs, J.G.M., 2003. Mechanisms involved in the re-establishment of *Sphagnum*-dominated vegetation in rewetted bog remnants. Wetlands Ecology and Management 11, 403–418.

Snow, C.R.S., Marrs, R.H., 1997. Restoration of *Calluna* heathland on a bracken *Pteridium* infested site. Biological Conservation 81, 35–42.

Soro, A., Sundberg, S., Rydkin, H., 1999. Species diversity, niche metrics and species associations in harvested and undisturbed bogs. Journal of Vegetation Science 10, 549–560.

Sottocornola, M., Boudreau, S., Rochefort, L., 2007. Peat bog restoration: effects of phosphorus on plant re-establishment. Ecological Engineering 31, 29–40.

Tallowin, J.R.B., Smith, R.E.N., 2001. Restoration of a *Cirsio-Molinietum* fen meadow on agricultural improved pasture. Restoration Ecology 9, 167–178.

Taylor, J.A., 1983. The peatlands of Great Britain and Ireland. In: Gore, A.J.P. (Ed.), Ecosystems of the World 4B. Mires: Swamp, Bog, Fen, and Moor. Elsevier Scientific Publishing Company, Amsterdam, The Netherlands, pp. 1–46.

Thormann, M.N., Currah, R.S., Bayley, S.E., 1999. The mycorrhizal status of the dominant vegetation along a peatland gradient in southern boreal Alberta. Wetlands 19, 438–450.

Timmerman, T., Margoczi, K., Takacs, G., Vegelin, K., 2006. Restoration of peat-forming vegetation by rewetting species-poor fen grassland. Applied Vegetation Science 9, 241–250.

Tomassen, H.B.M., Smolders, A.J.P., van Herk, J.M., Lamers, L.P.M., Roelofs, J.G.M., 2003. Restoration of cut-over bogs by floating raft formation: an experimental feasibility study. Applied Vegetation Science 6, 141–152.

Tomassen, H.B.M., Smolders, A.J.P., Lamers, L.P.M., Roelofs, J.G.M., 2004. Development of floating rafts after the rewetting of cut-over bogs: the importance of peat quality. Biogeochemistry 71, 69–87.

Triisberg, T., Karofeld, E., Paal, J., 2013. Factors affecting the revegetation of abandoned extracted peatlands in Estonia: a synthesis from field and greenhouse experiments. Estonian Journal of Ecology 62, 192–211.

Troxler, T.G., 2007. Patterns of phosphorus, nitrogen, and $\Delta^{15}N$ along a peat development gradient in a coastal mire, Panama. Journal of Tropical Ecology 23, 683–691.

Tuittila, E.-S., Komulainen, V.M., Vasander, H., Laine, J., 1999. Restored cut-away peatland as a sink for atmospheric CO_2. Oecologia 120, 563–574.

Tuittila, E.-S., Vasander, H., Laine, J., 2000a. Impact of rewetting on the vegetation of a cut-away peatland. Applied Vegetation Science 3, 205–212.

Tuittila, E.-S., Rita, H., Vasander, H., Laine, J., 2000b. Vegetation patterns around *Eriophorum vaginatum* L. tussocks in a cut-away peatland in southern Finland. Canadian Journal of Botany 78, 47–58.

Tuittila, E.-S., Komulaienen, V.M., Vasander, H., Nykanen, H., Martikainen, P.J., Laine, J., 2000c. Methane dynamics of a restored cutaway peatland. Global Change Biology 6, 569–581.

Turunen, J., Tolonen, K., 1996. Rate of carbon accumulation in boreal peatlands and climate change. In: Lappalainen, E. (Ed.), Global Peat Resources. International Peat Society, Jyska, Finland, pp. 21–28.

USDA (U.S. Department of Agriculture), Natural Resources Conservation Service, 1999. Soil Taxonomy. Agricultural Handbook No. 436. U.S. Government Printing Office, Washington, DC.

Vasander, H., Tuittila, E.-S., Lode, E., Lundein, L., Ilomets, M., Sallantaus, T., Heikkila, R., Pitkanen, M.L., Laine, J., 2003. Status and restoration of peatlands in northern Europe. Wetlands Ecology and Management 11, 51–63.

Venterink, H.O., Davidsson, T.E., Kiehl, K., Leonardson, L., 2002. Impact of drying and rewetting on N, P and K dynamics in a wetland soil. Plant and Soil 243, 119–130.

Verhagen, R., Klooker, J., Bakker, J.P., van Diggelen, R., 2001. Restoration success of low production plant communities on former agricultural soils after topsoil removal. Applied Vegetation Science 4, 73–82.

Verry, E.S., Boelter, D.H., Paivanen, J., Nichols, D.S., Malterer, T., Gafni, A., 2011. Physical properties of organic soils. In: Kolka, R., Sebestyen, S., Verry, E.S., Brooks, K. (Eds.), Peatland Biogeochemistry and Watershed Hydrology at the Marcell Experimental Forest. CRC Press, Taylor and Francis Group, Boca Raton, FL, pp. 135–176.

Waddington, J.M., Warner, K.D., 2001. Atmospheric CO_2 sequestration in restored mined peatlands. Ecoscience 8, 359–368.

Waddington, J.M., Warner, K.D., Kennedy, G.W., 2002. Cutover peatlands: a persistent source of atmospheric CO_2. Global Biogeochemical Cycles 16, 1–7.

Waddington, J.M., Rochefort, L., Campeau, S., 2003. *Sphagnum* production and decomposition in a restored cutover peatland. Wetlands Ecology and Management 11, 85–95.

Watts, C.H., Clarkson, B.R., Didham, R.K., 2008. Rapid beetle community convergence following experimental habitat restoration in a mined peat bog. Biological Conservation 141, 568–579.

Wein, R.W., MacLean, D.A., 1973. Cotton grass (*Eriophorum vaginatum*) germination requirements and colonization potential in the Arctic. Canadian Journal of Botany 51, 2509–2513.

Whinam, J., Hope, G.S., Clarkson, B.R., Buxton, R.P., Alspach, P.A., Adam, P., 2003. *Sphagnum* in peatlands of Australasia: the distribution, utilization, and management. Wetlands Ecology and Management 11, 37–49.

Wind-Mulder, H.L., Vitt, D.H., 2000. Comparisons of water and peat chemistries of a post-harvested and undisturbed peatland with relevance to restoration. Wetlands 20, 616–628.

Wind-Mulder, H.L., Rochefort, L., Vitt, D.H., 1996. Water and peat chemistry comparisons of natural and post-harvested peatlands across Canada and their relevance to peatland restoration. Ecological Engineering 7, 161–181.

Yu, Z.C., 2012. Northern peatland carbon stock and dynamics: a review. Biogeosciences 9, 4071–4085.

Zak, D., Gelbrecht, J., Steinberg, C.E.W., 2004. Phosphorus retention at the redox interface of peatlands adjacent to surface waters in northeast Germany. Biogeochemistry 70, 357–368.

Zak, D., Gelbrecht, J., Wagner, C., Steinberg, C.E.W., 2008. Evaluation of phosphorus mobilization potential in rewetted fens by an improved sequential chemical extraction procedure. European Journal of Soil Science 59, 1191–1201.

Zak, D., Wagner, C., Payer, B., Augustin, J., Gelbrecht, J., 2010. Phosphorus mobilization in rewetted fens: the effect of altered peat properties and implications for their restoration. Ecological Applications 20, 1336–1349.

Part Three

Restoration of Estuarine Wetlands

Tidal Marshes

Introduction

Tidal marshes, salt, brackish, and tidal fresh marshes, are among the most productive ecosystems in the world. They are characterized by the pulsing of the astronomical tides and the variable mixing of freshwater and seawater. They differ from nontidal wetlands in that inundation is regular and frequent, oftentimes diurnal or twice daily. Tidal inundation is driven by the gravitational pull of the moon and, secondarily, the sun. Approximately every 24 h, two tidal cycles occur, with one high tide and one low tide every 12 h and 25 min (Figure 8.1). On the high tide, the duration of inundation is about 3–6 h, depending on marsh elevation within the tidal frame. The diurnal tide is superimposed on a lunar tidal cycle of 26 days corresponding to the waxing and waning of the moon. During the lunar cycle, two *spring* tides, separated by approximately 2 weeks and with higher than average high tides and lower than average low tides, occur. *Spring* tides occur during the full and new phases of the moon. On alternating 2-week intervals, *neap* tides, corresponding to half waxing and waning phases of the moon, occur. The amplitude of *neap* tides is muted as compared to the average and spring tides (Figure 9.1). The highest and lowest tides of the year occur around the time of the spring and fall equinox (Figure 8.1). At this time, the angle of the earth's inclination is perpendicular to the moon and sun, resulting in greater gravitational pull than that occurs at other times of the year. The highest (and lowest) tides occur when a spring tide is superimposed on the *equinox* tides. In spite of the varying inundation

Creating and Restoring Wetlands. http://dx.doi.org/10.1016/B978-0-12-407232-9.00008-7

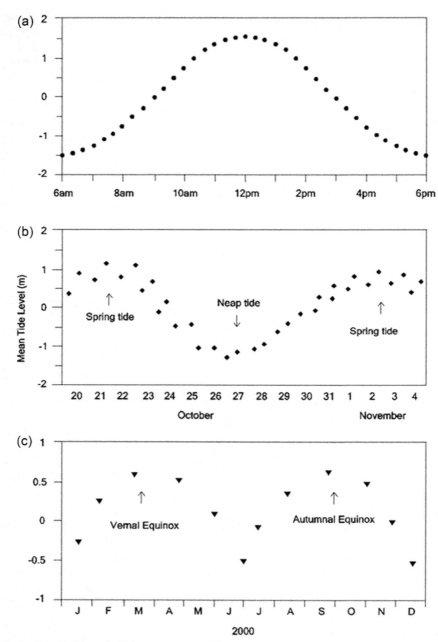

Figure 8.1 (a) Diurnal, (b) lunar, and (c) equinox tidal cycles of saline tidal marshes.

throughout the year, the tides are predictable and inundation is frequent but of short duration. Plants and animals of tidal marshes are well adapted to the predictable inundation, recession, and cycles of the tides.

Tidal marshes contribute to estuarine food webs by providing habitat for invertebrates and shellfish and nursery grounds for finfish. They transfer plant production directly to the estuary by *outwelling* of detritus (Odum, 2000) and indirectly through the *trophic relay* of finfish from the intertidal waters of the marsh to tidal creeks and estuaries (Kneib, 2000). Because of their high productivity and value as nursery habitat, much effort is devoted to creating and restoring tidal marshes.

Salt and Brackish Water Marshes

The chemistry of tidal floodwaters varies, depending on the relative proportion of seawater versus freshwater. Seawater contains much higher levels of dissolved salts, especially sodium chloride (NaCl) (Table 8.1) that is stressful to most plant species (Chapman, 1974) and, in part, determines the patterns of zonation of vegetation (Bertness and Pennings, 2000). Dissolved salts in seawater also contribute to pronounced differences in soil chemical properties, especially sulfur (S) chemistry between saline and fresh marshes. High concentrations of S in seawater support a microbial consortia that reduce sulfate (SO_4) to hydrogen sulfide that is toxic to plants and animals. Elements such as iron (Fe), manganese (Mn), and phosphorus (P), a limiting nutrient for plant growth, are greater in seawater. Because of the abundance of acid neutralizing bicarbonate (HCO_3^-) and alkalinity of seawater, pH is higher in saline tidal marshes than in freshwater marshes (Table 8.1).

The composition of dissolved salts in seawater differs from that of inland saline marshes. The salinity of seawater consistently is dominated by Na, Cl, and other halides (bromide, iodide, fluoride) whereas the salinity of inland saline marshes is more variable with SO_4^{2-}, Ca, and Mg present in greater concentrations (Chapman, 1974). In addition to the anaerobic soil conditions found in all wetlands, plants

Table 8.1 Concentration of Selected Ions (mg/L) in Freshwater Versus Seawater

	Seawater[a]	Fresh River Water[b]
Sodium	10,760	5
Potassium	399	1
Calcium	412	13
Magnesium	1294	3
Chloride	19,350	6
Sulfate	2712	8
Bicarbonate	145	52
Bromide	67	<1

[a]From Holland (1978).
[b]From Meybeck (1979).

Table 8.2 **Common Plant Species of Saline Tidal Marshes**

Cordgrass	*Spartina alterniflora*[a] (North America (NA)), *Spartina patens* (NA), *Spartina foliosa* (Western NA), *Spartina anglica* (Western Europe, Asia), *Spartina towsendii* (Europe)
Needlerush	*Juncus roemerianus* (NA), *Juncus gerardii* (NA), *Juncus maritimus* (Europe, Australia)
Salt grass	*Distichlis spicata* (NA), *Distichlis distichopylla* (AU)
Glasswort	*Salicornia virginica* (NA), *Salicornia europaea* (Europe), *Salicornia australis* (Australia)
Saltwort	*Batis maritima* (NA), *Batis argillicola* (Austalia, Asia)
Sea ox-eye daisy	*Borrichia frutescens* (NA)
Sea lavender	*Limonium carolinianum* (NA), *Limonium humile* (Europe), *Limonium vulgare* (Europe)
Arrow grass	*Triglochin maritima* (Europe), T. *spp.* (NA)
Alkali grass	*Puccinellia maritima* (Europe, NA)
Seepweed	*Suaeda linearis* (NA), *Suaeda maritima* (Europe), *Salicornia australis* (Australia)
Sea purslane	*Halimione (Atriplex) portulacoides* (Europe)

[a]*Spartina alterniflora* is an invasive species in many parts of the world, including western North America and Eastern Asia. A detailed description of saline marsh plant communities around the world is given by Chapman (1974).

and animals of salt and brackish water marshes must adapt to salinity that creates an osmotic imbalance between the internal (cell) and external medium. Many saline marsh plants exclude Na and Cl at the root interface or they take up Na and Cl and excrete NaCl salt through specialized salt glands on the leaves (Phleger, 1971). Other mechanisms include guttation (water secreting structures that exude salt), removal of salt saturated organs (i.e., leaf dropping), and salt hairs (Waisel, 1972). Hydrogen sulfide, a product of microbial sulfate reduction in saline marsh soils, also is toxic to wetland plants. Saline marsh species such as *Spartina* are much more tolerant to H_2S than freshwater vegetation such as, for example, *Phragmites australis* (Bart and Hartman, 2000; Chambers et al., 2002). Animal adaptations to salinity include the ability to migrate elsewhere to lower salinity waters. However, sessile species, such as the mollusks (oysters, clams, mussels), are unable to migrate and, for many of them, their internal osmotic concentration varies corresponding to floodwater salinity (Gross, 1964).

Because salt and brackish marshes are stressful environments, the species that inhabit them are few but widely distributed throughout the world (Chapman, 1974). Cosmopolitan genera of saline marshes include *Spartina, Juncus, Salicornia, Distichlis*, and others (Table 8.2). In cooler climates of the eastern Pacific (US and Canada) and northern Atlantic (Canada, Western Europe), *Puccinellia, Tryglochin*, and *Salicornia* are important saline marsh species (Chapman, 1974; McDonald, 1977). Generally, saline marsh vegetation grows within the range between mean low water and mean high water (McKee and Patrick, 1989) such that the width of the zone of vegetated habitat increases with increasing tide range. Across this elevation/inundation gradient, individual species exhibit strong patterns of zonation (Figure 8.2).

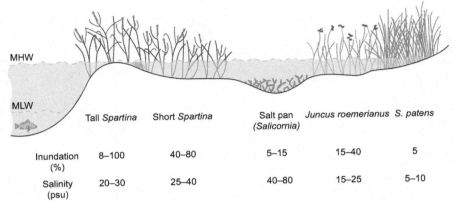

	Tall *Spartina*	Short *Spartina*	Salt pan (*Salicornia*)	*Juncus roemerianus*	*S. patens*
Inundation (%)	8–100	40–80	5–15	15–40	5
Salinity (psu)	20–30	25–40	40–80	15–25	5–10

Figure 8.2 Zonation of saline tidal marsh vegetation in relation to tide levels.
MHW: Mean High Water; MLW: Mean Low Water

Often, plant growth is more vigorous along the edge of the marsh, adjacent to the tidal creek, and decreases with distance inland (Gallagher et al., 1980). Enhanced growth of vegetation in the streamside marsh is attributed to greater soil drainage (King et al., 1982) and aerobic conditions (Howes et al., 1986) that lead to lower concentrations of toxins, salts, and sulfides, in the rooting zone (Mendelssohn and Morris, 2000).

In mid-elevations of the marsh, known as the marsh platform, vegetation is of shorter stature and is less productive than along the water's edge. Tidal flushing that removes toxins, salts, and sulfides also is less than along the marsh edge. Depressions within the marsh platform retain water brought in by the high tide. As water evaporates from them, salts are concentrated, creating salt pans with salinities well in excess of seawater. Vegetation of depressions and salt pans is sparse or absent and salt-tolerant halophytes such as *Salicornia* and *Batis* grow. At the landward edge of the marsh, tidal inundation is infrequent, occurring only during spring tides and storm tides. This "high" marsh is dominated by plant species adapted to lower salinity and less anaerobic soil conditions than species lower in the tidal frame.

Salinity also structures plant communities as the tides propagate up estuaries and tidal portions of rivers (Figure 8.3). At the seaward end of the gradient, saline tidal marshes dominated by *Spartina* and *Salicornia* exist. Inland of these are brackish marsh communities where salinity ranges from 30psu down to 5psu (Cowardin et al., 1992). Plant species diversity is greater as compared to saline marshes. Upstream of brackish marshes, oligohaline marshes with salinities of 5psu down to 0.5psu exist. Plant species diversity is greatest in these marshes (Odum, 1988). Upstream of oligohaline marshes are tidal freshwater marshes and tidal forests where salinity does not penetrate except during times of low river flow or drought. Fauna of tidal marshes also varies along the longitudinal, seaward to landward, axis of the estuary. Organisms that evolved from the marine environment including crabs, snails, and mollusks are present in greater numbers and diversity in saline tidal marshes whereas organisms that evolved from terrestrial environments such as insects are more abundant and diverse in oligohaline and fresh marshes.

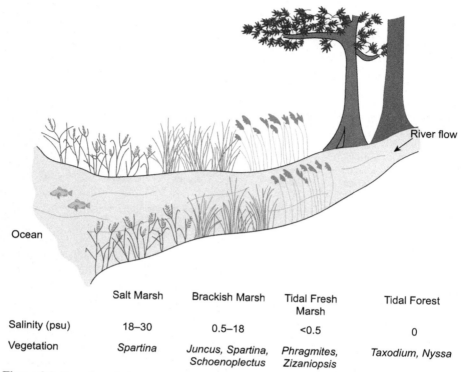

Figure 8.3 Zonation of plant communities in relation to salinity along the longitudinal gradient of an estuary, southeastern Georgia (USA). Salinity classification is from Cowardin et al. (1992).

	Salt Marsh	Brackish Marsh	Tidal Fresh Marsh	Tidal Forest
Salinity (psu)	18–30	0.5–18	<0.5	0
Vegetation	*Spartina*	*Juncus, Spartina, Schoenoplectus*	*Phragmites, Zizaniopsis*	*Taxodium, Nyssa*

Soils of saline marshes generally are minerogenic, dominated by sand, silt, or clay particles. Salinity promotes flocculation of clay particles and stems of emergent vegetation promote their settling (Morris et al., 2002). In cooler climates and in marshes with low sediment inputs, saline marsh soils contain more organic matter, and in some cases organic-rich soils or even peats may develop (Craft, 2007). Marshes with low (i.e., microtidal) tide range also may develop organic-rich soils (Pennings et al., 2012). Salinity also affects organic matter enrichment of marsh soils as tidal fresh and brackish marsh soils often contain more organic matter than saline tidal marshes (Craft, 2007).

Tidal Freshwater Marshes

Tidal freshwater marshes differ from salt and brackish water marshes in that they are inundated with freshwater only. They typically are located in the upper reaches of estuaries of large rivers. In the US, tidal freshwater marshes are found along all three coasts (Odum, 1988). In Europe, they are found along the Rhine, Thames, and Danube River deltas (Odum, 1988; Pringle et al., 1993; Barendregt, 2005). The

Table 8.3 Common Species/Genera of Tidal Freshwater Marshes

	Scientific Name	Common Name
Low marsh	*Alternanthera philoxeriodes*	Alligator weed
	Ludwigia spp.	Primrose
	Nuphar spp.[b]	Spatterdock, cow lily
	Nymphaea spp.	Water lily
Mid marsh	*Alisma* spp.	Water plantain
	Calamagrostis canadensis	Reed grass
	Carex spp.	Sedges
	Eleocharis spp.	Spikerush
	Iris spp.	Wild iris
	Juncus sp.[b]	Rushes
	Leersia oryzoides	Rice cutgrass
	Panicum spp.	Switchgrass
	Peltandra virginica[b]	Arrow arum
	Phalaris arundinacea[a]	Reed canary grass
	Phragmites australis[a]	Common reed
	Pontederia cordata	Pickerelweed
	Sagittaria spp.[b]	Duck potato
	Saururus cernuus	Lizard's tail
	Schoenoplectus spp.[b]	Bulrushes
	Sparganium spp.	Burr reed
	Typha spp.	Cattail
	Zizania aquatica	Wild rice
	Zizaniopsis miliacea	Giant cutgrass
High marsh	*Amaranthus cannabinus*	Water hemp
	Bidens spp.	Bur-marigold
	Hibiscus spp.	Mallow
	Lythrum salicaria[a]	Purple loosestrife
	Impatiens capensis	Jewelweed
	Polygonum spp.	Smartweed, tear thumb

[a]Invasive species.
[b]Species commonly planted in restoration projects. From Neff and Baldwin (2005).
From Odum et al. (1984), Bowers (1995), and Whigham et al. (2009).

Yellow and Ganges Rivers in Asia also contain extensive tidal freshwater marshes (Odum, 1988). Tidal freshwater marshes are much less common than saline tidal marshes, in large part because of their location in the upper part of the estuary, usually far removed from the ocean and at the head of tides where major ports and cities are located.

Tidal freshwater marshes contain much greater plant species diversity than saline tidal marshes. Common species include a number of grasses, sedges, and rushes, many of which also are found in inland freshwater marshes (Table 8.3). Succulent herbaceous vegetation including *Pontedaria* (pickerelweed), *Sagittaria* (duck potato), and others are common in the frequently inundated lower elevations.

Tidal Marsh Loss

Worldwide, there has been dramatic loss of tidal marshes, in large part because much of the world's population lives in the coastal zone (Culliton, 1998). Many were reclaimed to create new land. Others were ditched or drained for mosquito control. In the US, it is estimated that 27% (8000 km^2) of estuarine wetlands, tidal marshes, that were present in the early twentieth century have been lost to human activities (Mitsch and Gosselink, 2000), and probably much more were converted before that time.

In the future, tidal marshes will be threatened by accelerated rates of sea level rise linked to global warming (see Chapter 13, The Future of Wetland Restoration). Globally, sea level currently is rising at the rate of about 3 mm per year (Rahmstorf et al., 2007). In order to persist, marshes must maintain their elevation relative to the tidal frame by accreting mineral sediment and building soil organic matter (SOM) (McCaffrey and Thomson, 1981) or migrate inland (Craft et al., 2009), which will be increasingly difficult on coastlines where urbanization blocks their migration.

Theory

Development of tidal marshes is guided mostly by allogeneic succession, though, in cooler climates and in fresh regions of the estuary, accumulation of organic matter may be important. Because saline and brackish are such stressful environments, few species are present and they produce monocultures often separated by abrupt boundaries that correspond to changes in elevation and inundation (Johnson and York, 1915; Chapman, 1940; Ranwell, 1972). Establishment from seeds occurs rarely because of the dynamic hydrologic regime.

Saline and brackish marshes are exposed to periodic natural disturbance from wind and wave action that erodes edges of the marsh. Smaller scale disturbances include deposition of wrack, rafts of dead plant stems, that cover smaller (tens to hundreds of square meters) patches of marsh vegetation (Ranwell, 1972; Wiegert and Freeman, 1990).

Herbivory generally is low in saline and brackish marshes unless domesticated grazers such as cattle, horses, pigs, or sheep are introduced (Reimold et al., 1975; Bakker and Ruyter, 1981). Grazing denudes vegetation, reduces primary production, and alters epibenthic and benthic invertebrate communities (Reimold et al., 1975; Reader and Craft, 1999). There are cases where grazing has severely and perhaps irreversibly degraded tidal marshes. In marshes of the Hudson Bay lowlands of Canada, grazing by lesser snow geese (*Chen caerulescens carulescens*) decimated tidal marsh vegetation and led to an alternative stable state consisting of hypersaline, unvegetated mudflat (Abraham et al., 2005; Jefferies et al., 2006). In brackish and tidal freshwater marshes, herbivory by rodents (*Nutria* in Gulf Coast marshes) and waterfowl (geese) also strip vegetation (Mitsch and Gosselink, 2000; Smith and Odum, 1981; Kirwan et al., 2008).

In spite of these stressors, it generally is easier to restore salt, brackish and tidal freshwater marshes than many other types of wetlands. Frequent and predictable inundation by the tides ensures that the site will be flooded. Furthermore, since allogeneic processes predominate, restoration can be achieved relatively quickly compared to autogenically driven wetlands such as peatlands.

Practice

Tidal marshes are restored in a variety of ways. In the US, saline and brackish marshes were established to stabilize dredge material (Seneca et al., 1985; Broome et al., 1988b; LaSalle et al., 1991; Streever, 2000; Bolam et al., 2006) and eroding shorelines (Broome et al., 1986) following oil spills (Seneca and Broome, 1982, 1992) and for mitigation of wetland loss as proscribed by the Clean Water Act (Zedler, 1995; Hough and Robertson, 2008). The first projects consisted of planting salt marsh vegetation, *Spartina alterniflora* and *Spartina patens*, for shoreline erosion control in the Chesapeake Bay and date to 1928 (Knutson et al., 1981). Other projects for erosion control were established in 1946 (Sharp and Vaden, 1970) and 1956 (Phillips and Eastman, 1959). Some of the earliest tidal marsh experimental restoration projects were conducted by W.W. Woodhouse Jr. and colleagues Ernest D. Seneca and Stephen W. Broome at North Carolina State University, and they developed techniques that are widely used by tidal marsh restoration ecologists today (Woodhouse, 1982; Woodhouse and Knutson, 1982).

In areas where subsidence of land occurs, wetlands have been restored using thin layer placement of dredged material to maintain elevation (DeLaune et al., 1990; Croft et al., 2006; Schrift et al., 2008). Also in deltaic regions such as the Mississippi River, large-scale restorations using river diversions and crevasse splays are built to divert river water and sediment into shallow bays and submerging marshes to build land and restore wetlands (DeLaune et al., 2003; Lane et al., 2006; Turner and Streever, 2002).

In Western Europe, tidal marshes are restored using similar techniques, including dredged material (Bernhardt and Handke, 1992), accretion enhancement (Hofstede, 2003), dike removal or breaching (Bernhardt and Koch, 2003; Eertman et al., 2002), and depoldering (Van Staveren et al., 2014). In the United Kingdom and Western Europe, marshes are restored by managed realignment or "managed retreat" whereby coastal defenses such as clay banks are established landward of existing defenses to allow the sea to inundate formerly protected areas and allow them to convert to marsh (Emmerson et al., 1997; Miren et al., 2001; Crooks et al., 2002; Townsend and Pethick, 2002; Wolters et al., 2005b; French, 2006; Garbutt et al., 2006; Garbutt and Boorman, 2009; Esteves, 2014). These projects restore tidal inundation and hydrology but depend on natural colonization to reestablish vegetation. Managed realignment also poses legal hurdles since many areas that were reclaimed more than a century ago, now contain unique assemblages of vegetation and are protected as Special Areas of Conservation through the European Union's Habitats Directive (Pethick, 2002).

Tidal wetlands also were established by planting vegetation to build land and control erosion. The practice of transplanting vegetation in Europe extends to the early part of the twentieth century when *S. townsendii*, now known as *Spartina anglica*, was planted to reclaim land for agriculture, slow coastal erosion, and reduce channel silting (Ranwell, 1967). In the 1930s, *S. townsendii* was planted in Australia and New Zealand, and, in the 1950s, *Spartina alterniflora* was introduced to New Zealand (Ranwell, 1967). In China, saline tidal marshes were established by planting *S. anglica* and *Spartina alterniflora* on intertidal flats to reduce coastal erosion (Chung, 1989, 1994, 2006) and to reclaim land (Chung et al., 2004). Such projects have achieved these goals but, unfortunately, led to invasion and replacement of native marsh plant communities (Zhi et al., 2007; Li et al., 2009).

Restoring Hydrology

Restoration of tidal marshes first consists of reintroducing tidal inundation by removing dikes, levees, spoil banks, fill, or tide gates (Frenkel and Morlan, 1989; Turner et al., 1994; Roman et al., 1995; Niering, 1997; Craft, 2001; Warren et al., 2002; Williams and Orr, 2002; Orr et al., 2003). On sites that were filled, fill is removed by excavation to lower the elevation of the soil surface to within the tidal frame. Plugging mosquito ditches is a common restoration technique in the northeastern US (Konisky et al., 2006).

Restoration of tidal inundation has many benefits: (1) reintroduction of tidal pulsing that is critical for the reproduction and survival of plant and animal species; (2) introduction of salinity and other elements important in estuarine and marine environments; (3) restored connectivity to the estuary that enables organisms to access, colonize, and forage in the marsh; and (4) outwelling of detritus to the estuary.

In marshes that are diked, tidal inundation is restored by breaching (Figure 8.4) or removing the entire dike. On developed shorelines of the northeastern US coast, tide gates were installed decades ago to ameliorate flooding during spring tides and storms. Here, a common method of restoration involves removing them to reintroduce tidal inundation, increase salinity, reduce coverage of the brackish species *Phragmites*, and favor expansion of saline marsh vegetation (Buchsbaum et al., 2006; Konisky et al., 2006; Raposa, 2008; Anisfeld, 2012; Smith and Warren, 2012).

Figure 8.4 (a) Diked *Spartina alterniflora* marsh at Sapelo Island, Georgia in 1955. Note the bleached mussels and absence of vegetation. (b) The same site in 1998, 40 years after the dike was breached and tidal inundation was reintroduced.
Photo credit: Christopher Craft.

(a)

(b)

In the past, many saline tidal marshes were diked, then drained. The effects of diking include reduced sedimentation (Anisfeld et al., 1999), soil subsidence (Roman et al., 1995; Portnoy and Giblin, 1997b), acidification (Portnoy and Giblin, 1999), and release of heavy metals (Anisfeld and Benoit, 1997). All of these are problematic, especially subsidence since it leads to a lowering of the marsh soil surface within the tidal frame, making it difficult to reestablish vegetation (Portnoy and Giblin, 1997b, 1999; Anisfeld and Benoit, 1997). Another factor to consider when reintroducing tidal inundation to diked marshes is the reintroduction of salinity, especially sulfate (SO_4) used by sulfate-reducing bacteria to decompose organic matter. In marshes containing organic-rich soils, reintroducing salinity increases sulfate reduction and accumulation of sulfides, a toxin that inhibits reestablishment of vegetation (Portnoy and Giblin, 1997a).

Williams and Orr (2002) evaluated 15 sites reflooded by breaching dikes in San Francisco Bay. The sites varied in size from 18 to 220 ha and ranged in age from 2 to 29 years old. The sites naturally revegetated over time, but three factors (restricted tidal exchange, limited sediment supply, and erosion of deposited mud by wind tides) slowed the rate of colonization. Sites that were older, smaller in size, and higher in the tidal frame developed faster (Williams and Orr, 2002; Brand et al., 2012). Efforts to restore tidal marshes by breaching dikes result in different outcomes elsewhere. In Florida, impounded marshes historically kept pace with rising sea level though accretion of organic matter. However, because of limited sediment supply, breaching dikes in this region leads to submergence and conversion to open water unless hydrology is incrementally reintroduced or fill is added to raise it to the appropriate elevation prior to breaching (Parkinson et al., 2006).

Reestablishing Vegetation

Once tidal inundation is reestablished, vegetation may recolonize naturally through seeding if there is a nearby source of propagules (Frenkel and Morlan, 1989; Wolters et al., 2005a) or by clonal propagation if remnant vegetation persists, but the development of continuous cover of vegetation is slow. Elevation within the tidal frame is the single biggest determinant of natural colonization as sites that are too low will remain mudflats whereas sites that are too high will not be colonized by hydrophytic vegetation. Garbutt and Boorman (2009) found that on sites restored by managed realignment, vegetation colonized as long as the elevation was suitable and a seed source was nearby. After 5 years, species diversity of the restored site was similar to a local reference marsh (Wolters et al., 2008). In Belgium, a 14-ha restoration site also colonized quickly, within 5 years (Erfanzadeh et al., 2010). Early colonizers were annual species such as *Suaeda maritima*, *Salicornia maritime*, and *Salsola kali* whose seeds were shorter, weighed less, and were more buoyant than other potential colonizers.

The intensity of restoration activities also determines the speed at which vegetation colonizes the marsh. Gallego Fernandez and Garcia Novo (2007) compared the pace of vegetation colonization of two restored tidal marshes in southwestern Spain whose intensity of restoration activities differed. The high intensity restoration involved excavation of channels and surface reshaping whereas, at the adjacent low intensity restoration site, only a few channels were constructed. Within 3 years, mudflats at the high intensity restoration were covered with vegetation dominated by *Spartina densiflora*, while in the low intensity restoration, the plant community was dominated by *Sarcocornia perennis*. After 5 years, however, vegetation of the two sites was converging toward that of plant communities found in natural tidal marshes of the region.

Because marsh species have limited ability to disperse (Morzaria-Luna and Zedler, 2007; Wolters et al., 2008), on unvegetated sites, transplanting seedlings may be necessary. Seeding generally is not effective, especially in the lower half of the intertidal zone (Seneca et al., 1976; Webb et al., 1984; Broome et al., 1988b), because the seeds are washed out during ebb tide. Direct seeding also depends on storm-free periods (Seneca et al., 1976). Success is greater on sheltered or protected areas. Where seeding is successful, good coverage of vegetation develops within 1–3 years. After two growing seasons, aboveground biomass of seeded areas was comparable to that of transplanted areas (Seneca et al., 1976).

Transplanting in general is more successful over a wide range of conditions than is seeding (Broome et al., 1988b), and it is important to consider the location of the restoration within the larger landscape to ensure survival of planted vegetation. Avoid sites with large fetch and exposure to wind and wave action as well as sites with boat traffic since they produce waves that may uproot recently planted vegetation. Sites with a gentle (1–3%) slope and large tide range are ideal for establishing vegetation (Broome et al., 1988b). The gentle slope allows for adequate but not too much drainage as the tides recede, and the large tidal range maximizes the amount of surface where vegetation can be established.

Spacing of transplants is an important consideration since the closer the spacing, the more quickly vegetation will spread and cover the site (Broome et al., 1986). However, closer spacing necessitates more transplants and, hence, higher costs. When using transplants, it is important to obtain them from a nursery that propagates native species from the same geographic region or ecoregion, an ecotype, in which the project is located. This ensures that the transplants are matched to the environmental conditions of the site. The ideal transplants are those that have a well-developed root system. Such transplants typically are grown in potting soil/peat substrate, are well watered and fertilized, and are less likely to suffer from planting shock than bare root seedlings.

It is critical to match the appropriate species with the salinity of the tidal floodwaters. Salt-tolerant species such as *Spartina alterniflora*, *Spartina foliosa*, *S. anglica*, and others should be planted in areas with salinities of 20–35 psu. Brackish species such as *Juncus* spp., *Distichlis* spp., *Limonium* spp., and others should be planted where tidal floodwater salinity is 10–20 psu. Oligohaline species, of which there are many, should be planted where salinity is 5–10 psu or less. In arid regions and high salinity microsites, including depressions and salt pans, vegetation such as *Salicornia* spp., *Batis* spp., or *Suaeda* spp. are planted. Although slow growing as compared to other saline marsh vegetation, these species are able to survive better in hypersaline environments.

Matching transplants to the appropriate inundation regime is equally important. Different species are adapted to different hydroperiod and, hence, grow at different elevations within the marsh. Elevation and distance from the estuary/tidal creek, along with tide range, determine the depth, duration, and frequency of tidal inundation and, to some extent, salinity. In restored saline marshes of southern California, three distinct vegetation zones corresponded to differences in elevation and proximity to tidal creeks (Zedler et al., 1999). Cordgrass habitat dominated by *S. foliosa* occupied bayward sites and lower elevations of the marsh. The marsh plain, a 30-cm elevation range with heterogenous topography, was the most species-rich habitat. The high marsh with a 30–70 cm elevation range was dominated by *Salicornia subterminalis*. Waterlogged depressions supported annual species such as *Salicornia biglovii* that were not present otherwise (Varty and Zedler, 2008).

Generally, *Spartina* spp., a C_4 photosynthetic plant, is relatively easy to propagate. It grows quickly and can be transplanted in a matter of months after the seeds germinate. C_3 plants such as *Juncus* spp. grow more slowly and must be grown longer before they are large enough to transplant. The timing of planting should correspond to the onset of the growing season of the region in order to maximize the length of time the plants can become established, grow, and spread.

On many sites, N is applied to the planting hole to jumpstart plant growth and facilitate establishment and spread of vegetation. Many studies have shown that saline, brackish, and tidal freshwater marsh vegetation is N limited and additions of N increase aboveground biomass production (Sullivan and Daiber, 1974; Valiela and Teal, 1974; Jefferies and Perkins, 1977; Covin and Zedler, 1988; Kiehl et al., 1997; Frost et al., 2009; Ket et al., 2011). Usually a single application at the time of planting is sufficient to jumpstart the community though, in some marshes, annual fertilizer additions may be needed to support consistently productive stands of vegetation (Boyer and Zedler, 1998). A reduced form of N such as (polymer coated) urea or an ammonium-based fertilizer is preferred (Broome et al., 1988b) because NH_4^+ sorbs onto cation exchange sites in the soil while nitrate N readily leaches. In addition to N, P is sometimes limiting (Seneca and Broome, 1982), especially in coarse-textured sandy soils. In North Carolina, Broome et al. (1983) found that additions of 112 kg N/ha and 49 kg P/ha produced better growth of *Spartina alterniflora* than either N or P alone. As with seeding, broadcast application of fertilizer is inefficient since it will be washed out by the ebbing tide.

Restoring and Re-creating Soils

In marshes where tidal inundation was restricted or eliminated, reintroducing it usually is sufficient to restore soil properties. Anaerobic conditions develop quickly as frequent and regular tidal inundation fills soil pores with water and drives out air and oxygen. Chemical characteristics such as alkaline pH, abundance of base cations (Table 8.1), and ample phosphorus (P) characteristic of seawater develop quickly once salinity is reintroduced. Saline marsh soils created from dredged material also develop relatively quickly as the soil material already is chemically reduced though it is low in organic matter (Figure 8.5).

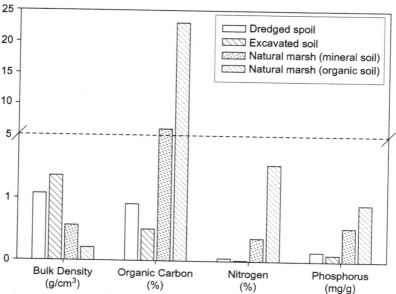

Figure 8.5 Comparison of soil properties of saline tidal marshes created on dredge spoil and excavated soil with mineral and organic soil natural marshes. Compiled from Craft et al. (1988, 1991b) and Pennings et al. (2012).

Where tidal marshes are created on terrestrial land, reducing conditions may be slow to develop because of the paucity of organic matter in the soil (Figure 8.6). Such excavations often remove the organic matter-rich surface layer, the topsoil, or A horizon, exposing dense mineral subsoil, the B or C horizons. Organic matter and its organic carbon are needed to fuel microbial reactions such as iron reduction, sulfate reduction, and denitrification that are characteristic of wetland soils. Accumulation of organic matter, added by plant growth and death, is a prerequisite for development of organic-rich wetland soils. Hydrology also determines the rate at which wetland soil properties form (Craft et al., 2002; Callaway et al., 2012). For example, in a created saline marsh, anaerobic conditions and organic-rich surface soil occurred more quickly under continuous inundation than under soil saturation alone (Craft et al., 2002).

Accumulation of sizable pools of SOM and nitrogen takes considerable time. Once vegetation becomes established, plant detritus, mostly roots and rhizomes (McCaffrey and Thomson, 1981), begins to accumulate. Nitrogen in the form of organic N (Craft et al., 1991a) also accumulates in the soil as soon as the plant community becomes established and the rate of N accumulation quickly becomes comparable to natural marshes (Craft et al., 2003). Over time, an organic- and nitrogen-rich surface layer develops (Craft et al., 1988). Organic matter and N are critical to supporting long-term growth and health of vegetation, microbial, and animal communities (Craft, 1999). Nitrogen pools generally are lower in created and restored tidal marsh soils than in natural marshes and often less than the 100 g N/m² needed to form a self-sustaining ecological community (Bradshaw, 1983; Craft et al., 2003).

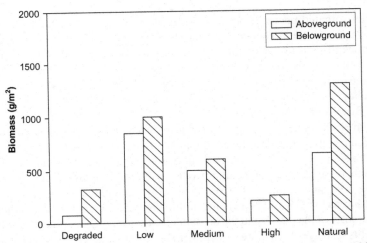

Figure 8.6 Effect of thin layer placement of dredged material at low, medium, and high rates on above- and belowground biomass.
Redrawn with permission of Stagg and Mendelssohn (2010). Reproduced with permission of John Wiley and Sons.

Organic matter also is needed to sustain heterotrophic microbial and benthic invertebrate communities. Decomposing plant detritus and associated microorganisms serve as the base of the tidal marsh food web. In saline tidal marshes, the invertebrate community is dominated by roundworms (nematodes, oligochaete, and polychaetes), burrowing crabs (*Uca* spp.), snails (*Littorina* spp., *Illyanassa* spp.), filter feeders (mollusks (oysters, *Crassostrea* spp.), clams (*Mercenaria* spp.), and mussels (*Geukensia* spp.)). Organic matter also improves soil physical properties by increasing porosity and air and water movement (Craft et al., 1991b).

Tidal marshes created by excavation have other problems in addition to nutrient limitations. Many excavated marsh soils consist of subsoil with high bulk density and low porosity (Craft et al., 1991b). Even with fertilizer additions, vegetation spreads slowly because of the difficulty in rooting. Excavation also may expose deposits of reduced S, such as pyrite (FeS, FeS_2), that upon contact with air are oxidized to acid sulfate minerals such as jarosite (Broome et al., 1988a). The pH of these soils, sometimes known as cat clays, decreases to 2–3, making it impossible to establish vegetation unless large amounts of lime are added and incorporated into the soil to neutralize acidity and raise pH.

Texture, the proportion of sand-, silt-, and clay-size particles, determines the ease and mechanism of planting the site. Clayey soils are very soft and need to be planted by hand. Sandy soils are firmer and can be planted by mechanical means such as a tractor. The ideal texture for restoration consists of a loamy soil, one with relatively equal proportions of sand-, silt-, and clay-size particles (Brady and Weil, 2002). The larger sand particles promote air and water movement and provide load-bearing capacity to support foot traffic. Clay particles, with their large surface area and negative charge, sorb essential cations such as ammonium, K, Ca, and Mg onto their cation exchange complex.

Amendments

Sometimes, organic matter is added to soils of excavated or dredged material sites to improve physical and chemical properties prior to planting. Amendments include topsoil stockpiled from the site during excavation; C-rich amendments such as peat, straw, leaf litter, wood compost, kelp compost; and N-rich amendments such as alfalfa and sewage compost (Broome et al., 1982, 2001; Gibson et al., 1994; Thompson et al., 1995; Kelley and Mendelssohn, 1995; Handa and Jefferies, 2000; Levin and Talley, 2002; O'Brien and Zedler, 2006).

Thin layer placement of dredge material is used to increase soil surface elevation to stem deterioration and subsidence of marshes (Ford et al., 1999). The primary benefit is to raise the elevation of the soil surface to reduce waterlogging and has been shown to increase soil redox potential (stem density and plant productivity) and nutrient uptake (DeLaune et al., 1990; Croft et al., 2006) and reduce porewater H_2S (Mendelssohn and Kuhn, 2003). The benefits of thin layer placement vary depending on the initial elevation of the marsh and the amount of sediment added. Some studies indicate that benefits last 7 years or more (Slocum et al., 2005) while others indicate a decline in plant performance after the first year (La Peyre et al., 2009).

The amount of dredge material added also affects the rate of ecosystem development. Tong et al. (2013) reported that thin layer placement of dredge material in large amounts to a Louisiana (USA) marsh increased surface elevation by 20 cm above that of the reference marsh. As a result, above- and belowground biomass and benthic macroinvertebrate communities were reduced compared to the reference marsh and negative impacts persisted for 7 years. Stagg and Mendelssohn (2010) added dredge material at low, medium, and high rates to a subsiding marsh. Low and intermediate rates of sediment addition had the greatest benefit (Figure 8.6), decreasing flood duration and frequency, increasing drainage, redox potential, and above- and belowground biomass compared to untreated areas and areas receiving high rates of sediment additions.

Ecosystem Development

With restoration of hydrology and establishment of vegetation, ecosystem processes of saline tidal marshes develop relatively quickly, aboveground biomass may achieve equivalence to natural marshes within a matter of years (Webb and Newling, 1985; Craft et al., 2003). In the Bay of Fundy, Canada, reintroduction of tidal inundation to a diked freshwater impoundment led to 100% cover of *Spartina alterniflora* within 3 years (Van Proosdj et al., 2010). Belowground biomass is slower to develop (Shafer and Streever, 2000; Craft et al., 2003), especially in semiarid and Mediterranean climates (Boyer et al., 2000). Boyer et al. (2000) measured belowground biomass in two constructed marshes planted 5 and 10 years earlier. Planted marshes contained less belowground biomass that was attributed to low fertility of the coarse sandy soil.

The development of structural characteristics such as species diversity takes longer than biomass. In a comparison of sites restored by managed realignment with natural marshes, Mossman et al. (2012) reported that restored sites contained more pioneer species and fewer perennials (e.g., *Limonium vulgare, Triglochin maritima, Plantago maritime, Armeria maritima*) characteristic of the natural marshes. Brooks et al. (2015) attributed the dominance of pioneer species to the compact soils and lack of topographic heterogeneity that limit colonization by characteristic marsh species. Garbutt and Wolters (2008) compared plant species composition along a chronosequence of saline tidal marshes restored by managed realignment. Eighteen sites, ranging in age from 2 to 107 years were sampled. Overall, sites less than 100 years old had fewer species per quadrat than 100+-year-old sites and reference marshes.

Succession of microphytobenthos, including diatoms and photosynthetic bacteria, proceeds quickly relative to emergent vegetation. Janousek et al. (2007) measured benthic algal communities following hydrologic restoration and planting with *S. foliosa* in a tidal marsh in southern California. Within 2 years, sediment chlorophyll concentrations, community composition, and species diversity were similar to a natural marsh.

Soil properties, especially organic matter, are slow to develop and created and restored marshes, especially excavated and dredge spoil marshes, contain less organic matter than older restored marshes and mature natural marshes (Cammen, 1976; Lindau and Hossner, 1981; Craft et al., 1988, 2002, 2003; Langis et al., 1991; Moy and Levin, 1991; Zedler and Langis, 1991; Simenstad and Thom, 1996; Streever, 2000; Havens et al., 2002; Morgan and Short, 2002; Edwards and Proffitt, 2003; Fearnley, 2008; Armitage et al., 2014). Marshes created from dredged material often contain more sand and less nitrogen and sometimes P than natural marshes (Poach and Faulkner, 1998; Craft et al., 2003; Fearnley, 2008). With increasing organic matter, microbial activity, including decomposition, methane production, and denitrification, increases (Craft et al., 2003). For full restoration of soil properties, especially subsurface layers, several hundred years may be needed to recreate conditions, especially organic C pools, that are similar to old, mature natural marshes (Craft et al., 2002, 2003; Burden et al., 2013).

Colonization by epibenthic and benthic invertebrates also depends on accumulating soil and macroorganic matter, the living and dead root and rhizome mat (Minello and Zimmerman, 1992; Scatolini and Zedler, 1996; Levin and Talley, 2002), and their ability to recruit to the site (Moseman et al., 2004). Moy and Levin (1991) compared infauna density and diversity in a 3-year-old saline marsh created by excavation and a natural marsh in North Carolina. The low organic matter-created marsh was dominated by surface deposit feeders whereas the natural marsh was dominated by subsurface deposit feeders. Sacco et al. (1994) also reported lower SOM and infauna density and species richness in six planted *Spartina alterniflora* marshes relative to six reference marshes. Invertebrates with planktonic (dispersing) larvae colonize the site within a matter of months to years, whereas invertebrates such as oligochaetes and other taxa with nondispersing larvae take much longer, perhaps a decade or more, to develop populations comparable to natural marshes (Levin et al., 1996; Posey et al., 1997; Craft and Sacco, 2003).

Macrofauna communities also may take a long time to develop. For example, abundance of the California horn snail (*Cerirthidea californica*) was less in a 3-year-old created marsh in southern California than in a nearby natural marsh (Armitage and Fong, 2004).

Studies of a restored impounded marsh indicate that, for some invertebrate species (e.g., some amphipods, ribbed mussel *Geukensia demissa*), populations may take a decade or more to become equivalent to natural marshes (Swamy et al., 2002). In Connecticut 12 years following tidal restoration of an impounded marsh, abundance of the mud snail, *Melampus bidentatus*, was comparable in the restored marsh and unimpounded marsh downstream (Fell et al., 1991).

Use of created and restored tidal marshes by nekton, motile epifauna, and finfish depends on enhancing connectivity by adding tidal creeks and maximizing the amount of edge habitat (West and Zedler, 2000; Minello and Rozas, 2002). Removing tidal restrictions such as tide gates or enlarging culverts quickly enables estuarine species to use restored marshes (Dionne et al., 1999; James-Pirri et al., 2001; Raposa, 2002). Restoration projects often incorporate large amounts of marsh–open water edge and grading to lower soil surface elevation to facilitate access (Minello and Webb, 1997; Dionne et al., 1999; Desmond et al., 2000; Wallace et al., 2005). In some cases, marsh terraces are constructed to maximize the amount of edge (Rozas and Minello, 2001) but this technique does not actually create much vegetated marsh habitat (Rozas et al., 2005).

Greater vegetated edge and subtidal habitat lead to increased use by finfish (Minello et al., 1994; Havens et al., 2002). Creating low-order tidal creeks and small ponds, which serve as refugia, also increase nekton utilization (Williams and Zedler, 1999; Jivoff and Able, 2003; Garbutt and Boorman, 2009). By reestablishing tidal connections with culverts, Gilmore et al. (1982) in Florida reported a dramatic increase in fish utilization of a marsh impounded for mosquito control. Studies have shown that with proper restoration of tidal flows and construction of tidal creeks, finfish utilization of created and restored marshes is comparable to natural reference marshes (Rulifson, 1991; Dionne et al., 1999; Raposa, 2008). For example, hydrologic restoration by removal of berms and tides increased abundance of the marsh killifish, *Fundulus*, to levels found in reference marshes though species richness was slower to increase (Figure 8.7). Migratory fish such as salmon also use restored marshes (Washington, USA) once tidal inundation is reintroduced (Shreffler et al., 1990; Miller and Simenstad, 1997; Gray et al., 2002).

Rozas et al. (2007) evaluated the effects of recent tidal marsh restoration in Galveston Bay, Texas, where 30% of tidal marsh was lost between the 1950s and 2002. In one cove, where losses reduced marsh area from 40 to 18 ha, restoration of 12 ha of marsh led to a rebound in populations of brown shrimp (*Farfantepenaeus aztecus*), white shrimp (*Litopenaeus setiferus*), and blue crab (*Callinectes sapidus*) in an area where populations were in decline. Rozas et al. (2005) compared blue crab and shrimp use of three salt marsh restoration projects consisting of terracing, mounds, and island construction, also in Galveston Bay. The three restoration methods increased fisheries use relative to unrestored habitat but usage still was less than in a natural reference marsh.

Not all studies indicate complete restoration of fisheries habitat following marsh creation and restoration. Zeug et al. (2007) compared nekton use of a 30-year-old created marsh with a natural marsh in south Texas. Overall, the natural marsh had greater biomass of shrimp and fish, including gulf toadfish, flatback mud crab, and snapping shrimp that were not found in the restored marsh. Lower use in the created marsh was attributed to the absence of oysters that increased habitat complexity, and low soil organic matter.

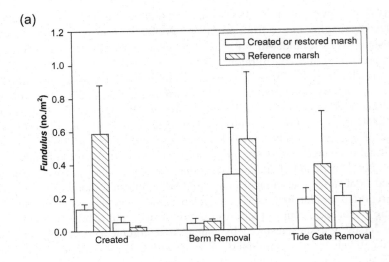

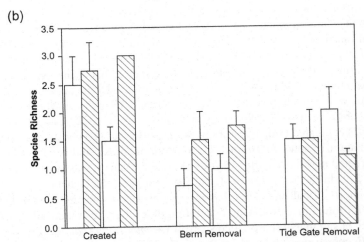

Figure 8.7 (a) *Fundulus* density and (b) species richness of created, restored, and reference saline tidal marshes.
Redrawn with permission from Dionne et al. (1999).

Analysis of food webs suggests that created and restored tidal saline marshes provide similar forage habitat relatively quickly following restoration. Stable isotope analysis of benthic macrofauna in young (<2 years old) constructed marshes indicated greater reliance on microalgae as a food source, whereas in reference marshes, emergent vegetation, *Spartina*, contributes proportionately more (Moseman et al., 2004). Wosniak et al. (2006) used stable isotopes to evaluate marsh utilization by the killifish, *Fundulus*, in three marshes restored by removing tidal restrictions such as culverts. Restoring tidal flow resulted in a shift in vegetation from the C_3 plant, *P. australis*, to the C_4 plant, *Spartina alterniflora*, with a concomitant increase in use of C_4 carbon by *Fundulus*. Llewellyn and La Peyre (2011) evaluated blue crab use of four created salt marshes, ranging in age from 5 to 24 years old and four paired natural marshes in Louisiana. Marshes older than 8 years were characterized by trophic diversity and breadth that was similar to natural marshes.

Waterfowl and wading birds also benefit from increasing marsh–open water edge, mudflat, and woody vegetation such as shrubs (Havens et al., 2002) and will quickly use restored marshes if these attributes are present (Brawley et al., 1998; Atkinson et al., 2004; Raposa, 2008; Garbutt and Boorman, 2009). Raposa (2008) evaluated bird use of a restricted tidal marsh 1 year before and 2 years following reintroduction of tidal exchange. Restoring tidal flows more than doubled the number of bird species after 2 years and abundance more than tripled. Greater use was attributed to increased tidal range that created more mudflat habitat for shorebirds. Seigel et al. (2005) reported similar increases in abundance and diversity, especially shorebirds, following *Phragmites* removal and restoration of intertidal habitat in urban marshes in New Jersey. On sites restored by managed realignment, waterbird populations achieved equivalence to natural intertidal areas within 3 years (Mander et al., 2007).

Other studies suggest that, while overall avian usage is high in created and natural marshes, different assemblages may frequent created and restored marshes. Melvin and Webb (1998) reported greater use of created marshes by gulls and terns and more use of natural marshes by marsh residents, rails and sparrows, migrating waterfowl, and shorebirds that was attributed to reduced habitat heterogeneity in the created marshes. Woody shrubs also increase abundance of some species but at the expense of others. A site restored by introduction of tidal exchange increased the numbers, species, and biomass of waders; waterfowl; shorebirds; and gulls (Slavin and Shisler, 1983) but, as shrub habitat decreased following restoration, passerine species declined. Havens et al. (2002) compared bird species richness and diversity in a constructed tidal marsh with two natural marshes in Virginia (USA). Both richness and diversity were lower in the constructed marsh even after 12 years (Figure 8.8) and it was attributed to the low abundance of shrub (*Iva frutescens* L, *Baccharis halimifolia* L) habitat.

Restoration of tidal inundation has variable effects on nesting habitat. DiQuinzio et al. (2002) found that, following restoration of a tidally restricted salt marsh, nesting density of salt marsh sharp-tailed sparrow (*Ammodramus caudacutus*) declined as *P. australis* marsh was replaced by *Spartina alterniflora* that grows at lower elevations and has shorter stems. They hypothesized that given time, nesting success would increase as high marsh vegetation (e.g., *Phragmites*) colonized and replaced shrub vegetation that grows at even higher elevations.

Studies indicate use of created and restored marshes by both generalist and specialist species. For example, Armitage et al. (2007) reported no difference in species diversity of wintering shorebirds in five 5- to 7-year-old restored tidal marshes and five paired reference marshes in southern California. Desrochers et al. (2008) compared avian communities of 11 created salt marshes with 11 reference marshes in Virginia. Marshes were created by excavation of uplands and reconnect with tidal creeks and the sites were 15 ± 4 years old at the time of sampling. Bird use by the obligate species, clapper rail (*Rallus longirostris*), did not differ between created and restored marshes. However, during the breeding season, created salt marshes contained lower densities and species richness, especially of wetland-dependent birds. Desrochers et al. (2008) attributed

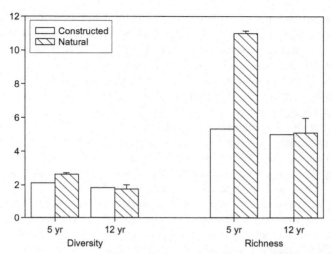

Figure 8.8 Bird species diversity and richness in a constructed and two natural marshes 5 and 12 years after marsh creation.
From Havens et al. (2002).

these differences to dominance by *Spartina alterniflora* in the created marshes in contrast to the natural marshes that contained a more diverse vegetation community.

Vegetation structure of created saline marshes often is different from natural marshes and it affects bird use. Planted marshes often contain stems that are short relative to vegetation in natural marshes (Zedler and Langis, 1991; Zedler, 1993; Craft et al., 2003). In southern California, the light-footed clapper rail (*Rallus longirostris levipes*), an endangered species, did not nest in a short-statured vegetation of a *S. foliosa* marsh in southern California because the nests were washed out during high tides. Additions of N increased stem height to levels found in natural marshes but the taller canopy was not maintained without annual fertilizer additions (Boyer and Zedler, 1998; Lindig-Cisneros et al., 2003).

Comparison of restored marshes of the (humid) eastern US with Mediterranean-type marshes in southern California suggests that the hypersaline marshes of southern California and similar Mediterranean climates develop at a slower rate and support slower growing and less vigorous plant species (Zedler et al., 2003). The slow rate of development is linked to the Mediterranean climate that is characterized by seasonal freshwater inflows with high interannual variability (Zedler, 2001). Another stressor is excessive sedimentation that accompanies infrequent but episodic flood events (Zedler et al., 2003). Such press (hypersalinity) and pulse (sedimentation) disturbances may explain why restored saline marshes of southern California do not exhibit clear trajectories of ecosystem development (Zedler and Callaway, 1999) whereas those on the east coast of the US do (Morgan and Short, 2002; Craft et al., 2003) (see Chapter 9, Performance Standards and Trajectories). A study of ecosystem development of restored brackish tidal marsh in the Pacific Northwest also did not show strong trajectories (Simenstad and Thom, 1996), but the time series of measurements, for 7 years following restoration, probably was not long enough to detect such trends.

Ecosystem processes such as N cycling also differ between created/restored and natural marshes especially during the early years. Soil N is less in created and restored marshes than in natural marshes (Craft et al., 1988, 2003; Langis et al., 1991; Zedler and Langis, 1991) and, as a result, microbial cycling of N also differs. Thompson et al. (1995) reported much lower denitrification in a 1-year-old restored marsh than in a natural marsh in North Carolina. Likewise, Broome and Craft (2009) also reported lower rates of denitrification in young constructed salt marshes than in older (>20 years) constructed marshes and mature natural marshes. Nitrogen fixation, in contrast, was comparable or greater in restored and constructed marshes than in natural marshes (Zedler and Langis, 1991; Currin et al., 1996; Piehler et al., 1998) because the developing canopy that is more open supports greater numbers of sediment cyanobacteria (Currin et al., 1996) and epiphytes (Currin and Paerl, 1998).

Restoring tidal inundation also facilitates outwelling of nutrients to the estuary. In a study of tidal exchange between two young (1–3 years old) created saline tidal marshes and the tidal creek, created marshes were a source of dissolved organic matter (C and N) to the estuary, comparable to what is observed in natural marshes (Craft et al., 1989). In contrast, restored marshes were sinks for inorganic N and P that served to augment low nutrient capital of these early successional soils.

Different ecosystem services of saline tidal marshes differ in their rate of development following creation and restoration (Table 8.4). With restoration of hydrology

Table 8.4 Development of Saline Tidal Marsh Ecosystem Services Following Creation or Restoration

	Time (years) to Equivalence to Mature Natural Marshes
Productivity and Habitat Functions	
Primary production	3–5
Plant species richness[a]	4–10
Benthic algae	<1–2
Microbial activity	5–15
Benthic invertebrates	5–20
Epifauna and finfish	2 to >15
Water and wading birds	3–10
Songbirds	10–15
Regulation Functions	
Sedimentation	1–3
Nutrient (N, P) retention	1–5
C Sequestration	3–5
N cycling	10–20
Outwelling of nutrients	1–5
Soil formation	10–100 s

[a]Tidal freshwater marshes.
Compiled from Broome and Craft (2009), Morgan and Short (2002), Garbutt and Boorman (2009), and others (see text).

and establishment of vegetation, the benthic algae community develops almost immediately while the plant community and its primary production becomes established in about 3–5 years. Finfish, other nekton, and wading birds begin to use the marsh as soon as hydrology is restored and vegetation begins to establish, especially when access is enhanced. Based on a review of nine salt marsh restoration projects in Connecticut, Warren et al. (2002) estimated that finfish assemblages and breeding bird populations achieve equivalence to reference marshes within 5 and 15 years, respectively. Microbial communities and benthic invertebrates that depend on accumulation of SOM may require 5–20 years to develop to levels found in mature natural marshes. Regulation functions such as sediment retention, nutrient storage, and C sequestration develop along with the plant community although soil formation takes much longer, decades to centuries (Table 8.4).

Restoration of Tidal Freshwater Marshes

As with saline tidal marshes, restoration of tidal freshwater marsh involves excavation, dredged material, or hydrologic restoration (Baldwin, 2009; Barendregt, 2009). Establishing freshwater vegetation on dredge spoil is common since these marshes typically are located in highly urbanized estuaries such as the Hudson River, Delaware Bay, and Chesapeake Bay along the US east coast and Scheldt River of Western Europe where dredging is frequent. Dredge spoil islands then are excavated to intertidal elevations to create restoration sites (Leck, 2003). In San Francisco Bay, tidal freshwater marshes are restored by breaching levees (Miller et al., 2008). These marshes were diked for agriculture nearly 150 years ago. Since then, the organic-rich peat soils subsided as much as 6 m, making them impossible to restore without water table management. By maintaining water tables at 25–55 cm, the marshes revegetated naturally with faster colonization at shallow water levels.

One of the first tidal freshwater marsh restoration projects in the US was on the Anacostia River, an urban tributary of the Potomac River in Washington, DC (Bowers, 1995; Niering, 1997). The Kenilworth marsh was established by placing fill on 13-ha mudflat and planting 340,000 plants of 16 species. Problems encountered during the restoration included retaining the fine-textured fill in place, dewatering it, and discouraging herbivory by waterfowl, Canada geese. A number of lessons were learned from this pioneering work, including understanding the environmental requirements of a number of freshwater species.

Baldwin (2004) and others evaluated vegetation development of Kingman marsh, also on the Anacostia River, that was restored in 2000 by placement of hydraulically dredged sediment. Seven species were planted and fencing was installed to deter herbivory. Plant cover developed quickly during the first growing season due in part to the planted species, but also to colonization by ruderal native and exotic species. Once the fencing was removed, plant cover declined as a result of herbivory by geese. New colonizers were species whose seeds dispersed by water (Neff and Baldwin, 2005). Species that were common in natural tidal freshwater marshes did not readily disperse to the site and included annuals such as *Bidens* and perennials, *Peltandra*, *Pontederia*, and *Sagittaria*, that occupy lower, wetter areas of the marsh (Neff and Baldwin, 2005).

The seedbank of Kingman marsh developed quickly. Between 2000, the year of restoration, and 2004, the number of taxa in the seedbank increased from 2 to 10 per 90 cm² (Neff et al., 2009). Common taxa included *Carex, Juncus, Ludwigia,* and the invasive *Lythrum.* Emergence of seedlings of annual and native species was greater in the reference marsh than in a 7-year-old restored marsh whereas emergence of native seedlings in a 1-year-old restored marsh was comparable (Figure 8.9) (Neff et al., 2009).

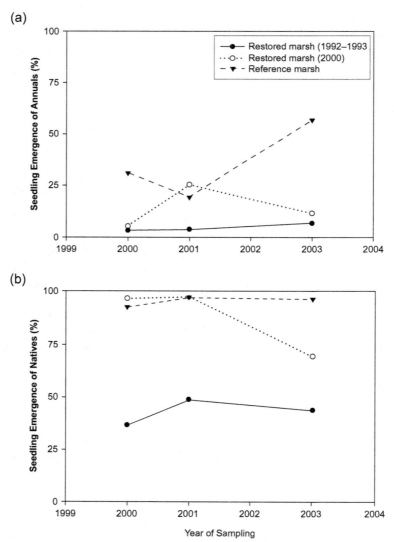

Figure 8.9 Seedling emergence of (a) annual and (b) native species from restored and reference tidal freshwater marshes.
From Neff et al. (2009).

Post-restoration monitoring of the older (10 years) Kenilworth and younger Kingman (2 years) marshes in the Anacostia River consisted of measurements of sediment accretion, soils, vegetation, benthic invertebrates, and birds (Baldwin et al., 2009). Both marshes accreted sediment to compensate for settling of dredged material and sea level rise. Soil organic matter, not surprisingly, was less than in reference marshes as has been reported for other tidal freshwater marshes in the region (Verhoeven et al., 2001). In the Anacostia River marshes, benthic invertebrate density and species richness were comparable among restored and natural tidal freshwater marshes in the area but were less relative to a high quality nonurban reference marsh outside of the city (Baldwin, 2009). Avian use of the restored marshes was considerable with 137 species observed in the younger marsh and 164 species in the older marsh.

Leck (2003) evaluated vegetation and seedbank development of a 32-ha tidal freshwater marsh mitigation wetland created by excavation in Delaware Bay. The site was planted with 14 herbaceous species and, within 1 year, vegetation completely covered the site. After 5 years, 92 species were present with 72 of them contributing to cover. A total of 177 species were present in the seedbank of the soil. Species richness was inversely related to inundation with the lowest number of species found along the channels. Overall, the created marsh contained many more species in the seedbank than a reference marsh including more seeds of invasive species such as purple loosestrife, reed canary grass, and *P. australis* (Leck and Leck, 2005). Greater numbers of species in the created marsh seedbank was attributed to soil disturbance caused during excavation of the site.

In the Pacific Northwest (Washington), a *Phalaris*-dominated tidal freshwater marsh that was restored by breaching a dike and reintroducing tidal inundation also revegetated quickly (Tanner et al., 2002). Restoration of tidal flows dramatically reduced coverage of *Phalaris*, and after 1 year, the number of marsh species increased from 5 to 14. Three years following restoration, 37 species were tallied. Finfish including juvenile Coho, Chum, and Chinook salmons utilized the restored marsh. A concern of this otherwise successful restoration project was colonization by purple loosestrife in some areas. Levings and Nishimura (1997) compared eight marshes planted with *Carex lyngbyei* and *Schoenoplectus americanus* with reference marshes in the Frazer River, British Columbia. Overall, cover of *Carex lynbyei* in the eight restored marshes was about 50% of that found in reference marshes. Abundance of invertebrates and finfish, however, was similar among transplanted and reference marshes.

In Europe, succession and ecosystem development of tidal freshwater marshes also occurs quickly following restoration. Within 3 years, vegetation covered the marsh and wading birds and ducks utilized the site (Barendregt, 2009). Successful restoration was achieved within 3–10 years. These studies from the US and Europe illustrate that tidal freshwater marshes can be successfully restored by restoring the proper elevation with respect to the tides, reintroducing tidal inundation, and providing necessary care such as control of herbivory during the early stages of plant establishment. Seeding or planting is not necessary if there is large, diverse, and viable seedbank or nearby source of seeds.

Keys to Ensure Success

Restoration of tidal marshes depends on identifying appropriate sites, proper construction and planting techniques, and maintenance (Table 8.5). Sites with large tide range contain more intertidal area where vegetation can be established. Sites protected from wind, those with short fetch, waves, and boat traffic, are more likely to be successful. A gentle (1–3%) slope ensures that the site will drain during low tide and will not become waterlogged. In terms of site preparation, it is important to maximize the ratio of marsh to open water, construct tidal creeks, and incorporate microtopographic features such as depressions to facilitate access to and utilization of the marsh by finfish and waterfowl.

In saline tidal marshes, the energetic tidal regime and lack of propagules in the seedbank necessitates planting, whereas, in tidal freshwater marshes, planting often is not needed because of the availability of seeds from nearby sources and in the seedbank. Planting considerations include using seedlings adapted to the appropriate inundation regime, salinity, and geographic region. It is important to fertilize with a slow-release source of N, incorporated into the soil at the time of planting, to accelerate growth when the transplants are placed in the ground. Following planting, it may be necessary to protect the tender young transplants from herbivory, especially by waterfowl. Once the plants are well established, it is necessary to monitor the site, remove invasive species as needed, and remove wrack after storms. Over the long term, rising sea level will threaten some restoration projects unless they are allowed to migrate inland. Tidal wetland restoration projects implemented today need to plan and allow for such migration in the future.

Table 8.5 Key Characteristics to Ensure Success

Site Selection
• Large tidal range
• Short fetch
• Low wave energy
• Gentle (1–3%) slope
• Ability to migrate (with sea level rise)
Construction Considerations
• Large marsh–open water ratio
• Construct tidal creeks
• Incorporate microtopographic features
Planting Considerations
• Use well-watered and well-fertilized greenhouse grown seedlings
• Plants selected for appropriate inundation regime, salinity, and geographic region
• Fertilize (incorporate) with N (and P)
• Protect from herbivores (e.g., waterfowl)
Maintenance
• Invasive species monitoring and removal
• Wrack removal

References

Abraham, K.F., Jefferies, R.L., Alisauskas, R.T., 2005. The dynamics of landscape change and snow geese in mid-continent North America. Global Change Biology 11, 841–855.

Anisfeld, S.C., 2012. Biogeochemical responses to tidal restoration. In: Roman, C.T., Burdick, D.M. (Eds.), Tidal Marsh Restoration: A Synthesis of Science and Management. Island Press, Washington, D.C., pp. 39–58.

Anisfeld, S.C., Benoit, G., 1997. Impacts of flow restrictions on salt marshes: an instance of acidification. Environmental Science and Technology 31, 1650–1657.

Anisfeld, S.C., Tobin, M.J., Benoit, G., 1999. Sedimentation rates in flow-restricted and restored salt marshes in Long Island Sound. Estuaries 22, 231–244.

Armitage, A.R., Fong, P., 2004. Gastropod colonization of a created coastal wetland: potential influences of habitat suitability and dispersal ability. Restoration Ecology 12, 391–400.

Armitage, A.R., Jensen, S.M., Yoon, J.E., Ambrose, R.F., 2007. Wintering shorebird assemblages and behavior in restored tidal wetlands in southern California. Restoration Ecology 15, 139–148.

Armitage, A.R., Ho, C.K., Madrid, E.N., Bell, M.T., Quigg, A., 2014. The influence of habitat construction technique on the ecological characteristics of a restored brackish marsh. Ecological Engineering 62, 33–42.

Atkinson, P.W., Crooks, S., Drewitt, A., Grant, A., Rehfisch, M.M., Sharpe, J., Tyas, C.J., 2004. Managed realignment in the UK–the first 5 years of colonization by birds. Ibis 146, 101–110.

Bakker, J.P., Ruyter, J.C., 1981. Effects of five years of grazing on salt-marsh vegetation. Vegetatio 44, 81–100.

Baldwin, A.H., 2004. Restoring complex vegetation in urban settings: the case of tidal freshwater marshes. Urban Ecosystems 7, 125–137.

Baldwin, A.H., 2009. Restoration of tidal freshwater wetlands in North America. In: Barendregt, A., Whigham, D., Baldwin, A. (Eds.), Tidal Freshwater Wetlands. Backhuys Publishers, Leiden, The Netherlands, pp. 207–222.

Baldwin, A.H., Hammerschlag, R.S., Cahoon, D.R., 2009. Evaluation of restored tidal freshwater wetlands. In: Perillo, G.M.E., Wolanski, E., Cahoon, D.R., Brinson, M.M. (Eds.), Coastal Wetlands: An Integrated Ecosystem Approach. Elsevier, Amsterdam, The Netherlands, pp. 801–831.

Barendregt, A., 2005. The impact of flooding regime on ecosystems in a tidal freshwater area. Ecohydrology and Hydrobiology 8, 95–102.

Barendregt, A., 2009. Restoration of European tidal freshwater wetlands. In: Barendregt, A., Whigham, D., Baldwin, A. (Eds.), Tidal Freshwater Wetlands. Backhuys Publishers, Leiden, The Netherlands, pp. 223–232.

Bart, D., Hartman, J.M., 2000. Environmental determinants of *Phragmites australis* expansion in a New Jersey salt marsh: an experimental approach. Oikos 89, 59–69.

Bernhardt, K.G., Handke, P., 1992. Successional dynamics of newly created saline marsh soils. Ekologia 11, 139–152.

Bernhardt, K.G., Koch, M., 2003. Restoration of a salt marsh system: temporal changes of plant species diversity and composition. Basic and Applied Ecology 4, 441–451.

Bertness, M.D., Pennings, S.C., 2000. Spatial variation in process and pattern in salt marsh plant communities in eastern North America. In: Weinstein, M.P., Kreeger, D.A. (Eds.), Concepts and Controversies in Tidal Marsh Ecology. Kluwer Academic Publishers, Dordrecht, The Netherlands, pp. 39–57.

Bolam, S.G., Schatzberger, M., Whomersley, P., 2006. Macro- and meiofaunal recolonization of dredged material used for habitat enhancement: temporal patterns in community development. Marine Pollution Bulletin 52, 1746–1755.

Bowers, J.K., 1995. Innovations in tidal marsh restoration. Restoration and Management Notes 13, 155–161.

Boyer, K.E., Zedler, J.B., 1998. Effects of nitrogen additions on the vertical structure of a constructed cordgrass marsh. Ecological Applications 8, 692–705.

Boyer, K.E., Callaway, J.C., Zedler, J.B., 2000. Evaluating the progress of restored cordgrass (*Spartina foliosa*) marshes: belowground biomass and tissue nitrogen. Estuaries 23, 711–721.

Bradshaw, A.D., 1983. The reconstruction of ecosystems. Journal of Applied Ecology 20, 1–17.

Brady, N.C., Weil, R.R., 2002. The Nature and Properties of Soils, thirteenth ed. Prentice Hall, Upper Saddle River, New Jersey.

Brand, L.A., Smith, L.M., Takekawa, J.Y., Athearn, N.D., Taylor, K., Shellenbarger, G.G., Schoellhamer, D.H., Spenst, R., 2012. Trajectory of early tidal marsh restoration: elevation, sedimentation and colonization of breached salt ponds in northern San Francisco Bay. Ecological Engineering 42, 19–29.

Brawley, A.H., Warren, R.S., Askins, R.A., 1998. Bird use of restoration and reference sites within the Barn Island Wildlife Management Area, Stonington, Connecticut, USA. Environmental Management 22, 625–633.

Brooks, K.L., Mossman, H.L., Chitty, J.A., Grant, A., 2015. Limited vegetation development on a created salt marsh associated with over-consolidated sediments and lack of topographic heterogeneity. Estuaries and Coasts 38, 325–336.

Broome, S.W., Craft, C.B., 2009. Tidal marsh creation. In: Perillo, G.M.E., Wolanski, E., Cahoon, D.R., Brinson, M.M. (Eds.), Coastal Wetlands: An Integrated Ecosystem Approach. Elsevier, Amsterdam, The Netherlands, pp. 715–736.

Broome, S.W., Craft, C.B., Seneca, E.D., 1988a. Creation and development of brackish-water marsh habitat. In: Zelazny, J., Feierabend, J.S. (Eds.), Proceedings of the Conference on Increasing Our Wetlands Resources. National Wildlife Federation, Washington, D.C., pp. 197–205.

Broome, S.W., Craft, C.B., Toomey Jr., W.A., 2001. Soil organic matter effects on infaunal community structure in restored and created tidal marshes. In: Weinstein, M.P., Kreeger, D.A. (Eds.), Concepts and Controversies in Tidal Marsh Ecology. Kluwer Academic Publishers, Dordrecht, The Netherlands, pp. 737–747.

Broome, S.W., Seneca, E.D., Woodhouse Jr., W.W., 1982. Establishing brackish marshes on graded upland sites in North Carolina. Wetlands 2, 152–178.

Broome, S.W., Seneca, E.D., Woodhouse Jr., W.W., 1983. The effects of source, rate and placement of nitrogen and phosphorus fertilizers on growth of *Spartina alterniflora* transplants in North Carolina. Estuaries 6, 212–226.

Broome, S.W., Seneca, E.D., Woodhouse Jr., W.W., 1986. Long-term growth and development of transplants of the salt-marsh grass *Spartina alterniflora*. Estuaries 9, 63–74.

Broome, S.W., Seneca, E.D., Woodhouse Jr., W.W., 1988b. Tidal salt marsh restoration. Aquatic Botany 32, 1–22.

Buchsbaum, R.N., Catena, J., Hutchins, E., James-Pirri, M.-J., 2006. Changes in salt marsh vegetation, *Phragmitess australis* and nekton in response to increased tidal flushing in a New England salt marsh. Wetlands 26, 544–557.

Burden, A., Garbutt, R.A., Evans, C.D., Jones, D.L., Cooper, D.M., 2013. Carbon sequestration and biogeochemical cycling in a salt marsh subject to coastal managed realignment. Estuarine. Coastal and Shelf Science 120, 12–20.

Callaway, J.C., Borgnis, E.L., Turner, R.E., Milan, C.S., 2012. Carbon sequestration and sediment accretion in San Francisco Bay tidal wetlands. Estuaries and Coasts 35, 1163–1181.

Cammen, L.M., 1976. Macroinvertebrate colonization of *Spartina* marshes artificially established on dredge spoil. Estuarine and Coastal Marine Science 4, 357–372.

Chambers, R.M., Osgood, D.T., Kalapasev, N., 2002. Hydrologic and chemical control of *Phragmites* growth in tidal marshes of SW Connecticut, USA. Marine Ecology Progress Series 239, 83–91.

Chapman, V.J., 1940. Studies in salt marsh ecology: comparisons with marshes on the east coast of North America. Journal of Ecology 28, 118–152.

Chapman, V.J., 1974. Salt Marshes and Salt Deserts of the World. Interscience Press, New York.

Chung, C.H., 1989. Ecological engineering of coastlines with salt-marsh plantations. In: Mitsch, W.J., Jorgensen, S.E. (Eds.), Ecological Engineering: An Introduction to Ecotechnology. John Wiley and Sons, Chichester, UK, pp. 255–289.

Chung, C.H., 1994. Creation of *Spartina* plantations as an effective measure for reducing coastal erosion in China. In: Mitsch, W.J. (Ed.), Global Wetlands: Old World and New. Elsevier, New York, pp. 443–452.

Chung, C.H., 2006. Forty years of ecological engineering with *Spartina* plantations in China. Ecological Engineering 27, 49–57.

Chung, C.H., Zhuo, R.Z., Xu, G.W., 2004. Creation of *Spartina* plantations for reclaiming Dongtai, China, tidal flats and offshore sands. Ecological Engineering 23, 135–150.

Covin, J.D., Zedler, J.B., 1988. Nitrogen effects on *Spartina foliosa* and *Salicornia virginica* in the salt marsh of Tijuana Estuary, California. Wetlands 8, 51–65.

Cowardin, L.M., Carter, V., Golet, F.C., LaRoe, E.T., 1992. Classification of Wetlands and Deepwater Habitats of the United States. United States Fish and Wildlife Service, Biological Services Program. Washington, D.C., FWS/OBS-79/31.

Craft, C.B., 1999. Co-development of wetland soils and benthic invertebrate communities following salt marsh creation. Wetlands Ecology and Management 8, 197–207.

Craft, C.B., 2001. Soil organic carbon, nitrogen and phosphorus as indicators of recovery in restored *Spartina* marshes. Ecological Restoration 19, 87–91.

Craft, C.B., 2007. Freshwater input structures soil properties, vertical accretion and nutrient accumulation of Georgia and United States (U.S.) tidal marshes. Limnology and Oceanography 52, 1220–1230.

Craft, C.B., Clough, J., Ehman, J., Joye, S., Park, D., Pennings, S., Guo, H., Machmuller, M., 2009. Forecasting the effects of accelerated sea level rise on tidal marsh ecosystem services. Frontiers in Ecology and the Environment 7, 73–78.

Craft, C.B., Sacco, J.N., 2003. Long-term succession of benthic infauna communities on constructed *Spartina alterniflora* marshes. Marine Ecology – Progress Series 257, 45–58.

Craft, C.B., Broome, S.W., Seneca, E.D., 1988. Nitrogen, phosphorus and organic carbon pools in natural and transplanted marshes. Estuaries 11, 272–280.

Craft, C.B., Broome, S.W., Seneca, E.D., 1989. Exchange of nitrogen, phosphorus and organic carbon between transplanted marshes and estuarine waters. Journal of Environmental Quality 18, 206–211.

Craft, C.B., Broome, S., Campbell, C., 2002. Fifteen years of vegetation and soil development after brackish-water marsh creation. Restoration Ecology 10, 248–258.

Craft, C.B., Megonigal, J.P., Broome, S.W., Cornell, J., Freese, R., Stevenson, R.J., Zheng, L., Sacco, J., 2003. The pace of ecosystem development of constructed *Spartina alterniflora* marshes. Ecological Applications 13, 1417–1432.

Craft, C.B., Seneca, E.D., Broome, S.W., 1991a. Loss on ignition and kjeldahl digestion, for estimating organic carbon and total nitrogen in estuarine marsh soils: calibration with, dry combustion. Estuaries 14, 175–179.

Craft, C.B., Seneca, E.D., Broome, S.W., 1991b. Porewater chemistry of natural and, created marsh soils. Journal of Experimental Marine Biology and Ecology 152, 187–200.

Croft, A.L., Leonard, L.A., Alphin, T.D., Cahoon, L.B., Posey, M.H., 2006. The effects of thin layer sand renourishment on tidal marsh processes: Masonboro Island, North Carolina. Estuaries and Coasts 29, 737–760.

Crooks, S., Schutten, J., Sheem, G.D., Pye, K., Davy, A.J., 2002. Drainage and elevation as factors in the restoration of salt marsh in Britain. Restoration Ecology 10, 591–602.

Culliton, T.J., 1998. Population, Distribution, Density, and Growth. NOAA's State of the Coast Report. National Oceanic and Atmospheric Administration (NOAA), Silver Spring, Maryland.

Currin, C.A., Paerl, H.W., 1998. Epiphytic nitrogen fixation associated with standing dead shoots of smooth cordgrass, *Spartina alterniflora*. Estuaries 21, 108–117.

Currin, C.A., Joye, S.B., Paerl, H.W., 1996. Diel rates of N_2–fixation and denitrification in a transplanted *Spartina alterniflora* marsh: implications for N flux dynamics. Estuarine, Coastal, and Shelf Science 42, 597–616.

DeLaune, R.D., Jugsujinda, A., Peterson, G.W., Patrick Jr., W.H., 2003. Impact of Mississippi River freshwater reintroduction on enhancing marsh accretionary processes in a Louisiana estuary. Estuarine Coastal and Shelf Science 58, 653–662.

DeLaune, R.D., Pezeshki, S.R., Pardue, J.H., Whitcomb, J.H., Patrick Jr., W.H., 1990. Some influences of sediment addition to a deteriorating salt marsh in the Mississippi River deltaic plain: a pilot study. Journal of Coastal Research 6, 181–188.

Desmond, J.S., Zedler, J.B., Williams, G.D., 2000. Fish use of tidal creek habitats in two southern California salt marshes. Ecological Engineering 14, 233–252.

Desrochers, D.W., Keagy, J.C., Cristol, D.A., 2008. Created versus natural wetlands: Avian communities in Virginia salt marshes. Ecoscience 15, 36–43.

Dionne, M., Short, F.T., Burdick, D.M., 1999. Fish utilization of restored, created and reference salt-marsh habitat in the Gulf of Maine. American Fisheries Society Symposium 22, 384–404.

DiQuinzio, D.A., Paton, P.W.C., Eddleman, W.R., 2002. Nesting success of salt marsh sharp-tailed sparrows in a tidally restricted salt marsh. Wetlands 22, 179–185.

Edwards, K.R., Proffitt, C.E., 2003. Comparison of wetland structural characteristics between created and natural salt marshes in southwest Louisiana, USA. Wetlands 23, 344–356.

Eertman, R.H.M., Kornman, B.A., Stikvoort, E., Verbeek, H., 2002. Restoration of the Sieperda tidal marsh in the Scheldt estuary, The Netherlands. Restoration Ecology 10, 438–449.

Emmerson, R.H.C., Manatunge, J.M.A., Macleod, C.L., Lester, J.N., 1997. Tidal exchange between Orplands managed retreat site and the Blackwater Estuary, Essex. Journal of Water and Environmental Management 11, 363–372.

Erfanzadeh, R., Garbutt, A., Petillon, J., Maelfait, J.P., Hoffman, M., 2010. Factors affecting the success of early salt-marsh colonizers: seed availability rather than site suitability and dispersal traits. Plant Ecology 206, 335–347.

Esteves, L.S., 2014. Managed Realignment: A Viable Long-term Coastal Strategy? Springer-Briefs in Environmental Science. Springer, New York. 19–34.

Fearnley, S., 2008. The soil physical and chemical properties of restored and natural back-barrier salt marshes on Isles Dernieres, Louisiana. Journal of Coastal Research 24, 84–94.

Fell, P.E., Murphy, K.A., Peck, M.A., Recchia, M.L., 1991. Re-establishment of *Melampus bidentatus* (Say) and other macroinvertebrates on a restored impounded tidal marsh: comparison of populations above and below the impoundment dike. Journal of Experimental Marine Biology and Ecology 152, 133–148.

Ford, M.A., Cahoon, D.R., Lynch, J.C., 1999. Restoring marsh elevation in a rapidly subsiding salt marsh by thin-layer deposition of dredged material. Ecological Engineering 12, 189–205.

French, P.W., 2006. Managed realignment–the developing story of a comparatively new approach to soft engineering. Estuarine, Coastal and Shelf Science 67, 409–423.

Frenkel, R.E., Morlan, J.C., 1989. Can we restore our salt marshes? Lessons from the Salmon River, Oregon. The Northwest Environmental Journal 7, 119–135.

Frost, J.W., Schleicher, T., Craft, C., 2009. Effects of nitrogen and phosphorus additions on primary production and invertebrate densities in a Georgia (USA) tidal freshwater marsh. Wetlands 29, 196–203.

Gallagher, J.L., Reimold, R.J., Linthurst, R.A., Pfeiffer, W.J., 1980. Aerial production, mortality and mineral accumulation-export dynamics in *Spartina alterniflora* and *Juncus roemerianus* stands in a Georgia salt marsh. Ecology 61, 303–312.

Gallego Fernandez, J.B., Garcia Novo, F., 2007. High-intensity versus low-intensity restoration alternatives of a tidal marsh in Gaudalquivir estuary, SW Spain. Ecological Engineering 30, 112–121.

Garbutt, A., Boorman, L.A., 2009. Managed realignment: recreating intertidal habitats on formerly reclaimed land. In: Perillo, G.M.E., Wolanski, E., Cahoon, D.R., Brinson, M.M. (Eds.), Coastal Wetlands: An Integrated Ecosystem Approach. Elsevier, Amsterdam, The Netherlands, pp. 763–795.

Garbutt, A., Wolters, M., 2008. The natural regeneration of salt marsh on formerly reclaimed land. Applied Vegetation Science 11, 335–344.

Garbutt, R.A., Reading, C.J., Wolters, M., Gray, A.J., Rothery, P., 2006. Monitoring the development of intertidal habitats on former agricultural land after managed realignment of coastal defenses at Tollesbury, Essex, UK. Marine Pollution Bulletin 53, 155–164.

Gibson, K.D., Zedler, J.B., Langis, R., 1994. Limited response of cordgrass (*Spartina foliosa*) to soil amendments in a constructed marsh. Ecological Applications 4, 757–767.

Gilmore, R.G., Cooke, D.W., Donohoe, C.J., 1982. A comparison of the fish populations and habitat in open and closed salt marsh impoundments in east-central Florida. Northeast Gulf Science 5, 25–37.

Gray, A., Simenstad, C.A., Bottom, D.L., Cornwell, T.J., 2002. Contrasting functional performance of juvenile salmon habitat in recovering wetlands of the Salmon River Estuary, Oregon, U.S.A. Restoration Ecology 10, 514–526.

Gross, W.J., 1964. Trends in water and salt regulation among aquatic and amphibious crabs. Biological Bulletin 127, 447–466.

Handa, I.T., Jefferies, R.L., 2000. Assisted revegetation trials in degraded salt-marshes. Journal of Applied Ecology 37, 944–958.

Havens, K.J., Varnell, L.M., Watts, B.D., 2002. Maturation of a constructed tidal marsh relative to two natural reference marshes over 12 years. Ecological Engineering 18, 305–315.

Hofstede, J.L.A., 2003. Integrated management of artificially created salt marshes in the Wadden Sea of Schleswig-Holstein, Germany. Wetlands Ecology and Management 11, 183–194.

Holland, H.D., 1978. The Chemistry of the Atmosphere and Oceans. John Wiley and Sons, New York.

Hough, P., Robertson, M., 2008. Mitigation under section 404 of the Clean Water Act: where it comes from, what it means. Wetlands Ecology and Management 17, 15–33.

Howes, B.L., Dacey, J.W.H., Goehringer, D.D., 1986. Factors controlling the growth form of *Spartina alterniflora*: feedbacks between above-ground production, sediment oxidation, nitrogen and salinity. Journal of Ecology 74, 881–898.

James-Pirri, M.J., Raposa, K.B., Catena, J.G., 2001. Diet composition of mummichogs, *Fundulus heteroclitus*, from restoring and unrestricted regions of a New England (U.S.A) salt marsh. Estuarine, Coastal and Shelf Science 53, 205–213.

Janousek, C.N., Currin, C.A., Levin, L.A., 2007. Succession of microphytobenthos in a restored coastal wetland. Estuaries and Coasts 30, 265–276.

Jefferies, R.L., Perkins, N., 1977. The effects on the vegetation of the additions of inorganic nutrients to salt marsh soils at Stiffkey, Norfolk. Journal of Ecology 65, 867–882.

Jefferies, R.L., Jano, A.P., Abraham, K.M., 2006. A biotic agent promotes large-scale catastrophic change in the coastal marshes of Hudson Bay. Journal of Ecology 94, 234–242.

Jivoff, P.R., Able, K.W., 2003. Evaluating salt marsh restoration in Delaware Bay: the response of blue crabs, Callinectes sapidus, at former salt hay farms. Estuaries 26, 709–719.

Johnson, D.S., York, H.L., 1915. The Relation of Plants to Tide Levels. Carnegie Institute of Washington, Washington, D.C.

Kelley, S., Mendelssohn, I.A., 1995. An evaluation of stabilized, water-based drill cuttings and organic compost as potential sediment sources for marsh restoration and creation in coastal Louisiana. Ecological Engineering 5, 497–517.

Ket, W.A., Schubauer, J.P., Craft, C.B., 2011. Effects of five years of nitrogen and phosphorus additions on a Zizaniopsis miliacea tidal freshwater marsh. Aquatic Botany 95, 17–23.

Kiehl, K., Esselink, P., Bakker, J.P., 1997. Nutrient limitation and plant species composition in temperate salt marshes. Oecologia 111, 325–330.

King, G.M., Klug, M.J., Wiegert, R.G., Chalmers, A.G., 1982. Relation of soil water movement and sulfide concentration to Spartina alterniflora production in a Georgia salt marsh. Science 218, 61–63.

Kirwan, M.L., Murray, A.B., Boyd, W.S., 2008. Temporary vegetation disturbance as an explanation for permanent loss of tidal wetlands. Geophysical Research Letters 35, L05403 http://dx.doi.org/1029/2007GL032681.

Kneib, R.T., 2000. Salt marsh ecoscapes and production transfers by estuarine nekton in the southeastern United States. In: Weinstein, M.P., Kreeger, D.A. (Eds.), Concepts and Controversies in Tidal Marsh Ecology. Kluwer Academic Publishers, Dordrecht, The Netherlands, pp. 267–291.

Knutson, P.L., Ford, J.C., Inskeep, M.R., 1981. National survey of planted salt marshes (vegetative stabilization and wave stress). Wetlands 1, 129–157.

Konisky, R.A., Burdick, D.M., Dionne, M., Neckles, H.A., 2006. A regional assessment of salt marsh restoration and monitoring in the Gulf of Maine. Restoration Ecology 14, 516–525.

Lane, R.R., Day Jr., J.W., Day, J.N., 2006. Wetland surface elevation, vertical accretion and subsidence at three Louisiana estuaries receiving diverted Mississippi River water. Wetlands 26, 1130–1142.

Langis, R., Zalejko, M., Zedler, J.B., 1991. Nitrogen assessments in a constructed and a natural salt marsh of San Diego Bay. Ecological Applications 1, 40–51.

La Peyre, M.K., Gossman, B., Piazza, B.P., 2009. Short- and long-term response of deteriorating brackish marsh and open-water ponds to sediment enhancement by thin-layer dredge disposal. Estuaries and Coasts 32, 390–402.

LaSalle, M.W., Landin, M.C., Sims, J.G., 1991. Evaluation of the flora and fauna of a Spartina alterniflora marsh established on dredged material in Winyah Bay, South Carolina. Wetlands 11, 191–208.

Leck, M.A., 2003. Seed-bank and vegetation development in a created tidal freshwater wetland on the Delaware River, Trenton, New Jersey, USA. Wetlands 23, 310–343.

Leck, M.A., Leck, C.F., 2005. Vascular plants of a Delaware River tidal freshwater wetland and adjacent terrestrial areas: seed bank and vegetation comparisons of reference and constructed marshes and annotated species list. Journal of the Torrey Botanical Society 132, 323–354.

Levin, L.A., Talley, T.S., 2002. Natural and manipulated sources of heterogeneity controlling early faunal development of a salt marsh. Ecological Applications 12, 1785–1802.

Levin, L.A., Talley, D., Thayer, G., 1996. Succession of macrobenthos in a created salt marsh. Marine Ecology Progress Series 141, 67–82.

Levings, C.D., Nishimura, D.H.J., 1997. Created and restored marshes in the lower Frazer River, British Columbia: summary of their functioning as fish habitat. Water Quality Research Journal of Canada 32, 599–618.

Li, B., Liao, C., Zhang, X., Chen, H., Wang, Q., Chen, Z., Gan, X., Wu, J., Zhao, B., Ma, Z., Cheng, X., Jiang, L., Chen, J., 2009. Spartina alterniflora invasions in the Yangtze River Estuary, China: an overview of current status and ecosystem effects. Ecological Engineering 35, 511–520.

Lindau, C.W., Hossner, L.R., 1981. Substrate characteristics of an experimental marsh and three natural marshes. Soil Science Society of America Journal 45, 1171–1176.

Lindig-Cisneros, R., Desmond, J., Boyer, K.E., Zedler, J.B., 2003. Wetland restoration thresholds: can a degradation transition be reversed with increased effort? Ecological Applications 13, 193–203.

Llewellyn, C., La Peyre, M., 2011. Evaluating ecological equivalence of created marshes: comparing structural indicators with stable isotope indicators of blue crab support. Estuaries and Coasts 34, 172–184.

Mander, L., Cutts, N.D., Allen, J., Masik, K., 2007. Assessing the development of newly created habitat for wintering estuarine birds. Estuarine, Coastal and Shelf Science 75, 163–174.

McCaffrey, R.J., Thomson, J., 1981. A record of the accumulation of sediment and trace metals in a Connecticut salt marsh. Advances in Geophysics 22, 165–236.

McDonald, K.B., 1977. Plant and animal communities of Pacific North American salt marshes. In: Chapman, V.J.H. (Ed.), Ecosystem of the World 1: Wet Coastal Ecosystems. Elsevier Scientific Publishing Company, Amsterdam, The Netherlands, pp. 167–191.

McKee, K.L., Patrick, W.H., 1989. The relationship of smooth cordgrass (Spartina alterniflora) to tidal datums: a review. Estuaries 11, 143–151.

Melvin, S.L., Webb Jr., J.W., 1998. Differences in the avian communities of natural and created Spartina alterniflora salt marshes. Wetlands 18, 59–69.

Mendelssohn, I.A., Kuhn, N.L., 2003. Sediment subsidy: effects on soil-plant responses in a rapidly submerging coastal salt marsh. Ecological Engineering 21, 115–128.

Mendelssohn, I.A., Morris, J.T., 2000. Eco-physiological controls on the productivity of Spartina alterniflora Loisel. In: Weinstein, M.P., Kreeger, D.A. (Eds.), Concepts and Controversies in Tidal Marsh Ecology. Kluwer Academic Publishers, Dordrecht, The Netherlands, pp. 59–80.

Meybeck, M., 1979. Concentrations des eaux fluviales en elements majeurs et apports en solution aux oceans. Review de Geologie Dynamique et de Geographie Physique 21, 215–246.

Miller, J.A., Simenstad, C.A., 1997. A comparative assessment of a natural and created estuarine slough as rearing habitat for juvenile Chinook and Coho salmon. Estuaries 20, 792–806.

Miller, R.L., Fram, M., Fugii, R., Wheeler, G., 2008. Subsidence reversal in a re-established wetland in the Sacramento-San Joaquin Delta, California, USA. San Francisco Estuary and Watershed Science 6, 1–20.

Minello, T.J., Rozas, L.P., 2002. Nekton in Gulf coast wetlands: fine-scale distributions, landscape patterns, and restoration implications. Ecological Applications 12, 441–455.

Minello, T.J., Webb Jr., J.W., 1997. Use of natural and created Spartina alterniflora salt marshes by fishery species and other aquatic fauna in Galveston Bay, Texas, USA. Marine Ecology Progress Series 151, 165–179.

Minello, T.J., Zimmerman, R.J., 1992. Utilization of natural and transplanted Texas salt marshes by fish and decapod crustaceans. Marine Ecology Progress Series 90, 273–285.

Minello, T.J., Zimmerman, R.J., Medina, R., 1994. The importance of edge for natant macrofauna in a created salt marsh. Wetlands 14, 184–198.

Mirin, O., Albizu, I., Amezaga, I., 2001. Effect of time on the natural regeneration of salt marsh. Applied Vegetation Science 4, 247–256.

Mitsch, W.J., Gosselink, J.G., 2000. Wetlands, third ed. John Wiley and Sons, New York.

Morgan, P.A., Short, F.C., 2002. Using functional trajectories to track constructed salt marsh development in the Great Bay Estuary, Maine/New Hampshire, U.S.A. Restoration Ecology 10, 461–473.

Morris, J.T., Sundareshwar, P.V., Nietch, C.T., Kjerfve, B., Cahoon, D.R., 2002. Responses of coastal wetlands to rising sea level. Ecology 83, 2869–2877.

Morzaria-Luna, H.N., Zedler, J.B., 2007. Does seed availability limit plant establishment during salt marsh restoration? Estuaries and Coasts 30, 12–25.

Moseman, S.M., Levin, L.A., Currin, C., Forder, C., 2004. Colonization, succession, and nutrition of macrobenthic assemblages in a restored wetland at Tijuana Estuary, California. Estuarine, Coastal and Shelf Science 60, 755–770.

Mossman, H.L., Davy, A.J., Grant, A., 2012. Does managed coastal realignment create salt marshes with 'equivalent biological characteristics' to natural reference sites? Journal of Applied Ecology 49, 1446–1456.

Moy, L.D., Levin, L.A., 1991. Are *Spartina* marshes a replaceable resource? A functional approach to evaluation of marsh creation efforts. Estuaries 14, 1–16.

Neff, K.P., Baldwin, A.H., 2005. Seed dispersal into wetlands: techniques and results for a restored tidal freshwater marsh. Wetlands 25, 392–404.

Neff, K.P., Rusello, K., Baldwin, A.H., 2009. Rapid seed bank development in restored tidal freshwater wetlands. Restoration Ecology 17, 539–548.

Niering, W.A., 1997. Tidal wetlands restoration and creation along the east coast of North America. In: Urbanska, K.M., Webb, N.R., Edwards, P.J. (Eds.), Restoration Ecology and Sustainable Development. Cambridge University Press, New York, pp. 259–285.

O'Brien, E.L., Zedler, J.B., 2006. Accelerating the restoration of vegetation in a southern California salt marsh. Wetlands Ecology and Management 14, 269–286.

Odum, E.P., 2000. Tidal marshes as outwelling/pulsing systems. In: Weinstein, M.P., Kreeger, D.A. (Eds.), Concepts and Controversies in Tidal Marsh Ecology. Kluwer Academic Publishers, Dordrecht, The Netherlands, pp. 3–7.

Odum, W.E., 1988. Comparative ecology of tidal freshwater and salt marshes. Annual Review of Ecology and Systematics 19, 147–176.

Odum, W.E., Smith III, T.J., Hoover, J.K., McIvor, C.C., 1984. The Ecology of Tidal Freshwater Marshes of the Eastern United States. A Community Profile. U.S. Fish and Wildlife Service, Washington, D.C. FWS/OBS-83/17.

Orr, M., Crooks, S., Williams, P.B., 2003. Will restored tidal marshes be sustainable? San Francisco Estuary and Water Science 6, 1–33.

Parkinson, R.W., DeLaune, R.D., Hutcherson, C.T., Stewart, J., 2006. Tuning surface water management and wetland restoration programs with historic sediment accumulation rates: Merritt Island National Wildlife Refuge, East-central Florida, USA. Journal of Coastal Research 22, 1268–1277.

Pennings, S.C., Alber, M., Alexander, C.R., Booth, M., Burd, A., Cai, W., Craft, C., DePratter, C.B., Di Iorio, D., Hopkinson, C.S., Joye, S.B., Meile, C.D., Moore, W.S., Silliman, B.R., Thompson, V., Wares, J.P., 2012. South Atlantic tidal wetlands. In: Batzer, D., Baldwin, A. (Eds.), Wetland Habitats of North America: Ecology and Conservation Concerns. University of California Press, Berkeley, CA, pp. 45–62.

Pethick, J., 2002. Estuarine and tidal wetland restoration in the United Kingdom: policy versus practice. Restoration Ecology 10, 431–437.

Phillips, W.A., Eastman, F.D., 1959. Riverbank stabilization in Virginia. Journal of Soil and Water Conservation 14, 257–259.

Phleger, C.R., 1971. Effects of salinity on growth of salt marsh grass. Ecology 52, 908–911.

Piehler, M.F., Currin, C.A., Cassanova, R., Paerl, H.W., 1998. Development and N_2-fixing activity of benthic microbial community in transplanted *Spartina alterniflora* marshes in North Carolina. Restoration Ecology 6, 290–296.

Poach, M.E., Faulkner, S.P., 1998. Soil phosphorus characteristics of created and natural wetlands in the Atchafalaya Delta, LA. Estuarine, Coastal and Shelf Science 46, 195–203.

Portnoy, J.W., Giblin, A.E., 1997a. Biogeochemical effects of seawater restoration to diked salt marshes. Ecological Applications 7, 1054–1063.

Portnoy, J.W., Giblin, A.E., 1997b. Effects of historic tidal restrictions on salt marsh sediment chemistry. Biogeochemistry 36, 275–303.

Portnoy, J.W., Giblin, A.E., 1999. Salt marsh diking and restoration: biogeochemical implications of altered wetland hydrology. Environmental Management 24, 111–120.

Posey, M.H., Alphin, T.D., Powell, C.M., 1997. Plant and infaunal communities associated with a created marsh. Estuaries 20, 42–47.

Pringle, C., Vellidis, G., Heliotis, F., Bandacu, D., Cristofor, S., 1993. Environmental problems of the Danube delta. American Scientist 31, 350–361.

Rahmstorf, S., Cazenave, A., Church, J.A., Hansen, J.E., Keeling, R.F., Parker, D.E., Somerville, R.C.J., 2007. Recent climate observations compared to projections. Science 316, 709.

Ranwell, D.S., 1967. World resources of *Spartina townsendii (sensu lato)* and economic use of *Spartina* marshland. Journal of Applied Ecology 4, 239–256.

Ranwell, D.S., 1972. Ecology of Salt Marshes and Sand Dunes. Chapman and Hall, London.

Raposa, K., 2002. Early responses of fishes and crustaceans to restoration of a tidally restricted New England salt marsh. Restoration Ecology 10, 665–676.

Raposa, K.B., 2008. Early ecological responses to hydrologic restoration of a tidal pond and salt marsh complex in Narragansett Bay, Rhode Island. Journal of Coastal Research 55, 180–192.

Reader, J., Craft, C., 1999. Comparison of wetland structure and function on grazed and ungrazed salt marshes. Journal of the Elisha Mitchell Scientific Society 115, 236–249.

Reimold, R.J., Linthurst, R.A., Wolf, P.L., 1975. Effects of grazing on a salt marsh. Biological Conservation 8, 105–125.

Roman, C.T., Garvine, R.W., Portnoy, J.W., 1995. Hydrological modeling as a basis for ecological restoration of salt marshes. Environmental Management 19, 559–566.

Rozas, L.P., Minello, T.J., 2001. Marsh terracing as a wetland restoration tool for creating fishery habitat. Wetlands 21, 327–341.

Rozas, L.P., Caldwell, P., Minello, T.J., 2005. The fishery value of salt marsh restoration projects. Journal of Coastal Research 40, 37–50.

Rozas, L.P., Minello, T.J., Zimmerman, R.J., Caldwell, P., 2007. Nekton populations, long-term wetland loss, and the recent effect of habitat restoration in Galveston Bay, Texas. Marine Ecology Progress Series 344, 119–130.

Rulifson, R.A., 1991. Finfish utilization of man-initiated and adjacent natural creeks of South Creek Estuary, North Carolina using multiple gear types. Estuaries 14, 447–464.

Sacco, J.N., Seneca, E.D., Broome, S.W., 1994. Infauna community development of artificially established salt marshes in North Carolina. Estuaries 17, 489–500.

Scatolini, S.R., Zedler, J.B., 1996. Epibenthic invertebrates of natural and constructed marshes of San Diego Bay. Wetlands 16, 24–37.

Schrift, A.M., Mendelssohn, I.A., Materne, M.D., 2008. Salt marsh restoration with sediment slurry amendment following a drought-induced large-scale disturbance. Wetlands 28, 1071–1085.

Seigel, A., Hatfield, C., Hartman, J.M., 2005. Avian response to restoration of urban tidal marshes in the Hackensack Meadowlands, New Jersey. Urban Habitats 3, 87–116.

Seneca, E.D., Broome, S.W., 1982. Ecological study of the Amoco Cadiz oil spill. In: Report of the NOAA-CNEXO Joint Scientific Commission. National Oceanic and Atmospheric Administration. U.S. Department of Commerce, Washington, D.C.

Seneca, E.D., Broome, S.W., 1992. Restoring tidal marshes in North Carolina and France. In: Thayer, G.W. (Ed.), Restoring the Nation's Marine Environment. Maryland Sea Grant, College Park, Maryland, pp. 53–78.

Seneca, E.D., Broome, S.W., Woodhouse Jr., W.W., 1985. The influence of duration-of-inundation on development of a man-initiated *Spartina alterniflora* Loisel marsh in North Carolina. Journal of Experimental Marine Biology and Ecology 94, 259–268.

Seneca, E.D., Broome, S.W., Woodhouse Jr., W.W., Cammen, L.M., Lyon III, J.T., 1976. Establishing *Spartina alterniflora* marsh in North Carolina. Environmental Conservation 3, 185–188.

Shafer, D.J., Streever, W.J., 2000. A comparison of 28 natural and dredged material salt marshes in Texas with an emphasis on geomorphic variables. Wetlands Ecology and Management 8, 353–366.

Sharp, W.C., Vaden, J., 1970. 10-year report on sloping techniques used to stabilize eroding tidal river banks. Shore and Beach 38, 31–35.

Shreffler, D.K., Simenstad, C.A., Thom, R.M., 1990. Temporary residence by juvenile salmon in a restored estuarine wetland. Canadian Journal of Fisheries and Aquatic Sciences 47, 2079–2084.

Simenstad, C.A., Thom, R.M., 1996. Functional equivalency trajectories of the restored Gog-Le-Hi-Te estuarine wetland. Ecological Applications 6, 38–56.

Slavin, P., Shisler, J., 1983. Avian utilization of a tidally restored salt hay farm. Biological Conservation 26, 271–285.

Slocum, M.G., Mendelssohn, I.A., Kuhn, N.L., 2005. Effects of sediment slurry enrichment on salt marsh rehabilitation: plant and soil responses over seven years. Estuaries 28, 519–528.

Smith, S.M., Warren, R.S., 2012. Vegetation responses to tidal restoration. In: Roman, C.T., Burdick, D.M. (Eds.), Tidal Marsh Restoration: A Synthesis of Science and Management. Island Press, Washington, D.C., pp. 59–80.

Smith, T.J., Odum, W.E., 1981. The effects of grazing by snow geese on coastal salt marshes. Ecology 62, 98–106.

Stagg, C.L., Mendelssohn, I.A., September 2010. Restoring ecological function to a submerged salt marsh. Restoration Ecology 18 (Suppl. S1), 10–17.

Streever, W.J., 2000. *Spartina alterniflora* marshes on dredged material: a critical review of the ongoing debate over success. Wetlands Ecology and Management 8, 295–316.

Sullivan, M.J., Daiber, F.C., 1974. Response in production of cord grass, *Spartina alterniflora*, to inorganic nitrogen and phosphorus fertilizer. Chesapeake Science 14, 121–123.

Swamy, V., Fell, P.E., Body, M., Keaney, M.B., Nyaku, M.W., Mcilvain, E.C., Keen, A.L., 2002. Macroinvertebrate and fish populations in a restored impounded salt marsh 21 years after the reestablishment of tidal flooding. Environmental Management 29, 516–530.

Tanner, C.D., Cordell, J.R., Rubey, J., Tear, L.M., 2002. Restoration of freshwater intertidal habitat functions at Spencer Island, Everett, Washington. Restoration Ecology 10, 564–576.

Thompson, S.P., Paerl, H.W., Go, M.C., 1995. Seasonal patterns of nitrification and denitrification in a natural and a restored salt marsh. Estuaries 18, 399–408.

Tong, C., Baustian, J.J., Graham, S.A., Mendelssohn, I.A., 2013. Salt marsh restoration with sediment-slurry application: effects on benthic macroinvertebrates and associated soil-plant variables. Ecological Engineering 51, 151–160.

Townsend, I., Pethick, J., 2002. Estuarine flooding and managed retreat. Philosophical Transactions of the Royal Society of London 360, 477–495.

Turner, R.E., Streever, B., 2002. Approaches to Coastal Wetland Restoration: Northern Gulf of Mexico. SPB Academic Publishing BV, The Hague, The Netherlands.

Turner, R.E., Swenson, E.M., Lee, J.M., 1994. A rationale for coastal wetland restoration through spoil bank management in Louisiana, USA. Environmental Management 18, 271–282.

Valiela, I., Teal, J.M., 1974. Nutrient limitation in salt marsh vegetation. In: Reimold, R.J., Queen, W.H. (Eds.), Ecology of Halophytes. Academic Press, New York.

Van Proosdij, D., Lundholm, J., Neatt, N., Bowron, T., Graham, J., 2010. Ecological re-engineering of a freshwater impoundment for salt marsh restoration in a hypertidal system. Ecological Engineering 36, 1314–1332.

Van Staveren, M.F., Warner, J.F., van Tatenhove, J.P.M., Wester, P., 2014. Let's bring in the floods: depoldering in the Netherlands as a strategy for long-term delta survival? Water International 39, 686–700.

Varty, A.K., Zedler, J.B., 2008. How waterlogged microsites help an annual plant persist among salt marsh perennials. Estuaries and Coasts 31, 300–312.

Verhoeven, J.T.A., Whigham, D.F., van Logtestijn, R., O'Neill, J., 2001. A comparative study of nitrogen and phosphorus cycling in tidal and non-tidal riverine wetlands. Wetlands 21, 210–222.

Waisel, Y., 1972. Biology of Halophytes. Academic Press, New York.

Wallace, K.J., Callaway, J.C., Zedler, J.B., 2005. Evolution of tidal creek networks in a high sedimentation environment: a 5-year experiment at Tijuana Estuary, California. Estuaries 28, 795–811.

Warren, R.S., Fell, P.E., Rozra, R., Brawley, A.H., Orsted, A.C., Olson, E.T., Swamy, V., Niering, W.A., 2002. Salt marsh restoration in Connecticut: 20 years of science and management. Restoration Ecology 10, 497–513.

Webb, J.W., Newling, C.J., 1985. Comparison of natural and man-made salt marshes in Galveston Bay complex, Texas. Wetlands 4, 75–86.

Webb, J.W., Dodd, J.D., Koerth, B.H., Weichert, A.T., 1984. Seedling establishment of *Spartina alterniflora* and *Spartina patens* on dredged material in Texas. Gulf Research Reports 7, 325–329.

West, J.M., Zedler, J.B., 2000. Marsh-creek connectivity: fish use of a tidal salt marsh in southern California. Estuaries 23, 699–710.

Whigham, D.F., Baldwin, A.H., Barendregt, A., 2009. Tidal freshwater wetlands. In: Perillo, G.M.E., Wolanski, E., Cahoon, D.R., Brinson, M.M. (Eds.), Coastal Wetlands: An Integrated Ecosystem Approach. Elsevier, Amsterdam, The Netherlands, pp. 515–533.

Wiegert, R.G., Freeman, B.J., 1990. Tidal Salt Marshes of the Southeastern Atlantic Coast: A Community Profile. U.S. Department of Interior. Fish and Wildlife Service, Washington, D.C. Biological Report 85(7.29).

Williams, G.D., Zedler, J.B., 1999. Fish assemblage composition in constructed and natural tidal marshes of San Diego Bay: relative influence of channel morphology and restoration history. Estuaries 22, 702–716.

Williams, P.B., Orr, M.K., 2002. Physical evolution of restored breached levee salt marshes in the San Francisco Bay Estuary. Restoration Ecology 10, 527–542.

Wolters, M., Garbutt, A., Bakker, J.P., 2005a. Plant colonization after managed realignment: the relative importance of diaspore dispersal. Journal of Applied Ecology 42, 770–777.

Wolters, M., Garbutt, A., Bakker, J.P., 2005b. Salt-marsh restoration: evaluating the success of de-embankments in north-west Europe. Biological Conservation 123, 249–268.

Wolters, M., Garbutt, A., Bekker, R.M., Bakker, J.P., Carey, P.D., 2008. Restoration of salt-marsh vegetation in relation to site suitability species pool and dispersal traits. Journal of Applied Ecology 45, 904–912.

Woodhouse Jr., W.W., 1982. Coastal sand dunes of the U.S. In: Lewis III, R.R. (Ed.), Creation and Restoration of Coastal Plant Communities. CRC Press, Boca Raton, Florida, pp. 1–44.

Woodhouse Jr., W.W., Knutson, P.L., 1982. Atlantic coastal marshes. In: Lewis III, R.R. (Ed.), Creation and Restoration of Coastal Plant Communities. CRC Press, Boca Raton, Florida, pp. 45–70.

Wosniak, A.S., Roman, C.T., Wainright, S.G., McKinney, R.A., Jane-Pirr, M., 2006. Monitoring food web changes in tide-restored salt marshes: a stable carbon isotope approach. Estuaries and Coasts 29, 568–578.

Zedler, J.B., 1993. Canopy architecture of natural and planted cordgrass marshes: selecting habitat evaluation criteria. Ecological Applications 3, 123–138.

Zedler, J.B., 1995. Salt marsh restoration: lessons from California. In: Cairns Jr., J. (Ed.), Reha-bilitating Damaged Ecosystems. Lewis Publishers, Boca Raton, Florida, pp. 75–95.

Zedler, J.B., 2001. Introduction. In: Zedler, J.B. (Ed.), Handbook for Restoring Tidal Wetlands. CRC Press, Boca Raton, Florida, pp. 1–37.

Zedler, J.B., Callaway, J.C., 1999. Tracking wetland mitigation: do mitigation sites follow desired trajectories? Restoration Ecology 7, 69–73.

Zedler, J.B., Langis, R., 1991. Comparisons of constructed and natural salt marshes of San Diego Bay. Restoration and Management Notes 9, 21–25.

Zedler, J.B., Callaway, J.C., Desmond, J.S., Vivian-Smith, G., Williams, G.D., Sullivan, G., Brewster, A.E., Bradshaw, B.K., 1999. Californian salt-marsh vegetation: an improved model of spatial pattern. Ecosystems 2, 19–35.

Zedler, J.B., Morzaria-Luna, H., Ward, K., 2003. The challenge of restoring vegetation on tidal, hypersaline substrates. Plant and Soil 253, 259–272.

Zeug, S.C., Shervette, V.R., Hoeinghaus, D.J., Davis III, S.E., 2007. Nekton assemblage struc-ture in natural and created marsh edge habitats of the Guadalupe Estuary, Texas, USA. Estuarine, Coastal and Shelf Science 71, 457–466.

Zhi, Y., Li, H., An, S., Zhao, L., Zhou, C., Deng, Z., 2007. Interspecific competition: *Spartina alterniflora* is replacing *Spartina anglica* in coastal China. Estuarine, Coastal and Shelf Science 74, 437–448.

Mangroves

Chapter Outline

Introduction

Mangroves are forested wetlands of estuaries and coastal shorelines of the tropics and subtropics. Similar to tidal marshes, they are inundated with a mixture of seawater and freshwater of varying salinity. Mangroves provide important ecosystem services such as storm protection and wave dissipation, biological productivity that provides a fuel source for cooking and heating in developing countries, and habitat/nursery grounds for juvenile species of finfish. The word mangrove is used to describe salt- and flood-tolerant trees and shrubs that live in the intertidal zone. The word "mangrove" is of Portuguese origin, from the word "mangue" for tree and from the English word "grove" (Dawes, 1998). Trees that comprise mangrove forests are not necessarily closely related taxonomically. However, they share common ecophysiological traits that enable them to live in anaerobic and saline soils. According to Tomlinson (1986), mangrove species have most or all of the following features: (1) trees are a major component of the community; (2) species exhibit morphological adaptations of aerial roots and vivipary; and (3) they exhibit adaptations to exclude salt. Duke (1992) defines mangroves as "a tree, shrub, palm or ground fern, generally exceeding one half meter in height and which normally grows above mean sea level in the intertidal zone of marine coastal environments or estuarine margins."

Mangrove forests contain a variety of tree species. Paleotropical forests contain at least 24 genera and 58 species (Table 9.1), including those belonging to the genus *Rhizophora, Avicennia, Sonneratia, Bruguiera,* and *Kandelia*. Mangrove forests of the Neotropics are less diverse with 5 genera and 12 species. Common neotropical mangroves include *Rhizophora* and *Avicennia* (Table 9.1). Spalding et al. (2010) identified 73 mangrove species worldwide including several Pteridaceae or ferns of the genus

Table 9.1 Common Mangrove Genera and Species of the Neotropics and Paleotropics

Neotropics (5 Genera, 12 Species)	Paleotropics (24 Genera, 58 Species)
Avicennia germinans	*Acanthus* sp.
Avicennia bicolor	*Aegialitis* sp.
Avicennia schauriana	*Aegicercus* sp.
Conocarpus erectus	*Aglaia* sp.
Laguncularia racemosa	*Avicennia marina*
Nypa fruticans	*Avicennia officinalis*
Pelliciera rhizophorae	*Avicennia alba*
Rhizophora mangle	*Bruguiera gymnorhiza*
Rhizophora racemosa	*Camptostemon* sp.
Rhizophora x harrisonii	*Ceriops decandra*
	Ceriops tagal
	Diospiros sp.
	Dolichandrone sp.
	Exocoecaria sp.
	Heritiera sp.
	Kandelia candel
	Lumnizera sp.
	Mora sp.
	N. fruticans
	Osbornia sp.
	Pelliciera sp.
	Pemphis sp.
	Rhizophora mucronata
	Rhizophora apiculata
	Rhizophora stylosa
	Scyphiphora sp.
	Sonneratia apetala
	Sonneratia caseolaris
	Xylocarpus sp.

Note the greater diversity of species in the Paleotropics.
From Tomlinson (1986), Dawes (1998), Field (1998a,b), Mendelssohn and McKee (2000).

Acrostichum, highlighting the evolving addition of species to this diverse forest community. Rhizophoraceae is the largest family, with 16 genera, including *Rhizophora*, *Bruguiera*, *Kandelia*, and *Ceriops* (Dawes, 1998).

In developing countries, mangroves are an important source of firewood, charcoal, construction/building materials, and medicines (Bandaranayake, 1998; Walters et al., 2008). The role of mangroves as nursery and feeding grounds for consumable finfish and shellfish is well documented (Odum and Heald, 1972; Lewis et al., 1985; Primavera, 1998; Shervette et al., 2007), although the exact benefits (Manson et al., 2005) and precise valuation of the resource (Barani and Hambrey, 1999) is difficult to ascertain. Many species of juvenile shrimp and finfish are found in mangroves, supporting their function as a nursery (Hogarth, 2007). There also are strong linkages existing between fish of coral reefs whose juveniles use mangrove habitat (Nagelkerken et al., 2000; Mumby et al., 2004). Overall, it appears that the benefits of mangroves to higher level trophic groups stem from an abundance of food, shelter, and low predation pressure (Sheridan and Hays, 2003; Nagelkerken et al., 2008). Outwelling of plant detritus and nutrients from mangrove forests occurs, but the fate of this material in the nearshore environment is unclear (Lee, 1995).

The contribution of mangrove forests to shoreline stabilization (Davis, 1940) and protection from hurricanes, tsunamis, and wave energy is increasingly recognized (Othman, 1994; Dahdouh-Guebas et al., 2005) and mangroves have been planted in many areas to protect shorelines from erosion and to attenuate storm waves (Teas, 1977). On the southeast coast of India, villages where mangroves were present along the shore suffered much less damage and mortality from the December 26, 2004 Indiana Ocean Tsunami than villages without this protective buffer (Danielson et al., 2005; Kathiresan and Rajendran, 2005). The presence of mangrove forests also is associated with the reduction in casualties from land falling cyclones (Das and Vincent, 2009). Danielson et al. (2005) and Alongi (2008) suggested that mangrove forest buffers of at least 100 m in width significantly reduce wave flow pressure of tsunamis. Likewise, Mazda et al. (1997) reported that a 100-m wide mangrove forest reduces wave energy by as much as 20%.

In aggregate, mangroves are estimated to be worth $13,837–$15,182/ha/year, based on services related to raw materials/food, coastal protection/erosion control, maintenance of fisheries, and C sequestration (Barbier et al., 2011). Most studies, however, value mangroves at lesser though not inconsequential amounts. Brander et al. (2012) synthesized 130 value estimates for mangrove forests mostly in Southeast Asia. The mean value was US$4185/ha/year with a median value of $237/ha/year. Walton et al. (2006b) valued the direct benefits—timber, fisheries, and tourism—of mangroves at US$315/ha/year whereas Tognella-de-Rosa et al. (2006) estimated direct benefits of less than US$8/ha/year.

Mangrove forests are disappearing at an alarming rate as humans cut the forests for firewood and convert them to agriculture such as rice, aquaculture, especially shrimp ponds, palm oil plantations, salt pans, and development for tourism (Ellison and Farnsworth, 1996; de Graaf and Xuan, 1998; Semesi, 1998; Barbier and Cox, 2003; Spalding et al., 2010). Other causes of mangrove loss include urban development, mining, and overexploitation for timber, fish, crustaceans, and shellfish (Alongi, 2002). Climate change, especially rising sea level, is expected to accelerate the loss of mangroves, especially in areas where sediment needed to support continued soil accretion is limited (Field, 1995; Gilman et al., 2008). In particular, island mangroves may have a more difficult time keeping pace with sea level rise because they lack the sediment supply of deltaic regions (Ellison, 2003). Furthermore, coastal development will limit their ability to migrate inland as sea level rises. Overall, the rate of mangrove loss is estimated at 1–2% per year worldwide, but the rate of loss is even greater in developing countries where more than 90% of the world's mangroves are located (Duke et al., 2007).

Because of the widespread loss of mangroves, many countries have developed strategies to protect them. The most widely used strategy is the establishment of protected areas for conservation purposes including protection of adjacent terrestrial habitat (Spalding et al., 2010) though lack of oversight and poor enforcement of regulations often limits their effectiveness (Chong, 2006). International conventions such as World Heritage sites, the Ramsar convention on wetlands (see Chapter 2, Definitions), UNESCO Man and Biosphere program, and local community involvement also are used to protect mangroves (Spalding et al., 2010). Restoration also is a key strategy. It is undertaken by planting seedlings and saplings on degraded lands and on mudflats. In developing countries, extensive acreage of mangrove is planted to restore stocks of firewood, rehabilitate degraded areas, and protect against wind and wave action from hurricanes, typhoons and even tsunamis. In the US, mangrove habitat is created and restored to compensate for habitat loss under the Clean Water Act.

Ecology of Mangroves

Mangroves are adapted to waterlogged, saline soils. They possess prop or stilt roots, pneumatophores, knee roots, and buttressed roots to live in anaerobic soils (Dawes, 1998; Spalding et al., 2010) (Figure 9.1). Many species have salt-secreting glands and also exclude salt at the roots (Waisel, 1972). They store salt in bark and in older leaves and produce organosaline compounds to compensate for salinity in the external environment (Saenger, 2002).

Figure 9.1 Photos illustrating mangrove adaptations to anaerobic conditions. Note the (a) prop roots (Daintree River, Queensland, Australia), and (b) pneumatophores (Mekong River Delta, Vietnam).
Photo credit: Chris Craft.

Mangroves produce water-dispersed propagules that exhibit vivipary, whereby the seed germinates on the parent plant, begins to grow out of the seed coat, and then out of the fruit while still attached to the parent (Tomlinson, 1986). This feature is found in all Rhizophoraceae, including *Rhizophora*, *Bruguiera*, *Kandelia*, and *Ceriops* (Tomlinson, 1986). Viviparous seeds differ from other seeds in that there is no dormant period. After a time, the spindle-shaped seedling drops from the parent where it floats horizontally. Over time, its orientation gradually shifts to the vertical axis, improving its chances for intercepting and taking root in the muddy substrate. Other species such as *Avicennia* and *Aegiceras* also produce seeds that germinate on the parent plant (Hogarth, 2007).

Like tidal marsh vegetation, mangrove species form zones along gradients of inundation and salinity (Figure 9.2). In the Neotropics, red mangrove, *Rhizophora mangle*,

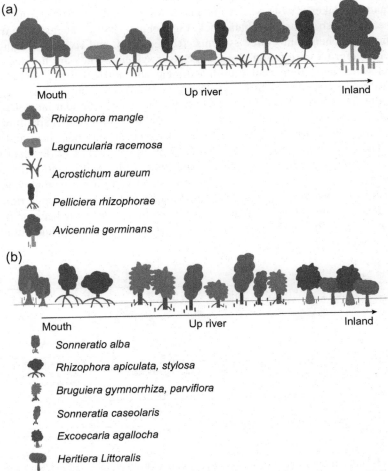

Figure 9.2 Simplified zonation of mangrove species in the Neotropics (Panama–Caribbean Sea) and Paleotropics (far North Queensland) mangrove forests.
After Davis (1940) and Duke et al. (1998).

grows seaward of other species (Figure 9.2(a)). It is adapted to frequent inundation with water of high salinity. Inland of *Rhizophora*, black mangrove, *Avicennia germinans*, is the dominant species. It grows in areas with less frequent inundation. At the upland edge, in areas that are infrequently inundated, white mangrove, *Laguncularia racemosa*, and gray mangrove, *Conocarpus erectus*, are found. In the Paleotropics, similar patterns of zonation exist (Figure 9.2(b)). At the water's edge, *Sonneratia alba* and *Avicennia marina* dominate (Thampanya et al., 2002). Inland of this zone, *Rhizophora* spp., *Bruguiera gymnorhiza*, and others are found. At the most inland locations, *Ceriops* sp. is found along with many other species. These patterns of zonation vary with distance from the water's edge but also longitudinally along riverine estuaries where a number of low salinity mangrove species, including *Exoecaris*, *Heritiera*, are found.

Mangrove forests are classified on the basis of geomorphology and hydrology. Lugo and Snedaker (1974) parsed mangrove forests into six types: fringe, basin, riverine, overwash, dwarf, and hammock. Lugo et al. (1989) condensed the original classification scheme into three systems: fringe, basin, and riverine (Figure 9.3). Riverine mangroves are found along large river systems. They are flooded mostly by river water and often receive high sediment and nutrient loads, making them the most productive mangrove forests (Semesi, 1998). Riverine mangroves are the most extensive geomorphic type and examples include the Sundarbans of Bangladesh, Niger delta of West Africa, and Orinoco delta of Venezuela (Spalding et al., 2010). Fringe mangroves exist at the immediate coast, on shorelines and island edges. They are subject to tidal inundation with mostly seawater. Productivity is less than in riverine mangroves. Basin mangroves are located inland of fringe and riverine mangroves. Inundation is infrequent, mostly from storm tides and wind tides. They often are more saline than fringe mangroves because evaporation in the basin condenses salts. Basin mangroves often are more anaerobic because they are not well flushed. Productivity is low because of the high salinity and sulfide and lack of connectivity.

	Fringe	Riverine	Basin
Species	*Rhizophora mangle*	*Rhizophora mangle*	*Avicennia germinans*
Hydrology	Tidal;frequent	Tidal–river pulsing;infrequent	Spring-storm tide; Infrequent
Salinity (psu)	20-35	0-20	>35
Canopy Height (m)	8-10	10-12	<5-7
Litterfall (g/m²/yr)	900	1200	<700

Figure 9.3 Diagram illustrating geomorphic position; hydrology; porewater salinity and sulfide; and productivity (canopy height, litterfall) of fringe, riverine, and basin mangrove forests. After Mendelssohn and McKee (2000).

Based on their landscape position, Ewel et al. (1998) suggested that different mangrove systems provide different ecosystem services. Riverine mangroves, being the most productive, are important for plant and animal productivity. Fringe mangroves are important for shoreline protection. Basin mangroves are important sinks for nutrients both from natural sources and from anthropogenic sources from the surrounding terrestrial landscape. All systems are a source of wood products (Ewel et al., 1998).

Global Distribution and Loss through Time

Mangrove forests are mostly tropical, with some excursions into subtropical regions such as Florida, South Africa, Victoria Australia, and southern Japan where warm tropical currents transport propagules (Tomlinson, 1986). In the Neotropics, the distribution of mangroves ranges from 32° N to 28° S (Mendelssohn and McKee, 2000). At the edge of its range in the Neotropics, black mangrove (*Avicennia*) dominates but in a stunted form caused by periodic freezes (Tomlinson, 1986). It is more tolerant of cold than other species and is found in the northern Gulf of Mexico whereas the other species are not (Sherrod and McMillan, 1985; Sherrod et al., 1986). In the Paleotropics, mangroves extend from 32° N to 38° S (Krauss et al., 2008). On the Neotropical Pacific and the African coasts, mangroves are limited by arid climate (Saenger, 2002).

Global estimates of mangrove area vary. The Food and Agricultural Organization (FAO) of the United Nations inventoried mangroves and compared their estimates with previous inventories (FAO, 2007). Worldwide, they estimated mangrove area at 157,050 km^2 in 2005 (Table 9.2). By far, the largest areas were in Southeast Asia. Spalding et al. (2010) pegged mangrove area at 152,361 km^2, slightly less than the FAO estimate. Giri et al. (2011) estimated total mangrove area of 137,760 km^2. The largest extent of mangroves occurred in Asia (42%) followed by Africa (20%), North and Central America (15%), Oceania (12%), and South America (11%) (Table 9.2). Many of the world's largest areas of mangroves are located in the deltas of the Bay of Bengal, the Ganges, Irrawaddy, and Mekong Rivers (Blasco et al., 2001). Among countries in the

Table 9.2 **Worldwide Area of Mangroves by Region**

	Area (km^2)	Percentage of Total
North and Central America	22,402	14.7
South America	23,882	15.7
East and South Africa	7917	5.2
Central and West Africa	20,040	13.2
Middle East	624	0.4
East Asia	215	0.1
Southeast Asia	51,049	33.5
South Asia	10,344	6.8
Australasia	10,171	6.7
Pacific Ocean	5717	3.7

From Spalding et al. (2010).

Table 9.3 Recent Area of Mangroves in the 15 Most Mangrove-Rich Countries

	Country	Area (km^2) Giri et al. (2011)	Area (km^2) Spalding et al. (2010)	(%)
Paleotropics	Indonesia	31,139	31,894	20.9
	Australia	9780	9910	6.5
	Malaysia	5054	7097	4.7
	Myanmar	4946	5029	3.3
	Papua New Guinea	4801	4265	2.6
	Bangladesh	4366	4951	3.2
	India	3683	4326	2.8
	Madagascar	2781	–	–
	Philippines	2631	–	–
	Nigeria	6.537	7356	4.8
	Guinea Bissau	3387	–	–
	Mozambique	3189	–	–
Neotropics	Brazil	9627	13,000	8.5
	Mexico	7419	7701	5.0
	Cuba	4215	4944	3.3
	Columbia	–	4079	2.7

From Giri et al. (2011) and Spalding et al. (2010). Percent of total area is derived from Spalding et al. (2010).

Paleotropics, Indonesia contained the largest area of mangroves, followed by Australia (Table 9.3). Nigeria contained the largest area in Africa. In the Neotropics, Brazil, Mexico, and Cuba contain the most extensive mangrove forests.

There has been widespread loss of mangrove habitat as it was cut for fuel wood and converted to aquaculture, notably shrimp ponds. It is estimated that 20% of mangrove area was lost between 1980 and 2005 (FAO, 2007) though the rate of loss slowed in recent years. About 1850 km^2 were lost annually in the 1980s or 1.4% of the total area and declined to 1185 km^2/year (0.72%) in the 1990s. Between 2000 and 2005, losses declined to 1020 km^2/year (0.66%). Australia is a notable exception owing to its small population and remote location of its forests in the northern and northeastern regions.

Theory

Mangrove ecosystems are defined broadly, but because of their proximity to the sea their development is driven by inundation, salinity, changes in sea level that is rising in most parts of the world. Other disturbances, hurricanes and typhoons, tsunamis, frosts and freezes, also structure mangrove communities over shorter temporal and spatial scales. Mangroves possess characteristics of early successional r-strategists or pioneer species in their reproductive biology, including broad tolerance to environmental factors, rapid growth and maturity, high propagule output, and mechanisms for short- and long-distance dispersal (Tomlinson, 1986) that are required for survival in their habitat (Cintron-Molero, 1992). On the other hand, their community structure is similar to the mature stage of succession. Like saline tidal marshes, mangroves are considered steady state rather than successional communities owing to the stresses associated with inundation and salinity (Lugo, 1980). In spite of these constraints, mangroves are easier to restore than other forested wetlands due to the low pool of species that inhabit this environment (Lugo, 1998).

Mangroves are common in deltaic areas with high sediment inputs (Field, 1998b) that support their elevation gain in the face of rising sea level. Many mangroves in the Caribbean and Indo-Pacific grow on coral sands or on coralline rock (Mendelssohn and McKee, 2000). In these areas, mangroves maintain their elevation as sea level rises by building peat fueled by root production and incomplete decomposition (McKee et al., 2007a). This probably also is true for mangroves in areas of low salinity or freshwater where conditions are more favorable for accumulation of soil organic matter (SOM).

The role of nurse plants in mangrove seedling establishment and growth is well documented. McKee et al. (2007b) reported that the herbaceous species *Sesuvium portulacastrum* L. and *Distichlis spicata* L. (Greene) enhanced recruitment of red mangrove (*R. mangle*) in the Caribbean. Both species trapped dispersing propagules, enhancing survival of seedlings. *Distichlis* provided structural support for mangrove seedlings and it enhanced their growth by increasing aeration and ameliorating high temperatures. In Florida, saline tidal marsh species smooth cordgrass (*Spartina alterniflora*) and saltwort (*Batis maritima*) initially colonize bare soil and facilitate establishment of mangroves (Lewis and Dunstan, 1975; Lewis and Gilmore, 2007). Lewis et al. (2005) found that planting *S. alterniflora* facilitated natural recruitment by trapping mangrove seeds. With *Batis*, the mechanism for improved seedling success was a slight increase in elevation provided by its dense root network (Milbrant and Tinsley, 2006). Intraspecific facilitation also is recognized as a means to increase mangrove recruitment. Increasing planting density using either single or mixed species increased survival of mangrove seedlings in Sri Lanka and Kenya, respectively, by trapping sediment and increasing elevation, especially at low sites (Huxham et al., 2010). Planting in clumps increased the survival rate by ameliorating abiotic stress and reducing herbivory (Gedan and Silliman, 2009). Planting at higher densities not only increased surface accretion and elevation but also increased N content of the soil (Kumara et al., 2010).

Practice

The first efforts to restore mangroves were in the early part of the twentieth century and involved establishing large plantations to produce timber and fuel wood (Walton et al., 2006b) and for erosion control (Teas, 1977; Lewis, 1982). Mangroves also were planted over large areas in the Philippines beginning in the 1930s for wood supply and to provide protection from typhoons (Primavera and Esteban, 2008). Plantations of *Nypa* were established on a smaller scale to produce sugar and alcohol (Primavera and Esteban, 2008). Since the 1980s, mangroves have been planted for a variety of restoration goals, including coastal protection from storm surges, shoreline erosion control, fisheries enhancement (Field, 1996b), production of livestock fodder (Sato et al., 2005), biodiversity (Chen et al., 2009), and visual and aesthetic appeal (Saenger, 2002; Lacerda, 2003). In the US and elsewhere, mangrove restoration is used to mitigate for loss of habitat resulting from coastal development (Snedaker and Biber, 1996), urbanization (Saenger, 1996), and oil spills (Duke, 1996).

Rehabilitation of mangrove forests is widely used in developing countries where the local community has a say in how the mangrove resource is managed and utilized (Field, 1998b). These communities rely in part on the forest for wood products and aquaculture. Rehabilitation may take the form of: (1) conservation of the natural system, such as establishment of protected areas, and (2) sustainable production of natural resources, such as plantations. Since 1966, extensive areas in Bangladesh have been planted with mangroves for land building, shoreline protection, and timber (Saenger and Siddiqi, 1993). These plantings cover more than 148,000 ha, a size unmatched anywhere else in the world (Iftekhar and Islam, 2004) though extensive areas have been planted in the Philippines (Samson and Rollon, 2008) and Cuba (Suman, 2003).

Restoration, on the other hand, is commonly used in the US where laws were enacted to protect wetland resources (see Chapter 2, Definitions). Here, the restoration project strives to re-create the structure and functional characteristics of the forest and is driven by compensatory mitigation under the Clean Water Act (see Chapter 2, Definitions) (Lewis, 1979; Lewis and Haines, 1981) or by government-sponsored restoration programs (Milano, 1999). Restoration also is used to compensate for habitat loss in Australia (Saenger, 1996) and some other countries.

Most mangrove restoration and rehabilitation projects rely on planting (Figure 9.4), and over time many restorations come to resemble forest plantations (Kaly and Jones, 1998; Ellison, 2000). Before 1982, plantings were primarily for wood production and coastal stabilization (Ellison, 2000). Since that time, there has been more focus on the suite of ecosystem services they provide, including forestry, coastal protection, fisheries, and wildlife habitat.

Many projects fail because of planting at inappropriate elevation within the tidal frame (Primavera and Esteban, 2008) or on sites where mangroves did not exist before (Lewis, 2005), such as on lower elevation mudflats and sandbars (Samson and Rollon, 2008; Walters et al., 2008). Planting easy-to-propagate but inappropriate species such as *Rhizophora* instead of natural pioneer species like *Avicennia* or *Sonneratia* is another cause of failure (Primavera and Esteban, 2008). Only a fraction of the known species, about 30%, is used in plantings (Field, 1998b) and many are nonnative. Plantings, especially with *Rhizophora*, often produce monospecific stands that are vulnerable to infestation by pests (De Leon and White, 1999). Ellison (2000) pointed out that restoration projects that incorporate aquaculture or mariculture produce systems that more closely resemble natural mangrove forests. They also provide greater economic return than plantations. Over the long-term, the intense pressure to harvest timber or coastal development of planted forests leads to failure (Spalding et al., 2010).

Figure 9.4 (a) High school students planting *Rhizophora mucronata* seedlings at Thousand Islands, northern Jakarta, Indonesia; (b) The site 2 years after planting; and (c) 4 years later in 2009. Photo credit: Beginer Subhan, Department of Marine Science and Technology, Bogor Agricultural University (IPB).

Planting

Knowledge regarding planting techniques for mangroves is well documented and dates back a century or more in Malaysia (Watson, 1928) and the Philippines (Walters, 2000) and perhaps longer in the Sundarbans region of Bangladesh and

India (Chowdhury and Ahmed, 1994; Siddiqi, 2001). J.G. Watson, who pioneered mangrove silviculture nearly a century ago in Malaysia, first described the planting technique.

> the seeds should be stuck into a depth of a few inches only, so that they will not fall over; deep insertion is not recommended. The seedlings are thrust into the mud at intervals from 40 to 100 centimeters... young plantations are protected from damage by floating object.
>
> *From Watson (1928)*

In the Paleotropics, a limited number of mangrove species are used in restorations including *Rhizophora apiculata*, *Rhizophora mucronata*, *A. marina*, and *Sonneratia apetala* in the Paleotropics even though many are available (Table 9.4). These species have high survival rates and grow well in mixtures and monocultures (Figure 9.5) (Das et al., 1997). *Rhizophora* and *S. alba* grow well at lower elevations whereas *Ceriops tagal*, *Bruguiera cylindrica*, and *Xylocarpus granatum* grow better at higher elevations (Kitaya et al., 2002). In Bangladesh, a number of species have been tested for large-scale plantings but mostly *S. apetala* and *Avicennia officinalis* are planted (Saenger and Siddiqi, 1993). *Rhizophora mangle* is commonly planted in the Neo-tropics (Lewis, 1979; Ellison, 2000). It is easy to propagate (Walters et al., 2008), has a high survival rate (Elster, 2000), and grows well on sheltered shorelines and river-banks (Imbert et al., 2000). Mature propagules are planted by burying them about one-third of their length into the mud at densities of less than 5 per m^2 in protected sites to 15 per m^2 to build land (Imbert et al., 2000) or to compensate for greater exposure and wave energy (Saenger, 2002).

Table 9.4 Common Mangrove Species Planted in Different Countries of the World

Australia	*Avicennia marina, Aegiceras corniculatum*
Bangladesh	*Sonneratia apetala, Avicennia officinalis, Heritiera fomes*
China	*Kandelia candel, S. apetala, Sonneratia caseolaris*
Cuba	*Rhizophora mangle, Avicennia germinans, Laguncularia racemosa, Conocarpus erectus*
India	*A. marina, A. officinalis, S. caseolaris, Rhizophora mucronata, Rhizophora apiculata*
Indonesia	*Bruguiera gymnorrhiza, Rhizophora stylosa, R. apiculata, R. mucronata*
Malaysia	*R. apiculata, R. mucronata*
Mexico	*A. germinans*
Philippines	*R. apiculata, R. mucronata, R. stylosa, Ceriops tagal, Nypa fruticans*
USA	*Rhizophora mangle, A. germinans, L. racemosa, C. erectus*
Vietnam	*R. apiculata, R. mucronata, R. stylosa, K. candel, Avicennia alba, Ceriops decandra, S. caseolaris, N. fruticans*

From Field (1996a,b, 1998a,b), Li et al. (1998), Han et al. (2001), Zan et al. (2001), Toledo et al. (2001).

Planting nonnative species may lead to invasion of native mangrove stands. In southern China, mangrove forests are invaded by the introduced exotic *S. apetala* that was used to reclaim mudflats beginning in 1985 (Ren et al., 2008). *Sonneratia apetala* has traits such as rapid growth and reproductive capacity and high salt tolerance characteristic of competitor species (see Chapter 3, Ecological Theory and Restoration) (Ren et al., 2009). It grows taller than native species, shading them, and altering nutrient cycles by enriching soils with organic matter and nitrogen. It has been likened to *S. alterniflora*, salt marsh cordgrass, that was introduced to eastern China to reclaim land but spread and became a problem.

Propagules are collected from the wild, nursery-raised seedlings and shrubs, and stem cuttings (Saenger, 2002) (Figure 9.6). Propagules are planted at the onset of the rainy season. On exposed shorelines, *Rhizophora* is planted, usually in winter, when barometric pressure is high and sea level is low. On high energy or eroding shorelines where natural recruitment is not possible, mangroves are planted by encasing the seedling in a tube made of plastic (Riley and Salgado Kent, 1999). However, the technique is expensive and is not applicable to large-scale plantings. Propagules generally are planted in the upper third of the tidal range (Saenger, 2002). Lewis and Gilmore (2007) suggest that mangroves be established in areas that are inundated less than 30% of the time as more frequent flooding causes mortality. Saenger (2002) and Saenger and Siddiqi (1993) offer detailed reviews of mangrove seed collection, propagation and planting techniques, forest management, and problems encountered when establishing large-scale mangrove forests.

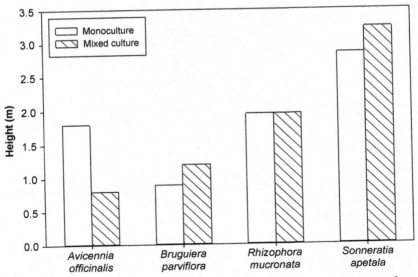

Figure 9.5 Height of four mangrove species planted in monocultures and mixtures after 2 years.
Adapted from Das et al. (1997).

Figure 9.6 (a) Viviparous seedlings; (b) mangrove nursery; and (c) a 32-year-old planted mangrove forest at Can Gio World Biosphere Reserve, Mekong Delta, Vietnam. Note the planted rows still evident in the 32-year-old planted forest.
Photo credit: Chris Craft.

Natural Colonization

Where a seed source is available, mangrove propagules will recruit into planted areas. Lewis et al. (2005) reported that seedlings of black (*A. germinans*) and white mangrove (*L. racemosa*) recruited into a mitigation wetland in southwestern Florida that had been planted with *S. alterniflora* as a nurse plant. In Kenya, Bosire et al. (2003) reported a number of mangrove species, including *Bruguiera gymnorrhiza* and *C. tagal*, colonized young 5-year-old stands planted with *R. mucronata*, *A. marina*, and *S. alba*. Likewise, Chen et al. (2004) observed recruitment of native *Kandelia candel* and *Aegiceras corniculata* into stands planted with nonnative *S. apetala* and *Sonneratia caseolaris* in China. On sites where the forest was clear cut, natural recruitment does not occur even after 25 years because of changes to local hydrodynamics caused by felling of the trees (Bosire et al., 2003). Long-term (50-year-old) plantings in the Philippines also indicate little recruitment of new species, suggesting that restoration of a species-rich mangrove forest requires planting a diverse assemblage of mangrove species (Walters, 2000).

Lewis (2005) recommends a five-step procedure to achieve successful mangrove restoration (see Keys to Ensure Success) in which planting seedlings is employed only if natural colonization does not occur. Careful site selection, removal of stressors and disturbances, and a nearby source of propagules are needed to make this technique work (Lewis, 2005). Removal of stressors such as wave action dramatically improves natural colonization by reducing wave energy and promoting sediment deposition. In Malaysia, a breakwater was constructed to create a sheltered area and then planted with seedlings of *A. marina* (Hashim et al., 2010; Kamali and Hashim, 2011). Initially, the breakwater increased sediment deposition that buried the seedlings, leading to high mortality of transplanted seedlings. Eight months later, however, natural seedlings recruited to the site and became established (Hashim et al., 2010). Blanchard and Prado (1995) evaluated natural regeneration of *R. mangle* in clear cut forests in Ecuador. They found that clear cutting in strips 20 m wide, leaving seed trees at 20-m spacing in the interstrips allowed for sufficient natural regeneration to reestablish the forest. Srivastava and Bal (1984) suggested that, for natural regeneration, at least 2500 well-distributed seedlings per hectare are needed.

Natural regeneration may be inhibited by early colonizers that quickly dominate the site. Following harvesting of mangroves in Southeast Asia, *Acrostichum* fern or "piai" often colonize higher elevations in densities that stymie natural recruitment and forest regeneration (Chan, 1989; Jawa and Srivastava, 1989). The sudden exposure of the site following clear cutting encourages the growth of *Acrostichum* and other fast growing species such as *Acanthus* and *Derris* in Malaysia (Chan et al., 1988). Srivastava et al. (1987) reported that, in clear cut areas where coverage of piai was greater than 60%, recruitment of mangrove species was insufficient for natural regeneration of the forest. *Acrostichum* is not a problem everywhere though. In Ecuador, Blancard and Prado (1995) observed little effect of *Acrostichum* on natural recruitment in strip-felled clear cuts.

Natural regeneration also is slowed by predation on dispersing propagules (Lewis, 1979; Field, 1998a,b). In Kenya, Bosire et al. (2005a) found that predation by herbivorous sesarmid crabs varied among species with crabs preferring propagules of *B. gymnorrhiza* and *C. tagal* over *R. mucronata*. Also in Kenya, Dahdouh-Guebas et al. (1998) found that grapsid crabs and snails cleared propagules, favoring those of *Rhizophora* over other species.

Ecosystem Development

Ecosystem development of mangroves and other forested wetlands is slower relative to herbaceous wetlands, owing to the larger size and slower time to maturity. However, because of the long history of planting mangroves for silviculture, there are many studies of ecosystem development of restored forests, especially growth, biomass production and forest development. Typically, young mangrove plantations have canopy height and aboveground biomass comparable to natural stands of the same age, though young restored forests often have greater stem densities (Bosire et al., 2008; Osland et al., 2012). Bosire et al. (2006) found lower recruitment into 8-year-old *S. alba* and *R. mucronata* stands than in natural stands in Kenya but it increased with stand age (Bosire et al., 2008). Natural stands contained fewer but larger trees (Figure 9.7). Walters (2000), however, reported little recruitment into 50- to 60-year-old *R. mucronata* plantations in the Philippines. The low recruitment into plantations may be due in part to weeding of non-planted species by the local population (Bosire et al., 2008).

Development of mangrove forest structure and productivity following restoration is typical of aggrading forests, with high productivity early on and increasing biomass with time. In Kenya, the rate of biomass accumulation in a 12-year-old *Rhizophora* plantation was much greater (11 t/ha/year) than an 80-year-old plantation (5.1 t/ha/year) (Putz and Chan, 1986; Kairo et al., 2008). Stevenson et al. (1999) evaluated mangrove structure of a disused shrimp pond on the Pacific coast of Central America and found that, 10 years after abandonment, basal area (18 m²/ha) was 70% as compared to a natural forest (28 m²/ha). Osland et al. (2012) compared vegetation development

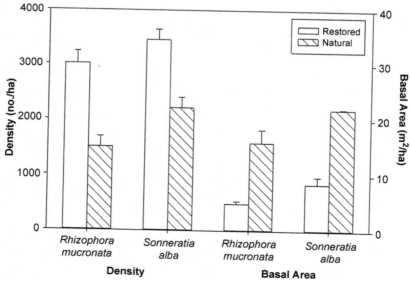

Figure 9.7 Density and basal area of two mangrove species in 8-year-old reforested and natural stands in Kenya.
Adapted from Bosire et al. (2006).

along a chronosequence of 1- to 20-year-old planted mangroves at nine sites in Florida. Mangrove height and diameter increased with age. After 20 years, density and diameter of trees still was less than in reference forests (Osland et al., 2012).

Shafer and Roberts (2008) assessed 17 mangrove forests in 2005 that were established in the 1980s for mitigation of mangrove loss in Florida. The sites, ranging in age from 17 to 25 years, were planted with red mangrove (*R. mangle*) or a combination of red mangrove and the nurse plant *S. alterniflora* and originally were sampled in 1988. Basal area and tree height increased with time, but after 25 years mitigation forests had less basal area and were shorter than natural stands. Mitigation forests also were denser than natural stands. Species composition changed over time as white mangrove (*L. racemosa*) colonized the sites. Factors limiting stand development included planting at the incorrect elevation leading to inappropriate hydrology (either too wet or too dry) and colonization by the invasive species, Brazilian pepper (*Schinus terebinthifolius*) at higher elevations (Shafer and Roberts, 2008).

In south China, Luo et al. (2010) compared stand structure and species composition of a 50-year-old planted forest with a natural mangrove forest dominated by *A. marina* and *Kandelia obovata*. The restoration site was planted with *K. obovata* (also known as *K. candel*) in the 1950s. Later, nonnative species *S. apetala* and *S. caseolaris* were introduced. The 50-year-old planted sites exhibited stand density and aboveground biomass that was similar to the natural forest. Even though the sites differed in the dominant species, overall species richness and evenness did not differ among stands.

Pneumatophores that provide additional stand complexity provide habitat epiphytic vegetation and other organisms. Crona et al. (2006) compared epiphytic algae in two 8-year-old planted *S. alba* forests with a natural forest in Kenya. Surface area available for colonization was greater in the natural stand. Algal biomass was positively correlated with pneumatophore area that increased with stand age. In spite of these differences, there was no difference in species richness among planted (10–23 species) and natural stands (18 species). Morrisey et al. (2003) compared benthic fauna of young (3–12 years old) naturally colonized mudflats and older (>60 years) stands of *A. marina* var. australasica in New Zealand. Greater numbers and diversity of benthic macrofauna were present in young stands, whereas in older stands, surface living organisms such as the gastropod *Potamopyrgus antipodarum* dominated. Morrisey et al. (2003) suggested that, as mangrove forests develop, the fauna community shifts from dominance by benthic organisms to organisms that live on the roots of the trees.

Pneumatophores also provide habitat and refuge for epibenthic organisms including crabs and mollusks. In Brazil, 5-year-old planted and naturally colonized stands contained comparable numbers and diversity of grapsid and ocypodoid crabs as a nearby natural stand (Ferreira et al., 2015). A 16-year-old plantation in the Philippines supported densities and productivity of mud crabs, a commercially exploited species, similar to natural stands (Walton et al., 2006a, 2007). Diversity of crabs, however, varied among young planted and mature natural mangroves. Some studies report lower diversity in planted stands (Al-Khayat and Jones, 1999) whereas other studies report greater (Macintosh et al., 2002) or no difference in diversity (Bosire et al., 2004). The number of crustacean and mollusk species in a 10-year-old replanted forest in Qatar was similar in planted (14–16 crustaceans, 8–9 mollusks) and natural stands

(14–15 crustaceans, 9–11 mollusks) (Al-Khayat and Jones, 1999). And, in Thailand, MacIntosh et al. (2002) recorded greater species richness of crabs and mollusks in 7-year-old replanted mangroves than in a 40-year-old stand. Other studies suggest that species richness of epibenthic organisms develops as the forest ages. In Malaysia, the density and diversity of epibenthic organisms was greatest in a mature mangrove forest, intermediate in a 15-year-old stand, and lowest in a 2-year-old stand of young *R. apiculata* seedlings (Sasekumar and Chong, 1998).

Density and diversity of benthic infauna generally are less in young planted forests. In Qatar, the number of polychaete species was lower in planted than in natural mangroves (Al-Khayat and Jones, 1999). Bosire et al. (2004) also reported lower density of benthic infauna in 5-year-old planted stands than in mature forests in Kenya though there was no difference in taxa richness. In Malaysia, Sasekumar and Chong (1998) reported lower species richness in a 2-year-old mangrove plantation than in a mature forest though infauna density was greater in the young forest. In China, Chen et al. (2007) reported similar macrobenthic communities in 19- and 43-year-old *K. candel* forests suggesting that 20 years is needed for infauna communities to become comparable to mature stands.

Nekton, such as shrimp and fish use planted mangrove stands to varying degrees. Crona and Ronnback (2005) compared shrimp use of two 8-year-old planted mangrove forests with a natural stand in Kenya. One planted forest was established in a clear cut (matrix plantation) and the second one was planted in a degraded forest (integrated plantation). Density of shrimp were highly variable among treatments and sampling years (2002 and 2003) (Figure 9.8), though natural stands and integrated

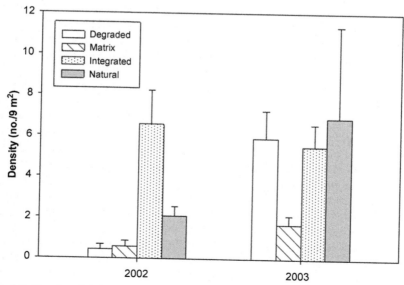

Figure 9.8 Density of shrimp caught in degraded, planted, and natural mangrove forests in Kenya.
Adapted from Crona and Ronnback (2005).

plantations harbored greater numbers of most taxa. Species diversity of shrimp taxa, including penaeid shrimp was similar in the integrated *S. alba* forest and the natural stand. Diversity of shrimp in the matrix plantation, however, was lower. Crona and Ronnback (2005) attributed differences among sites to greater root system structural complexity and longer inundation time in the integrated plantation and the natural forest that enhanced shrimp and fish use (Ronnback et al., 1999). Using hand nets and seine nets, Al-Kayat and Jones (1999) sampled planted and natural mangrove forests in Qatar and found lower diversity of juvenile and adult fish in planted mangroves. In contrast, Crona and Ronnback (2007) found no differences in finfish abundance and biomass among degraded, planted, and mature mangrove forests. There was also no difference in diversity and species composition. Of the species caught, 65% were associated with coral reefs and 75% were commercially important.

In the US, restoration of mangroves often involves reintroducing tidal inundation to impounded forests rather than planting. Impounded sites restored by hydrologic restoration show rapid and high levels of finfish use. Rey et al. (1990) reported that hydrologic restoration of impounded *R. mangle-* and *A. germinans*-dominated mangroves on the east coast of Florida resulted in a 100-fold increase in use by transient species. Also along the Florida Atlantic coast, Lewis and Gilmore (2007) reported a dramatic increase in both resident mangrove and transient species when tidal inundation was restored. In Tampa Bay, Florida, Vose and Bell (1994) compared macrobenthos and finfish utilization before and after removal of a berm. Following removal of the berm, there was increased use by polychaetes, amphipods, and finfish as measured by abundance, biomass, and community similarity, but after 22 months invertebrate and finfish use still was low compared to a reference mangrove forest.

Even though food resources may be less in young restored forests, fish feed in them. Llanso et al. (1998) compared foraging by two commercially important species, red drum and spotted seatrout, in the same Tampa Bay mangroves. They found that smaller fish fed on a variety of amphipods and benthic infauna whereas larger fish fed on polychaetes, xanthid crabs, palaemonid shrimps, and small fish.

A constraint for the development of ecosystem processes is low organic matter content of the soil. McKee and Faulkner (2000) compared soil properties of two sites restored by removing fill and two reference mangrove forests in southwestern Florida. The restored sites, 6 and 14 years old, had three to five times less SOM (10–12%) than the natural forests (38–56%). Bulk density was three times higher in the restored forest soils.

Mangroves planted on mudflats also contain less SOM, more coarse-textured particles, and lower moisture than natural mangrove soils (Al-Kayat and Jones, 1999) though SOM and fine-texture particles begin to increase once mangroves are planted and become established. Ren et al. (2008) compared SOM in 4- to 10-year-old stands planted with nonnative *S. apetala* in south China. SOM increased with stand age, from 1.14% in barren unplanted sites to 2.45% in the 10-year-old site but it still was less than natural mangrove forests (4.02%) in the region. Kairo et al. (2008) compared soil properties of a 12-year-old planted forest with an unplanted site. The planted forest contained more SOM (31%) and silt–clay (38%) than the unplanted site (22% SOM, 16% silt–clay). Osland et al. (2012) measured soil properties along a chronosequence of nine mangrove plantings, ranging in age from 1 to 20 years, in Tampa Bay, Florida. Soil organic C and N (0–10 cm) increased with site age while bulk density increased. The 20-year-old site exhibited surface soil properties that were within the range of nine reference forests.

Alteration of soil properties by restoration activities such as land clearing and refor-estation affects the development of ecosystem processes. Bosire et al. (2005b) com-pared litter decomposition of *S. alba* and *R. mucronata* in bare soil, 5- to 8-year-old planted mangroves and natural mangrove stands in Kenya. Natural stands exhibited the highest rate of decomposition while decomposition was lowest in the bare soil. Bosire et al. (2005b) attributed higher decomposition in forested stands to greater abundance of invertebrates, amphipods, nematodes, turbellarians, isopods, and polychaetes that facilitated litter breakdown. Alongi et al. (2000) measured benthic decomposition, C mineralization, and denitrification in 6-, 8- and 35-year-old planted mangroves in Vietnam. There was no difference in CO_2 flux with stand age. Denitrification was nil in the 6- and 8-year-old planted stands, but in the 35-year-old forest it was the largest contributor to C mineralization after aerobic respiration and sulfate reduction. A comparison of 2-, 15- and 60-year-old mangrove forests in Malaysia revealed that aerobic decomposition was greater and sulfate reduction was less in the 2-year-old site, leading to conditions that favor rapid seedling growth and forest regeneration (Alongi et al., 1998). Vovides et al. (2011) compared N fixation in restored, natural, and degraded mangrove forests in Baja, Mexico. Two sites were restored in 1995, one by planting with *A. germinans* and one by natural colonization. After 11 years, there was no difference in N fixation between restored and natural forests but it was lower in the degraded forest. In a greenhouse study, soil N increased fourfold in soils planted with *A. germinans*, *Rhizophora stylosa*, or *B. gymnorrhiza*, and it was linked to higher rates of N fixation in the planted soil (Inoue et al., 2011).

Complete restoration of mangrove structure and function requires anywhere from 25 to 50 years based on field observations (Lewis and Gilmore, 2007) (Table 9.5) and simulation modeling of forest structure (Twilley et al., 1998). Forest productivity develops to equivalent levels within 20 years, but stand structure and species richness take longer. Epiphytic organisms, algae, sponges, crabs, and mollusks, require ade-quate forest structure, especially pneumatophores, and require 5–20 years to develop. Benthic infauna requires a comparable amount of time to fully develop. Shrimp and finfish use restored forests at levels comparable to natural forests within 3–5 years fol-lowing reintroduction of tidal inundation (Lewis, 1992; Lewis and Gilmore, 2007). For plantings, 5 to more than 10 years are needed to provide structure for food resources and refuge from predators. And, for regulation functions such as decomposition and nutrient (N) cycling, 10–20 years may be needed to achieve equivalence to natural stands.

Keys to Ensure Success

Pioneering research on mangrove restoration and rehabilitation has been led by Lewis (US), Field (Australia), and others. Lewis (1982) edited *Creation and Restoration of Coastal Plant Communities*, perhaps the first book on coastal wetland restoration.

Table 9.5 Ecosystem Development of Mangrove Forests Following Planting or Restoration

	Time (years) to Equivalence to Natural Forests
Productivity and habitat functions	
Forest primary production	10–20
Forest stand structure and species richness	25–50
Algal production	5–10
Algal diversity	5–10
Epifauna density	10–20
Epifauna diversity	5–20
Benthic infauna density	5–20
Benthic infauna diversity	5–20
Shrimp and finfish	3–5 for hydrologic restorations, 5 to >10 for plantings
Regulation functions	
Soil organic matter and bulk density	>20[a]
Decomposition	>10
N fixation	5–10
Nutrient (N) cycling	>10
Denitrification	>10–20

See text for references.
[a]Silvicultural *restoration* projects require less time than plantings.

Field (1996b) edited *Restoration of Mangrove Ecosystems* that provides excellent case studies from a number of Paleotropical and Neotropical countries.

Field (1998a,b) and Lewis and Gilmore (2007) identified a suite of considerations for mangrove rehabilitation, including site selection, approaches to reestablish vegetation, monitoring, and maintenance (Table 9.6). The first consideration is placement in a sheltered area, partially protected from wind and wave action. Sometimes breakwaters are constructed. It also is important to restore hydrology by breaching dikes or levees or removing fill. Mangroves are land builders, so sites with fine-textured (mud) substrates and an adequate sediment supply are ideal. A gentle slope that provides for drainage during low tide is important as is selecting sites with predictable (tidal) inundation. The availability of freshwater, either from rainfall or river discharge, is important (Cintron-Molero, 1992; Field, 1998b). The presence of adjacent areas of mangroves and seagrass beds may be a good indicator that environmental conditions are favorable for establishment.

On impounded sites, it may be necessary to modify the physical environment by establishing a network of tidal creeks to provide for access (Lewis and Gilmore, 2007). Creating a heterogenous landscape, including basins for temporary refuge, contiguous vegetation, including terrestrial vegetation, and maintaining a source of freshwater inflow from the uplands is important for both flora and fauna (Lewis and Gilmore, 2007).

Table 9.6 Keys to Ensure Success When Restoring and Rehabilitating Mangroves

Site Selection
• Sheltered area, protected from wind and wave action
• Adequate sediment supply
• Gentle (1–3%) slope
• Appropriate hydrology with regular inundation and adequate freshwater
• Proximity to intact mangrove forests and seagrass beds (indicators of favorable conditions)
• Incorporate tidal creeks and landscape heterogeneity to provide access and refuge[a]
Revegetation
• Preferably natural regeneration; rely on planting in high(er) energy environments and in areas that lack dispersing propagules
• If planting, selection of appropriate species to match hydrology and salinity
• Nurse plants to facilitate recruitment, sedimentation and increase elevation
Maintenance
• Keep the site free of woody debris and garbage
• Replant if needed (replacement for mortality)
Social Considerations
• Community support is necessary
Constraints on Successful Restoration
• Hypersalinity
• High soil temperature
• Acid sulfate soils
• Low N and P
• Invasive/aggressive species (e.g., *Acrostichum*)
• Herbivory (e.g., crabs, monkeys)

[a]May be needed for impounded or filled sites.
From Cintron-Molero (1992), Field (1998a,b), Lewis (2005), Gilman and Ellison (2007), Lewis and Gilmore (2007).

Reestablishing vegetation relies on either natural colonization or planting. Recolonization depends on a nearby source of propagules to support seedling recruitment (Lewis, 2005). Cintron-Molero (1992) suggested that, in areas with abundant freshwater and nutrients, natural recruitment may be sufficient if there is a nearby source of propagules. In high-stress environments such as low freshwater or nutrients, or high salinity, fertilization and planting are needed. When planting, it is critical to match the appropriate native species to the proper environmental conditions, especially hydrology and salinity. Field (1996a, 1998a) offers a list of common species that are planted (see Table 9.4), characteristics to identify ripe propagules, and recommended spacing when planting. Nurse plants such as *S. alterniflora, D. spicata, B. maritima, S. portulacastrum* L., and *D. spicata* L. (Greene) may be planted beforehand or in tandem to facilitate recruitment of mangrove seedlings, trap sediment, and build elevation. Generally, most species are planted at a spacing of 2 m or less apart. While smaller spacing requires planting more individuals at greater cost to the project, it improves chances of successful revegetation. The greatest key to ensure success is to have the support of the local community. This is especially important in developing regions to minimize disturbance to the site (Gilman and Ellison, 2007). It also is important that the local community benefits economically from the restoration (Walters, 1997) by selectively harvesting trees and using the site for fishing and aquaculture while protecting the overall integrity of it.

Once the site is restored or rehabilitated, there may be unforeseen obstacles that require action. Soil conditions such as hypersalinity; acid sulfates; or low N, P, or Fe may need to be corrected (Field, 1998a,b; Sato et al., 2005). Aggressive or invasive species may need to be controlled. In Malaysia and Indonesia, many areas where mangroves were harvested became infested with the fern, *Acrostichum*, which proved difficult and expensive to remove (Field, 1998b). Other species such as *Acanthus* and *Derris* spp. also may be problematic (Chan et al., 1988). Barnacles and weeds may inhibit the growth of seedlings and herbivory by crabs (Lewis, 1979) and monkeys (Chan et al., 1988) may be a problem (Field, 1998a,b). The site also may need to be periodically cleared of woody debris and garbage that damages or covers saplings (Gilman and Ellison, 2007).

Finally, monitoring of the restored site and reference sites is necessary to document performance over time to determine if project goals are being met (Ellison, 2000; Lewis, 2005). For rehabilitation projects, take photographs and conduct aerial and on-the-ground reconnaissance (Field, 1998a). Monitor growth and losses due to seedling mortality, pests, and disease. Record impacts of human activities such as timber harvest, grazing, and aquaculture ponds. Be sure to measure the success of the project against the original criteria that were established. Monitoring should be conducted for an extended period of time. Most projects require monitoring for 5 years following restoration which is insufficient for mangroves where 30 years or more are needed to document complete restoration of structure and function (Field, 1998a; Lewis, 2005).

An excellent resource is *Ecological Mangrove Rehabilitation: A Field Manual for Practitioners* by Lewis and Brown (2014). The online publication offers detailed information on biophysical factors, assessment of resilience (including social and economic factors), design and planning implementation, monitoring, and case studies of successes and failures from Old and New World restoration projects. This publication is a must-read for anyone interested in implementing and maintaining mangrove restoration projects.

References

Al-Khayat, J.A., Jones, D.A., 1999. A comparison of the macrofauna of natural and replanted mangroves in Qatar. Estuarine, Coastal and Shelf Science 49 (Suppl. A), 55–63.
Alongi, D.M., 2002. Present state and future of the world's mangrove forests. Environmental Conservation 29, 331–349.
Alongi, D.M., 2008. Mangrove forest: resilience, protection from tsunamis, and responses to global change. Estuarine, Coastal and Shelf Science 76, 1–13.

Alongi, D.M., Sasekumar, A., Tirendi, F., Dixon, P., 1998. The influence of stand age on benthic decomposition and recycling of organic matter in managed mangrove forests of Malaysia. Journal of Experimental Marine Biology and Ecology 225, 197–218.

Alongi, D.M., Tirendi, F., Trott, L.A., Xuan, T.T., 2000. Benthic decomposition rates and pathways of the mangrove *Rhizophora apiculata* in the Mekong delta, Vietnam. Marine Ecology Progress Series 194, 87–101.

Bandaranayake, W.M., 1998. Traditional and medicinal uses of mangroves. Mangroves and Salt Marshes 2, 133–148.

Barani, E., Hambrey, J., 1999. Mangrove conservation and coastal management in Southeast Asia: what impact on fisheries resources? Marine Pollution Bulletin 37, 431–440.

Barbier, E.B., Cox, M., 2003. Does economic development lead to mangrove loss? A cross-country analysis. Contemporary Economic Policy 21, 418–432.

Barbier, E.B., Hacker, S.D., Kennedy, C., Koch, E.W., Stier, A.C., Silliman, B.R., 2011. The value of estuarine and coastal ecosystem services. Ecological Monographs 81, 169–193.

Blanchard, J., Prado, G., 1995. Natural regeneration of *Rhizophora mangle* in strip clearcuts in Northwest Ecuador. Biotropica 27, 160–167.

Blasco, F., Aizpuru, M., Gers, C., 2001. Depletion of the mangroves of continental Asia. Wetlands Ecology and Management 9, 245–256.

Bosire, J.O., Dahdouh-Guebas, F., Kairo, J.G., Koedam, N., 2003. Colonization of non-planted mangrove species into restored mangrove stands in Gazi Bay, Kenya. Aquatic Botany 76, 267–279.

Bosire, J.O., Dahdouh-Guebas, F., Kairo, J.G., Cannicci, S., Koedam, N., 2004. Spatial variations in macrobenthic fauna recolonization in a tropical bay mangrove. Biodiversity and Conservation 13, 1059–1074.

Bosire, J.O., Kairo, J.G., Kazunga, J., Koedam, N., Dahdouh-Guebas, F., 2005a. Predation on propagules regulates regeneration in a high-density reforested mangrove plantation. Marine Ecology Progress Series 299, 149–155.

Bosire, J.O., Dahdouh-Guebas, F., Kairo, J.G., Kazunga, J., Dehairs, F., Koedam, N., 2005b. Litter degradation and CN dynamics in reforested mangrove plantations at Gazi Bay, Kenya. Biological Conservation 126, 287–295.

Bosire, J.O., Dahdouh-Guebas, F., Kairo, J.G., Wartel, S., Kazunga, J., Koedam, N., 2006. Success rates of recruited tree species and their contribution to the structural development of reforested mangrove stands. Marine Ecology Progress Series 325, 85–91.

Bosire, J.O., Dahdouh-Guebas, F., Walton, M., Crona, B.I., Lewis III, R.R., Field, C., Kairo, J.G., Koedam, N., 2008. Functionality of restored mangroves: a review. Aquatic Botany 89, 251–259.

Brander, L.M., Wagtendonk, A.J., Hussain, S.S., McVittie, A., Verburg, P.M., de Groot, R.S., van der Ploeg, S., 2012. Ecosystem service value for mangroves in Southeast Asia: a meta-analysis and value transfer application. Ecosystem Services 1, 62–69.

Chan, H.T., 1989. A note on the eradication of *Acrostichum aureum* ferns in the Matang mangroves, Perak, Peninsular Malaysia. Journal of Tropical Forest Science 2, 171–173.

Chan, H.T., Chong, P.F., Ng, T.P., 1988. Silvicultural efforts in restoring mangroves in degraded coastal areas in Peninsular Malaysia. Galaxea 7, 307–314.

Chen, G.C., Ye, Y., Lu, C.Y., 2007. Changes of macro-benthic faunal community with stand age of rehabilitated *Kandelia candel* mangrove in Jiulongjiang Estuary, China. Ecological Engineering 31, 215–234.

Chen, L., Wang, W., Zhang, Y., Lin, G., 2009. Recent progresses in mangrove conservation, restoration and research in China. Journal of Plant Ecology 2, 45–54.

Chen, Y., Liao, B., Zheng, M.Li, S., Song, X., 2004. Dynamics and species diversities of artificial *Sonneratia apetala, S. caseolaris,* and *Kandelia candel* communities. Chinese Journal of Applied Ecology 15, 924–928.

Chong, V.C., 2006. Sustainable utilization and management of mangrove ecosystems of Malaysia. Aquatic Ecosystem Health and Management 9, 249–260.

Chowdhury, R.A., Ahmed, I., 1994. History of forest management. In: Hussain, Z., Acharya, G. (Eds.), Mangroves of the Sundarbans. Bangladesh, vol. II. International Union for Conservation of Nature and Natural Resources, Bangkok, Thailand, pp. 155–180.

Cintron-Molero, G., 1992. Restoring mangrove systems. In: Thayer, G.W. (Ed.), Restoring the Nation's Marine Resources. Maryland Sea Grant, College Park, Maryland, pp. 223–277.

Crona, B.I., Ronnback, P., 2005. Use of replanted mangroves as nursery grounds by shrimp communities in Gazi Bay, Kenya. Estuarine, Coastal and Shelf Science 65, 535–544.

Crona, B.I., Ronnback, P., 2007. Community structure and temporal variability of juvenile fish assemblages in natural and replanted mangroves, *Sonneratia alba* Sm., of Gazi Bay, Kenya. Estuarine, Coastal and Shelf Science 74, 44–52.

Crona, B.I., Holmgren, S., Ronnback, P., 2006. Re-establishment of epibiotic communities in reforested mangroves of Gazi Bay, Kenya. Wetlands Ecology and Management 14, 527–536.

Dahdouh-Guebas, F., Jayatissa, L.P., Di Nitto, D., Bosire, J.O., Lo Seen, D., Koedam, N., 2005. How effective were mangroves as a defence against the recent tsunami? Current Biology 15, 443–447.

Dahdouh-Guebas, F., Verneirt, M., Tack, J.F., Van Speybroeck, D., Koedam, N., 1998. Propagule predators in Kenyan mangroves and their possible effect on regeneration. Marine and Freshwater Research 49, 345–350.

Danielson, F., Sorensen, M.K., Olwig, M.F., Selvam, V., Parish, F., Burgess, N.D., Hiraishi, T., Karunagaran, V.M., Rasmussen, M.S., Hansen, L.B., Quarto, A., Suryadiputra, N., 2005. The Asian tsunami: a protective role for coastal vegetation. Science 310, 643.

Das, S., Vincent, J.R., 2009. Mangroves protected villages and reduced death toll during Indian super cyclone. Proceedings of the National Academy of Sciences 106, 7357–7360.

Das, P., Basak, U.C., Das, A.B., 1997. Restoration of the mangrove vegetation in the Mahanadi Delta, Orissa, India. Mangroves and Salt Marshes 1, 155–161.

Davis, J.H., 1940. The Ecology and Geologic Role of Mangroves in Florida. Publications of the Carnegie Institution of Washington 517. pp. 303–412.

Dawes, C.J., 1998. Marine Botany. John Wiley and Sons, New York.

De Leon, R.O.D., White, A.T., 1999. Mangrove rehabilitation in the Philippines. In: Streever, W. (Ed.), An International Perspective on Wetland Rehabilitation. Kluwer Academic Publishers, Dordrecht, The Netherlands, pp. 37–42.

Duke, N.C., 1992. Mangrove floristics and biogeography. In: Robertson, A.L., Alongi, D.M. (Eds.), Tropical Mangrove Ecosystems. American Geophysical Union, Washington, D.C, pp. 63–100.

Duke, N.C., 1996. Mangrove reforestation in Panama: an evaluation of planting in areas deforested by large oil spill. In: Field, C.D. (Ed.), Restoration of Mangrove Ecosystems. International Society for Mangrove Ecosystems, Okinawa, Japan, pp. 209–232.

Duke, N.C., Ball, M.C., Ellison, J.C., 1998. Factors influencing biodiversity and distributional gradients in mangroves. Global Ecology and Biogeography Letters 7, 27–47.

Duke, N.C., Meynecke, J.O., Dittman, S., Ellison, A.M., Anger, K., Berger, U., Cannicci, S., Diele, K., Ewel, K.C., Field, C.D., Koedam, N., Lee, S.Y., Marchland, C., Norhaus, I., Dahdouh-Guebas, F., 2007. A world without mangroves? Science 317, 41–42.

Ellison, A.M., 2000. Mangrove restoration: do we know enough? Restoration Ecology 8, 219–229.

Ellison, A.M., Farnsworth, E.J., 1996. Anthropogenic disturbance of Caribbean mangrove ecosystems: past impacts, present trends, and future predictions. Biotropica 28, 549–565.

Ellison, J.C., 2003. How South Pacific mangroves may respond to predicted climate change and sea-level rise. In: Gillespie, A., Burns, W.C.G. (Eds.), Climate Change in the South Pacific: Impacts and Responses in Australia, New Zealand and Small Island States. Kluwer Academic Publishers, Dordrecht, The Netherlands, pp. 289–301.

Elster, C., 2000. Reasons for reforestation success and failure with three mangrove species in Columbia. Forest Ecology and Management 131, 201–214.

Ewel, K.C., Twilley, R.W., Ong, J.E., 1998. Different kinds of mangrove forests provide different goods and services. Global Ecology and Biogeography Letters 7, 83–94.

Ferreira, A.C., Ganade, G., Luiz de Attayde, J., 2015. Restoration versus natural regeneration in a neotropical mangrove: effects on plant biomass and crab communities. Ocean & Coastal Management 110, 38–45.

Field, C.D., 1995. Impact of expected climate change on mangroves. Hydrobiologia 295, 75–81.

Field, C.D., 1996a. General guidelines for the restoration of mangroves. In: Field, C.D. (Ed.), Restoration of Mangrove Ecosystems. International Society for Mangrove Ecosystems, Okinawa, Japan, pp. 233–250.

Field, C.D., 1996b. Restoration of Mangrove Ecosystems. International Society for Mangrove Ecosystems, Okinawa, Japan.

Field, C.D., 1998a. Rationales and practices of mangrove afforestation. Marine and Freshwater Research 49, 353–358.

Field, C.D., 1998b. Rehabilitation of mangrove ecosystems: an overview. Marine Pollution Bulletin 37, 383–392.

Food and Agricultural Organization, 2007. The World's Mangroves: 1980–2005. Food and Agricultural Organization of the United Nations, Rome, Italy.

de Graaf, G.J., Xuan, T.T., 1998. Extensive shrimp farming, mangrove clearance and marine fisheries in southern provinces of Vietnam. Mangroves and Salt Marshes 2, 159–166.

Gedan, K.B., Silliman, B.R., 2009. Using facilitation theory to enhance mangrove restoration. Ambio 38, 109.

Gilman, E., Ellison, J., 2007. Efficiency of alternative low-cost approaches to mangrove restoration, American Samoa. Estuaries and Coasts 30, 641–665.

Gilman, E.L., Ellison, J., Duke, N.C., Field, C., 2008. Threats to mangroves from climate change and adaptation options: a review. Aquatic Botany 89, 237–250.

Giri, C., Ochieng, E., Tieszen, L.L., Zhu, Z., Singh, A., Loveland, T., Masek, J., Duke, N., 2011. Status and distribution of mangrove forests of the world using earth observation satellite data. Global Ecology and Biogeography 20, 154–159.

Han, W., Gao, X., Teunissen, E., 2001. Study on *Sonneratia apetala* productivity in restored forests in Leizhou Peninsula, China. Forest Research 12, 229–234.

Hashim, R., Kamali, B., Yamin, N.M., Zakaria, R., 2010. An integrated approach to coastal rehabilitation: mangrove restoration in Sungai Haji Dorani, Malaysia. Estuarine, Coastal and Shelf Science 86, 118–124.

Hogarth, P.J., 2007. The Biology of Mangroves and Seagrasses. Oxford University Press, Oxford, UK.

Huxham, M., Kumara, M.P., Jayatissa, L.P., Krauss, K.W., Kairo, J., Langat, J., Mencuccini, M., Skov, M.W., Kirui, B., 2010. Intra- and interspecific facilitation in mangroves may increase resilience to climate change. Philosophical Transactions of the Royal Society B 365, 2127–2135.

Iftekhar, M.S., Islam, M.R., 2004. Managing mangroves in Bangladesh: a strategy analysis. Journal of Coastal Conservation 10, 139–146.

Imbert, D., Rousteau, A., Scherrer, P., 2000. Ecology of mangrove growth and recovery in the Lesser Antilles: state of knowledge and basis for restoration projects. Restoration Ecology 8, 230–236.

Inoue, T., Nohara, S., Matsumoto, K., Anzai, Y., 2011. What happens to soil chemical properties after mangrove plants colonize? Plant and Soil 346, 259–273.

Jawa, R.R., Srivastava, P.B.L., 1989. Dispersal of natural regeneration in some piai-invaded areas of mangrove forests in Sarawak. Forest Ecology and Management 26, 155–177.

Kairo, J.G., Lang'at, J.K.S., Dahdouh-Guebas, F., Bosire, J., Karachi, M., 2008. Structural development and productivity of replanted mangrove plantations in Kenya. Forest Ecology and Management 255, 2670–2677.

Kaly, U.L., Jones, G.P., 1998. Mangrove restoration: a potential tool for coastal management in tropical developing countries. Ambio 27, 656–661.

Kamali, B., Hashim, R., 2011. Mangrove restoration without planting. Ecological Engineering 37, 387–391.

Kathiresan, K., Rajendran, N., 2005. Coastal mangrove forests mitigated tsunami. Estuarine, Coastal and Shelf Science 65, 601–606.

Kitaya, Y., Jintana, V., Piriyayotha, S., Jaijing, D., Yabuki, K., Izutani, S., Nishimiya, A., Iwasaki, M., 2002. Early growth of seven mangrove species planted at different elevations in a Thai estuary. Trees 16, 150–154.

Krauss, K.W., Lovelock, C.E., McKee, K.L., Lopez-Hoffman, L., Ewe, S.M.L., Sousa, W.P., 2008. Environmental drivers in mangrove establishment and early development: a review. Aquatic Botany 89, 105–127.

Kumara, M.P., Jayatissa, L.P., Krauss, K.W., Phillips, D.H., Huxham, M., 2010. High mangrove density enhances surface accretion, surface elevation change and tree survival in coastal areas susceptible to sea level rise. Oecologia 164, 545–553.

Lacerda, L.D., 2003. Mangrove Ecosystems: Function and Management. Springer, Berlin.

Lee, S.Y., 1995. Mangrove outwelling: a review. Hydrobiologia 295, 203–212.

Lewis III, R.R., 1979. Large scale mangrove restoration on St. Croix, U.S. Virgin Islands. In: Cole, D.P. (Ed.), Proceedings of the Sixth Annual Conference on Wetlands Restoration and Creation, Tampa, Florida, pp. 231–242.

Lewis III, R.R., 1982. Creation and Restoration of Coastal Plant Communities. CRC Press, Boca Raton, FL.

Lewis III, R.R., 1992. Coastal habitat restoration as a fishery management tool. In stemming the tide of coastal fish habitat loss. In: Proceedings of a Symposium on Conservation of Coastal Fish Habitat. March 7–9, 1991, Baltimore, MD, pp. 169–173.

Lewis III, R.R., 2005. Ecological engineering for successful management and restoration of mangrove forests. Ecological Engineering 24, 403–418.

Lewis III, R.R., Brown, B., 2014. Ecological Mangrove Rehabilitation. A Field Manual for Practitioners. http://www.mangroverestoration.com.

Lewis III, R.R., Dunstan, F.M., 1975. The possible role of *Spartina alterniflora* Loisel. in the establishment of mangroves in Florida. In: Lewis III, R.R. (Ed.), Proceedings of the Second Annual Conference on Restoration of Coastal Vegetation in Florida. Hillsborough Community College, Tampa, Florida, pp. 82–100.

Lewis III, R.R., Haines, K.C., 1981. Large scale mangrove restoration on Saint Croix, U.S. Virgin Islands – II. Second year. In: Proceedings of the Seventh Conference on Restoration of Coastal Vegetation in Florida. Hillsborough Community College, Tampa FL, pp. 137–148.

Lewis III, R.R., Gilmore, R.G., 2007. Important considerations to achieve successful mangrove forest restoration with optimum fish habitat. Bulletin of Marine Science 80, 823–837.

Lewis III, R.R., Hodgson, A.B., Mauseth, G.S., 2005. Project facilitates the natural reseeding of mangrove forests (Florida). Ecological Restoration 23, 276–277.

Lewis III, R.R., Gilmore Jr., R.G., Crewz, D.W., Odum, W.E., 1985. Mangrove habitat and fishery resource. In: Seaman Jr., W.J. (Ed.), Florida Aquatic Habitat and Fishery Resources of Florida. Florida Chapter of the American Fisheries Society, Kissimmee, Florida, pp. 281–336.

Li, Y., Zheng, D., Chen, H., Liao, B., Zheng, S., Cheng, X., 1998. Preliminary study on introduction of mangrove Sonneratia apetala Buch-Ham. Forest Research 11, 39–44.

Llanso, R.J., Bell, S.S., Vose, F.E., 1998. Food habits of red drum and spotted seatrout in a restored mangrove impoundment. Estuaries 21, 294–306.

Lugo, A.E., 1980. Mangrove ecosystems: successional or steady-state? Biotropica 12, 65–72.

Lugo, A.E., 1998. Mangrove forests: a tough system to invade but an easy one to rehabilitate. Marine Pollution Bulletin 37, 427–430.

Lugo, A.E., Snedaker, S.C., 1974. The ecology of mangroves. In: Johnston, R.F., Frank, P.W., Michener, C.D. (Eds.), Annual Review of Ecology and Systematics. Annual Reviews Inc, Palo Alto, California, pp. 39–64.

Lugo, A.E., Brinson, M.M., Brown, S., 1989. Forested Wetlands. Ecosystems of the World. Elsevier, Amsterdam, The Netherlands.

Luo, Z., Sun, O.J., Xu, H., 2010. A comparison of species composition and stand structure between planted and natural mangrove forests in Shenzhen Bay, South China. Journal of Plant Ecology 3, 165–174.

Macintosh, D.J., Ashton, E.C., Havanon, S., 2002. Mangrove rehabilitation and intertidal biodiversity: a study in the Ranong mangrove ecosystem, Thailand. Estuarine, Coastal and Shelf Science 55, 331–345.

McKee, K.L., Faulkner, P.L., 2000. Restoration of biogeochemical function in mangrove forests. Restoration Ecology 8, 247–259.

McKee, K.L., Cahoon, D.R., Feller, I.C., 2007a. Caribbean mangroves adjust to rising sea level through biotic controls on change in soil elevation. Global Ecology and Biogeography 16, 545–556.

McKee, K.L., Rooth, J.E., Feller, I.C., 2007b. Mangrove recruitment after forest disturbance is facilitated by herbaceous species in the Caribbean. Ecological Applications 17, 1678–1693.

Manson, F.J., Loneragan, N.R., Skilleter, G.A., Phinn, S.R., 2005. An evaluation of the evidence for linkages between mangroves and fisheries: a synthesis of the literature and identification of research directions. Oceanography and Marine Biology 43, 485–515.

Mazda, Y., Magi, M., Kogo, M., Hong, P.N., 1997. Mangroves as a coastal protection from waves in the Tong King Delta, Vietnam. Mangroves and Salt Marshes 1, 127–135.

Mendelsohn, I.A., McKee, K.L., 2000. Saltmarshes and mangroves. In: Barbour, M.G., Billings, W.D. (Eds.), North American Vegetation. Cambridge University Press, New York, pp. 501–536.

Milano, G.R., 1999. Restoration of coastal wetlands in southeastern Florida. Wetland Journal 11, 15–24.

Milbrandt, E.C., Tinsley, M.N., 2006. The role of saltwort (Batis maritima L.) in regeneration of degraded mangrove forests. Hydrobiologia 368, 369–377.

Morrisey, D.J., Skilleter, G.A., Ellis, J.I., Burns, B.R., Kemp, C.E., Burt, K., 2003. Differences in benthic infauna and sediment among mangrove (Avicennia marina var. australasica) stands of different ages in New Zealand. Estuarine, Coastal and Shelf Science 56, 581–592.

Mumby, P.J., Edwards, A.J., Arias-Gonzalez, J.E., Lindeman, K.C., Blackwell, P.G., Gall, A., Gorczynska, M.I., Harborne, A.R., Pascod, C.L., Renken, H., Wabnitz, C.C.C., Llewellyn, G., 2004. Mangroves enhance the biomass of coral reef fish communities in the Caribbean. Nature 427, 533–536.

Nagelkerken, I., van der Velde, G., Gorissen, M.W., Meijer, G.J., Van't Hof, T., den Hartog, C., 2000. Importance of mangroves, seagrass beds, and the shallow coral reef as a nursery for important coral reef fishes, using a visual census technique. Estuarine, Coastal and Shelf Science 51, 31–44.

Nagelkerken, I., Blaber, S.J.M., Bouillon, S., Green, P., Haywood, M., Kirton, L.G., Meynecke, J.-O., Pawlik, J., Penrose, H.M., Sasekumar, A., Somerfield, P.J., 2008. The habitat function of mangroves for terrestrial and marine fauna. Aquatic Botany 89, 155–185.

Odum, W.E., Heald, E.J., 1972. Trophic analysis of an estuarine mangrove community. Bulletin of Marine Science 22, 671–738.

Osland, M.J., Spivak, A.C., Nestlerode, J.A., Lessman, J.M., Almario, A.E., Heitmuller, P.T., Russell, M.J., Krauss, K.W., Alvarez, F., Dantin, D.D., Harvey, J.E., From, A.S., Cormier, N., Stagg, L., 2012. Ecosystem development after mangrove wetland creation: plant-soil change across a 20-year chronosequence. Ecosystems 15, 848–868.

Othman, M.A., 1994. Value of mangroves in coastal protection. Hydrobiologia 285, 277–282.

Primavera, J.H., 1998. Mangroves as nurseries: shrimp populations in mangrove and non-mangrove habitat. Estuarine, Coastal and Shelf Science 46, 457–464.

Primavera, J.H., Esteban, J.M.A., 2008. A review of mangrove rehabilitation in the Philippines – successes, failures and future prospects. Wetlands Ecology and Management 16, 345–358.

Putz, F.E., Chan, H.T., 1986. Tree growth, dynamics, and productivity in a mature mangrove forest in Malaysia. Forest Ecology and Management 17, 211–230.

Ren, H., Jian, S., Lu, H., Zhang, Q., Shen, W., Han, W., Yin, Z., Guo, Q., 2008. Restoration of mangrove plantations and colonization by native species in Leizhou Bay, South China. Ecological Restoration 23, 401–407.

Ren, H., Lu, H., Shen, W., Huang, C., Guo, Q., Li, Z., Jian, S., 2009. *Sonneratia aptela* Buch. Ham in the mangrove ecosystems of China: an invasive species or a restoration species? Ecological Engineering 35, 1243–1248.

Rey, J.R., Schaffer, J., Tremain, D., Crossman, R.A., Kain, T., 1990. Effects of re-establishing tidal connections in two impounded subtropical marshes on fishes and physical conditions. Wetlands 10, 27–45.

Riley, R.W., Salgado Kent, C.P., 1999. Riley encased methodology: principles and processes of mangrove habitat creation and restoration. Mangroves and Salt Marshes 3, 207–213.

Ronnback, P., Troell, M., Kautsky, N., Primavera, J.H., 1999. Distribution patterns of shrimps and fish among *Avicennia* and *Rhizophora* microhabitats in the Pagbilao mangroves, Philippines. Estuarine, Coastal and Shelf Science 48, 223–234.

Saenger, P., 1996. Mangrove restoration in Australia: a case study of Brisbane International Airport. In: Field, C.D. (Ed.), Restoration of Mangrove Ecosystems. International Society for Mangrove Ecosystems, Okinawa, Japan, pp. 36–51.

Saenger, P., 2002. Mangrove Ecology, Silviculture and Conservation. Kluwer Academic Publishers, Dordrecht, The Netherlands.

Saenger, P., Siddiqi, N.A., 1993. Land from the sea: the mangrove afforestation program of Bangladesh. Ocean and Coastal Management 20, 23–39.

Samson, M.S., Rollon, R.N., 2008. Growth performance of planted mangroves in the Philippines: revisiting forest management strategies. Ambio 37, 234–240.

Sasekumar, A., Chong, V.C., 1998. Faunal diversity in Malaysian mangroves. Global Ecology and Biogeography Letters 7, 57–60.

Sato, G., Fisseha, A., Gebrekiros, S., Karim, H.A., Negassi, S., Fisher, M., Yemain, E., Teclemariam, J., Riley, R., 2005. A novel approach to growing mangroves on the coastal mudflats of Eritrea with the potential for relieving poverty and hunger. Wetlands 25, 776–779.

Semesi, A.K., 1998. Mangrove management and utilization in eastern Africa. Ambio 27, 620–626.

Shafer, D.J., Roberts, T.H., 2008. Long-term development of tidal mitigation wetlands in Florida. Wetlands Ecology and Management 16, 23–31.

Sheridan, P., Hays, C., 2003. Are mangroves nursery habitat for transient fishes and decapods? Wetlands 23, 449–458.

Sherrod, C.L., McMillan, C., 1985. The distributional history and ecology of mangrove vegetation along the northern Gulf of Mexico Coastal region. Contributions in Marine Science 28, 129–140.

Sherrod, C.L., Hockaday, D.L., McMillan, C., 1986. Survival of red mangrove, *Rhizophora mangle*, on the Gulf of Mexico coast of Texas. Contributions in Marine Science 29, 27–36.

Shervette, V.R., Aquirre, W.E., Blasio, E., Cevallos, R., Gonzalez, M., Pozo, F., Gelwick, F., 2007. Fish communities of a disturbed mangrove wetland and an adjacent tidal river in Palmar, Ecuador. Estuarine, Coastal and Shelf Science 72, 115–128.

Siddiqi, N.A., 2001. Mangroves of Bangladesh. In: de Lacerda, L.D. (Ed.), Mangrove Ecosystems: Function and Management. Springer, Berlin, pp. 142–157.

Snedaker, S.C., Biber, P.D., 1996. Restoration of mangroves in the United States of America: a case study in Florida. In: Field, C.D. (Ed.), Restoration of Mangrove Ecosystems. International Society for Mangrove Ecosystems, Okinawa, Japan, pp. 170–188.

Spalding, M., Kainuma, M., Collins, L., 2010. World Atlas of Mangroves. Earthscan, London.

Srivastava, P.B.L., Bal, S., 1984. Composition and natural distribution pattern of natural regeneration after second thinning in Matang Mangrove Reserves, Perak, Malaysia. In: Proceedings of the Asian Symposium on Mangrove Environment, Research and Management, pp. 761–884.

Srivastava, P.B.L., Keong, G.B., Muktar, A., 1987. Role of *Acrostichum* species in natural regeneration of *Rhizophora* species in Malaysia. Tropical Ecology 26, 274–288.

Stevenson, N.J., Lewis, R.R., Burbridge, P.R., 1999. Disused shrimp ponds and mangrove rehabilitation. In: Streever, W.J. (Ed.), An International Perspective on Mangrove Rehabilitation. Kluwer Academic Publishers, Dordrecht, The Netherlands, pp. 277–297.

Suman, D., 2003. Can you eat a mangrove? Balancing conservation and development in the management of mangrove ecosystems in Cuba. Tulane Environmental Law Journal 16, 619–652.

Teas, H.J., 1977. Ecology and restoration of mangrove shorelines in Florida. Environmental Conservation 4, 51–58.

Thampanya, U., Vermaat, J.E., Duarte, C.M., 2002. Colonization success of common Thai mangrove species as a function of shelter from water movement. Marine Ecology Progress Series 237, 111–120.

Tognella-de-Rosa, M.M.P., Cunha, S.R., Soares, M.L.G., Schaeffer-Novelli, Y., Lugli, D.O., 2006. Mangrove evaluation: an essay. Journal of Coastal Research SI 39, 1219–1224.

Toledo, G., Rojas, A., Bashan, Y., 2001. Monitoring of black mangrove restoration with nursery-reared seedlings on an arid coastal lagoon. Hydrobiologia 444, 101–109.

Tomlinson, P.B., 1986. The Botany of Mangroves. Cambridge University Press, Cambridge.

Twilley, R.R., Rivera-Monroy, V.H., Chen, R., Botero, L., 1998. Adapting an ecological mangrove model to simulate trajectories in restoration ecology. Marine Pollution Bulletin 37, 404–419.

Vose, F.E., Bell, S.S., 1994. Resident fishes and macrobenthos in mangrove-rimmed habitats: evaluation of habitat restoration by hydrologic modification. Estuaries 17, 585–596.

Vovides, A.G., Bashan, Y., Lopez-Portillo, J.A., Guevara, R., 2011. Nitrogen fixation in preserved, reforested, naturally regenerated and impaired mangroves as an indicator of functional restoration in mangroves in an arid region of Mexico. Restoration Ecology 19, 236–244.

Waisel, Y., 1972. Biology of Halophytes. Academic Press, New York.

Walters, B.B., 1997. Human ecological questions for tropical restoration: experiences from planting native upland trees and mangroves in the Philippines. Forest Ecology and Management 99, 275–290.

Walters, B.B., 2000. Local mangrove planting in the Philippines: are fisherfolk and fishpond owners effective restorationists? Restoration Ecology 8, 237–246.

Walters, B.B., Ronnback, P., Kovacs, J.M., Crona, B., Hussain, S.A., Badola, R., Primavera, J.H., Barbier, E., Dahdouh-Geubas, F., 2008. Ethnobiology, socioeconomics and management of mangrove forests: a review. Aquatic Botany 89, 220–236.

Walton, M.E.M., Levay, L., Lebata, J.H., Binas, J., Primavera, J.H., 2006a. Seasonal abundance, distribution and recruitment of mud crabs (*Scylla* spp.) in replanted mangroves. Estuarine, Coastal and Shelf Science 66, 493–500.

Walton, M.E., Samonte, G., Primavera, J.H., Edwards-Jones, G., Levay, L., 2006b. Are mangroves worth replanting? the direct economic benefits of a community-based reforestation project. Environmental Conservation 33, 335–343.

Walton, M.E.M., Levay, L., Lebata, J.H., Binas, J., Primavera, J.H., 2007. Assessment of the effectiveness of mangrove rehabilitation using exploited and non-exploited indicator species. Biological Conservation 138, 180–188.

Watson, J.G., 1928. Mangrove Forests of the Malay Peninsula. Malayan Forest Records 6. Fraser and Neave, Singapore.

Zan, Q.J., Wang, Y.J., Liao, B.W., Zheng, D.S., Chen, Y.J., 2001. The structure of *Sonneratia apetala* + *S. caseolaris* – *Kandelia candel* mangrove plantations of Futian, Shenzhen. Forest Research 14, 610–615.

Part Four

From Theory to Practice

Measuring Success: Performance Standards and Trajectories of Ecosystem Development

10

Chapter Outline

Introduction

Successful restoration of wetlands requires measurement of key structural and functional properties with the expectation that they will eventually converge to those measured in natural reference wetlands of the same type. Webster's dictionary (http://www.merriam-webster.com) defines "success" as "the correct or desired result of an attempt" and so a successful restoration project is one where the preestablished goals are met within a reasonable amount of time. Kentula (2000) describes three types of success when it comes to restoring wetlands. Compliance success meets the terms of the agreement, similar to the definition above. Functional success requires that ecological functions, energy flow, and nutrient cycling are restored. Landscape success is a measure of restoration of structure and function at larger scales such as watersheds and landscapes. Vegetation traditionally has been used to gauge success of restoration projects, but, increasingly, other measures such as hydrology, soils, and fauna are used (Kentula, 2000).

Structural attributes describe the ecological community and include plant architecture such as height, density, biomass, and species richness. Density and diversity of fauna such as invertebrates, finfish, and birds also is measured. Functional attributes describe energy flow and nutrient cycling within the wetland and among adjacent terrestrial and aquatic ecosystems and measurements may include net primary production; decomposition; microbial processes such as denitrification, sulfate reduction, and methane production; and faunal productivity. Attributes describing the development of hydric soils such as reduction of ferric (Fe^{3+}) to ferrous (Fe^{2+}) iron and accumulation of organic matter also are important. Such measurements are referred to as

Creating and Restoring Wetlands. http://dx.doi.org/10.1016/B978-0-12-407232-9.00010-5

performance standards and are compared against measurements made prior to restoration or in comparable reference wetlands to determine whether the restored wetland has achieved equivalence or parity with it. Historical or prerestoration site information is extremely useful (White and Walker, 1997) but seldom is available because of the long-term and chronic alteration of ecosystems and landscapes by humans, so reference wetlands often are used for comparison.

Achieving equivalence takes time and the amount of time needed depends on the size, intensity, and duration of disturbance of the site (see Chapter 3, Ecological Theory and Restoration). If the environmental template, especially hydrology, is properly restored and appropriate plant propagules are available or introduced, community structure and ecosystem function will develop and follow a trajectory or pathway that proceeds toward that of the reference wetland. The shape and slope of the trajectories and the time it takes to achieve equivalence will vary depending on the attribute that is measured. However, if the proper environmental template and/or source of propagules are not established initially, trajectories may converge more slowly or they may proceed in a direction that never converges with the reference wetland.

Performance Standards

Performance standards consist of measurements that describe the environmental template as well as attributes of biotic and abiotic structures and function that develop over time. The environmental template consists of the physical characteristics, hydrology including water depth, duration, frequency, and timing/seasonality and chemical characteristics of the floodwaters. Postrestoration monitoring of structure and function enables one to determine whether the site is converging toward that of a natural wetland of the same type, a *reference wetland,* and to determine the amount of time needed to achieve complete reestablishment of structure and function.

A number of attributes have been suggested as performance standards to gauge success of restoration projects. The Society of Ecological Restoration (2004) proposed three categories of ecosystem attributes from which measurements should be made: (1) diversity, (2) vegetation structure, and (3) ecosystem processes (Ruiz-Jaen and Aide, 2005). Neckles et al. (2002) proposed a monitoring protocol for restored salt marshes, including performance standards describing hydrology, soil physicochemical properties, vegetation, nekton, and birds that were tested using 36 restoration projects in the Gulf of Maine (Konisky et al., 2006). Anisfeld (2012) proposed a framework for monitoring biogeochemical responses to tidal marsh restoration. The performance standards presented in this chapter include many attributes proposed by Neckles et al. and Anisfeld as well as additional ones describing ecological and microbial processes.

Reference Wetlands—A Yardstick for Comparison

In this book, a reference wetland is defined as a natural, mature wetland of the same type as the wetland to be restored, ideally located in the same landscape position and watershed. It exhibits minimal human disturbance and represents a high-quality, well-functioning system. The definition corresponds to Brinson and Rheinhardt's (1996) reference standard, "...wetlands that exhibit the highest level of functioning across the suite of functions," used in the hydrogeomorphic approach to assessing wetland function (see Chapter 2, Definitions). A reference wetland is used as a "yardstick" to determine whether performance standards measured in the restored wetland are moving along a trajectory that will eventually converge with the reference wetland. Performance standards of the restored site are compared with the reference wetland and the results are expressed as percent equivalence to it. This ratio is referred to as the response ratio and is used to compare restored versus reference wetlands as well as restored versus degraded sites (Suding, 2011). Ideally, multiple reference wetlands should be sampled, recognizing the inherent variability of natural systems (Findley et al., 2002).

When setting expectations for individual restoration projects, it is necessary to consider the existing landscape, scale stressors, and processes (Simenstad et al., 2006; Hobbs, 2007) that affect success. Sometimes it is not possible to restore a site to its reference condition, as is the case in many agricultural and urban landscapes where external stressors cannot be ameliorated. In this case, one should focus on restoring the functions that are most likely to succeed.

The Environmental Template—Hydrology and Water Chemistry

Successful restoration depends on establishing the appropriate hydrology and water chemistry for a given wetland type (Table 10.1). Hydrology—depth, duration, frequency, and seasonality of inundation or soil saturation—must be measured following restoration and compared against reference wetlands sampled at the same time. In addition to plotting water levels over daily to annual cycles, there are other ways to characterize hydrology with detailed measurements of water level. For example, water table during the growing season and monthly median water table level during onset

Table 10.1 Performance Standards of Hydrology and Surface Water Used to Evaluate Environmental Conditions Necessary for Successful Wetland Creation and Restoration

Hydrology	Water level, depth of inundation, and soil saturation
	Duration of inundation and saturation
	Frequency of inundation and saturation
	Timing and seasonality of inundation and saturation
Surface water	Salinity
	Conductivity
	pH
	Inorganic N (NH_4^+, NO_3^-) and P (PO_4^{3-})[a]
	(SO_4^{2-})[a]
	Base cations (Ca^{2+}, Mg^{2+}, K^+, Na^+)[a]

[a]Secondary standards.

of the growing season are useful predictors of community composition of forested wetlands (Johnson et al., 2014). Chemistry of the floodwaters, including conductivity, salinity, and pH, is measured to ensure that the restored site has the same chemical template as the reference wetlands (Table 10.1). Nutrient chemistry is important at sites where eutrophication or atmospheric deposition of N and sulfate occurs and involves measurement of inorganic nutrients, N and P. Base cations (Ca, Mg) may be measured in wetlands fed by groundwater.

The frequency at which measurements are made depends on the availability of financial resources or volunteer labor. Measurements of hydrology should at least be made on a monthly basis. For tidal wetlands, marshes, and mangroves, it probably is sufficient to measure depth and duration of inundation over a single tidal cycle during spring and neap cycles (see Chapter 8, Tidal Marshes). Measurement of water chemistry may involve in situ measurements using portable devices such as a refractometer for salinity and probes for pH and conductivity. At the least, measurements should be made monthly or seasonally to capture cyclic changes related to hydrologic patterns. The nutrient chemistry of flooding waters is highly variable in space and time and frequent measurements may be required. A number of standard methods are available for analysis of chemical constituents that describe water quality (APHA, 1998).

Community Structure and Ecosystem Function

Performance standards are selected to describe the ecological community and its functions set forth by the project goals. Standards of community structure describe the organization and arrangement of biota, vegetation, fauna and microorganisms, and soil properties. Functions such as productivity, decomposition, and denitrification describe how energy (carbon) flows and nutrients (N, P) cycle within and through the system. Primary performance standards are those that should be measured, whereas secondary standards are useful if additional resources are available. It is important to select standards that determine whether project goals are being met. For example, assessment of finfish utilization is important when restoring tidal wetlands whereas measurement of C sequestration is important in bogs.

Prospective performance standards are shown in Tables 10.2 and 10.3 and, though not inclusive, they provide a summary of key attributes of community structure and ecosystem function to consider when evaluating the success of wetland restoration projects.

Community Structure

Primary performance standards of community structure describe vegetation, fauna, microorganisms, and soil (Table 10.2). Percent cover of vegetation—stem height and density, flowering stems, and aboveground biomass—describes the physical structure or architecture of the plant community. The density and number of species of epibenthic invertebrates, such as crabs and snails that live on the soil surface, are measured. Crabs are sampled indirectly based on crab hole counts or directly by excavation (Teal, 1958; Taylor and Allanson, 1993). Snails are counted (Silliman et al., 2005).

Table 10.2 Primary and Secondary Performance Standards of Community Structure Used to Evaluate Ecosystem Development of Created and Restored Wetlands

	Primary	Secondary
Vegetation	Cover Stem height Stem density Flowering stems Aboveground biomass[a] Number of species[b] Species diversity[c]	Belowground biomass
Fauna	Epifauna density Epifauna number of species Epifauna species diversity	Epifauna biomass
		Infauna density Infauna diversity
	Nekton density[d] Nekton diversity Avian (bird) density Avian diversity	Nekton length/age structure Nekton biomass
Microorganisms		Microbial numbers/density Microbial biomass Microbial diversity
Soils	Bulk density Organic matter content Organic carbon Total nitrogen pH	Total phosphorus Available nitrogen Available phosphorus Organic matter quality (i.e., lignin)
Porewater		Salinity Conductivity pH Sulfide

[a]Total versus invasive species.
[b]Total, invasive, and rare species.
[c]See text for discussion of various methods.
[d]Includes finfish, shrimp, crabs, and others that swim.

Finfish and motile shellfish, such as shrimp and crabs, collectively are known as nekton. The density and number of species are counted during periods when floodwaters inundate the wetland and are sampled using nets or traps. Birds, the most charismatic of animals that rely on wetlands, are enumerated and identified by species. This is done by visual identification or identification of their call or song (Brittain et al., 2010).

Community structure also is described using various measurements of species number (richness) and indices of species diversity. Diversity indices take into account not only the number of species present but also their relative abundance. This should

Table 10.3 **Primary and Secondary Performance Standards of** *Ecosystem Function* **Used to Evaluate Ecosystem Development of Created and Restored Wetlands**

	Primary	Secondary
Vegetation	Stem height	Aboveground NPP
	Stem density	Belowground NPP
	Aboveground biomass[a]	
	Belowground biomass	
Fauna	Epifauna density	Infauna biomass
	Epifauna biomass	Epifauna trophic structure
	Infauna density	Infauna trophic structure
	Nekton density[b]	Nekton trophic structure
	Nekton biomass	
	Avian (bird) density	Avian trophic structure
Microorganisms	Microbial biomass	Microbial/soil respiration
		Denitrification
		Methanogenesis
		Sulfate reduction[c]
		Fe reduction
Soils	Bulk density	Total P
	Organic matter	Available N
	Organic C	Available P
	Total N	Organic matter quality
	pH	Accretion and sedimentation
		Carbon sequestration
		Nitrogen accumulation
		Phosphorus accumulation
Porewater		Salinity
		Conductivity
		pH
		Sulfide

[a]Total versus invasive species.
[b]Includes finfish, shrimp, crabs, and others that swim.
[c]Estuarine wetlands only.

be done for different groups of organisms, vegetation, invertebrates, finfish, and birds. Common measures of species diversity include Shannon's and Simpson's indices (Odum and Barrett, 2005). The number and abundance of rare and invasive species also should be noted since these species may contribute disproportionately, either positively or negatively, to the overall quality of the community. Matthews et al. (2009) employed a number of diversity standards for evaluating the vegetation of restored freshwater wetlands. These included native species richness, the Floristic Quality Index (FQI), Coefficient of Conservatism (C of C), *Carex* species richness, number of perennial species, and abundance of hydrophytic species. The FQI and C of C describe the overall quality of the plant community using regional lists (Hopple and Craft,

2013; Matthews et al., 2009). Individual species are ranked based on their quality, rare species being of high quality (C of C = 10), and weedy or invasive species being of low quality (C of C = 1). The FQI ranking is used to calculate a single numerical value for the plant community, with a higher value corresponding to a higher-quality community.

Soil properties are important standards of wetland structure since they provide physical and nutritional support for plants and fauna. Soil organic matter is key since it serves as a reservoir of nutrients, especially nitrogen, for vegetation and source of organic carbon (C) for detrital food webs. Organic matter by loss on ignition is measured by combusting a dried sample of known weight, then reweighing it (Nelson and Sommers, 1996). Measurement of soil N is important because it is the nutrient that typically limits plant growth. It is measured by Kjeldahl digestion or more commonly by dry combustion (Dumas method) using a CHN analyzer (Bremner, 1996). The advantage of using a CHN analyzer is that it measures carbon simultaneously. Measurement of C should focus on organic carbon, by far the dominant form of C in wetland soils. For soils that contain appreciable carbonate C, these should first be stripped out using dilute acid (HCl) (Hedges and Stern, 1984). Bulk density and pH describe fundamental, physical, and chemical properties of soils, respectively. Bulk density is determined gravimetrically using intact dried soil cores of known volume (Blake and Hartge, 1986). pH is determined using a hydrogen ion (pH) electrode placed in a slurry of soil and water (Thomas, 1996).

Oxidation–reduction potential, also known as redox potential, is used to determine the intensity of anaerobic conditions. It is a gauge of the relative abundance of oxidized (O_2) and reduced (hydrogen ions) chemical species and is measured using an electrical meter and platinum and reference electrodes. Measurement of redox potential in the field is difficult due to its great spatial and temporal variability. Rather, inferring the development of anaerobic condition in the field is better done using indicators of hydric soils such as low chroma or indicator of reduction in soils (IRIS) tubes that demonstrate the presence of reduced (ferrous) iron and organic matter enrichment of the surface soil (see Chapter 2, Definitions).

Secondary performance standards of community structure include belowground biomass of vegetation, epifauna, benthic infauna and microorganisms, and additional soil properties. Secondary standards, while more difficult to measure and requiring more resources than primary standards, provide more refined information of community structure. Belowground biomass (roots and rhizomes) is determined by collection of soil cores of known volume where the soil material is washed through a sieve and the roots and rhizomes are retained, dried, and weighed (Craft, 2013). Cores may be subdivided into different depths (e.g., 0–10, 10–30 cm) as most roots and rhizomes are concentrated near the soil surface. Epifauna biomass is determined by collecting, drying, and weighing crabs, snails, and other surface-feeding invertebrates from a plot of known area ranging from $0.25 \, m^2$ for crabs to $2 \, m^2$ for bivalves or by using a large diameter (1.8 m) drop net (Minello and Webb, 1997).

Benthic infauna are invertebrates that live in the soil. They create burrows and feed at the surface or on subsurface materials, usually organic matter and associated microorganisms. Infauna include nematodes and oligochaete and polychaete worms that are commonly found in estuarine wetlands, whereas larvae of flies (dipterans) dominate in freshwater wetlands (Craft, 2001). Infauna are sampled by collecting small diameter soil cores usually to a shallow depth, 5 cm (Sacco et al., 1994). Cores are fixed in the field with a preservative, formalin or alcohol, and stained with a dye such as Rose Bengal. Infauna are picked by hand. With proper expertise, they may be sorted into different taxonomic groups to characterize species richness and trophic (feeding) groups. Terrestrial insects are sampled using a vacuum suction sampler (McCall and Pennings, 2012). Sampling to characterize finfish biomass, length, and dry weight is done using seines, nets, or traps (Freeman et al., 1984; Bonar et al., 2009), and catch per unit effort between restored and reference wetlands is compared. The age structure of the population may be determined from length measurements.

Measurement of microbial numbers and diversity is time-consuming, expensive, and requires considerable expertise. It is easier and probably more useful to measure microbial contributions to wetland function (see *Ecosystem Functions* below). However, one measure of microbial structure is biomass that is relatively simple to measure. Microbial biomass is highly labile compared to other sources of organic matter and can be extracted from soils using chloroform (Jenkinson and Powlson, 1976). While this method does not distinguish between the various consortia of microorganisms, it does provide a measure of their relative abundance.

Secondary performance standards describing soils include total phosphorus (P), available forms of N (NH_4^+, NO_3^-) and P (PO_4^{3-}), and organic matter quality. Available N and P are extracted with various dilute salt solutions, 2N KCl for NH_4^+ and NO_3^- (Mulvaney, 1996), and NH_4F, NaH_2CO_3, or other salts for P (Kuo, 1996). Extractions are analyzed using the same colorimetric techniques used for N and P in surface waters. Measurement of total P involves digesting soil in hot, concentrated acid, then analyzing the diluted digest colorimetrically. The quality of organic matter is determined by measuring the lignin content of the soil. Lignin, a measure of the recalcitrant fraction of soil organic matter is extracted using concentrated sulfuric acid (Ryan et al., 1990). Another measure of organic matter quality is the lignin:nitrogen ratio that incorporates the relative abundance of both labile (nitrogen) and recalcitrant (lignin) components of organic matter.

Similar to surface waters, porewater salinity, conductivity, and pH provide useful information on physicochemical characteristics of aqueous soil environment. These are measured in situ with a refractometer (salinity) or electrical probes. Porewater sulfide is a measure of sulfate reduction in the soil. It is especially important in estuarine wetlands where sulfate from seawater is abundant and is reduced to sulfide by anaerobic microorganisms. Sulfide is toxic to many wetland plants as it inhibits nitrogen uptake (Bradley and Morris, 1990; Chambers, 1997; Chambers et al., 1998). It is measured using a sulfide electrode after the sulfide is fixed chemically (APHA, 1998).

Ecosystem Function

Performance standards describing ecosystem function are used to characterize development of processes such as productivity, energy flow, and nutrient cycling (Table 10.3). Primary standards such as stem height, stem density, and above- and belowground biomass, while measures of structure, also can serve as indices of productivity. Aboveground plant production is determined by nondestructive repeated measurements of stem height and density of herbaceous vegetation (Morris, 2007). For woody vegetation, litterfall is measured using litter traps while woody increment is measured with dendrometer bands (Conner and Cherry, 2013). Allometric equations using tree height and diameter as well as coring methods are used.

Belowground production is more difficult to measure. Techniques include root ingrowth into bags filled with peat or sand, sequential coring over time to sample roots and rhizomes, and mini-rhizotrons, small cameras that nondestructively photograph the root environment (Conner and Cherry, 2013; Craft, 2013). All of these methods have limitations. The ingrowth and sequential coring methods require much labor to sort and pick roots. The mini-rhizotron method is more commonly used in terrestrial soils (Tierney and Fahey, 2007) and has not been widely used in waterlogged or saturated soils. See Fahey and Knapp (2007) for a complete review of methods for measuring primary production of mosses, herbaceous vegetation, and woody vegetation.

Epifauna, benthic infauna, and nekton density and biomass also serve as proxies for productivity (Table 10.3). Using mesocosms placed within the vegetated edge of restoration and reference marshes, Rozas and Minello (2009) measured growth rates of key nekton species. They recommended that growth rate as well as mortality be used to assess tidal marsh restoration projects. Food webs of invertebrates and nekton may be measured by trophic or feeding group analysis to evaluate carbon flow through the system. Organisms are classified by feeding habitat, herbivore, carnivore, or detritivore, to evaluate the number and types of trophic linkages. Food web analysis requires a thorough knowledge of feeding habits of each taxa or species (see, for example, Sacco et al., 1994). Likewise, trophic structure of the avian community also can be inferred in this way.

Microorganisms are key agents of energy flow and nutrient cycles. They are essential for the conversion of labile and recalcitrant organic matter to CO_2 and CH_4 using a variety of terminal electron acceptors including Fe and SO_4. They mediate key processes in the N cycle including denitrification. They also produce enzymes that mineralize various organic C, N, and P compounds, including amylase (C), urease (N), and phosphatase (P).

Microbial function is estimated by measuring their biomass using chloroform extraction (Jenkinson and Powlson, 1976) and other methods (Paul et al., 1999; Rinklebe and Langer, 2013). Respiration, an indicator of microbial activity, is performed using laboratory incubations of soil (Bridgham and Ye, 2013) or in situ with chambers (Holland et al., 2007). With both methods, it is important to exclude CO_2 respired by plant roots by removing roots prior to incubation. Measurement of microbial processes such as Fe reduction, denitrification, and methanogenesis, is done using incubations and measurement of anaerobic decomposition products such as Fe^{2+} and Mn^{2+} (Pentrak et al., 2013), N_2O (Groffman et al., 1999; Burgin et al., 2013), and CH_4 (Inglett et al., 2013). Sulfate reduction, important in saline tidal marshes and mangroves, requires measurement of changes in reduced sulfur (e.g., H_2S, FeS, FeS_2—pyrite) pools (Ulrich et al., 1997) or by labeling soils with radioactive ^{35}S (Cornwell, 2013). Microbial communities are characterized by a number of other methods, including enzyme assays, substrate utilization profiles, lipid analysis, and nucleic acid analysis (Sinsabaugh et al., 1999). Standard methods for these and other soil-based analyses are available in Sparks (1996), Robertson et al. (1999), and DeLaune et al. (2013).

Measurement of vertical soil accretion, sediment deposition, C sequestration, and N and P accumulation in soil may be done. These responses typically occur slowly and over long time scales. As such, measurements are made less frequently (monthly, annually, or even longer) than measures of the biota. Plots containing marker layers such as feldspar may be established (Callaway et al., 2013) (Figure 10.3(b)). Periodically, cores may be collected and the volume (mm) and mass (g) of the dried sediment measured. Organic C and N and P of the dried sediment also may be measured to estimate rates of C sequestration and N and P accumulation in soil (Noe and Hupp, 2005).

Trajectories of Ecosystem Development

With removal of stressors, reestablishment of hydrology, and sometimes assisted colonization, ecosystem attributes will follow a trajectory or pathway toward a self-sustaining and fully functioning wetland. Models of succession and ecosystem development (see Chapter 3, Ecological Theory and Restoration) offer insight into the patterns, pathways, and progress toward the desired endpoint community. Different performance standards develop at different rates and sometimes along different trajectories. Life history traits such as size and age to reproduction determine how quickly an organism will arrive, propagate, and colonize a site. The availability of propagules and their ability to disperse to a site is important (Galatowitsch, 1996; Donath et al., 2003). Dispersal often is stochastic, occurring infrequently and randomly. On some sites, assisted colonization may be needed to introduce the desired species in a timely manner. Herbaceous vegetation, with their short (annual, biennial, perennial) life cycles, will reestablish and mature more quickly than woody vegetation (Figure 10.1(a)). Ecosystem development of wetlands with predictable, pulsing hydrology such as that occurs in tidal wetlands and floodplains is faster than in closed, nonpulsed wetlands (Figure 10.1(b)). Ecosystem functions such as productivity and nutrient cycling are likely to develop faster than community structure (habitat and biodiversity) (Figure 10.1(c)).

Once established, trajectories of measured performance standards may be linear or curvilinear, including asymptotic or rise to maximum (negative exponential) and peaked (double exponential) trajectories. Evidence to support such trajectories comes from studies of chronosequences of restoration sites including saline tidal marsh (Morgan and Short, 2002; Craft et al., 2003), freshwater forest (Meyer et al., 2008; Berkowitz, 2013), freshwater marsh (Jinbo et al., 2007; Matthews et al., 2009), and mangrove (Osland et al., 2012). Sometimes, performance standards follow undesirable trajectories. This is true in the case of invasive species such as *Phragmites* dominance of intertidal brackish water "high marshes" (see Chapter 11, Case Studies) and *Phalaris* or *Typha* dominance of freshwater marshes (Matthews

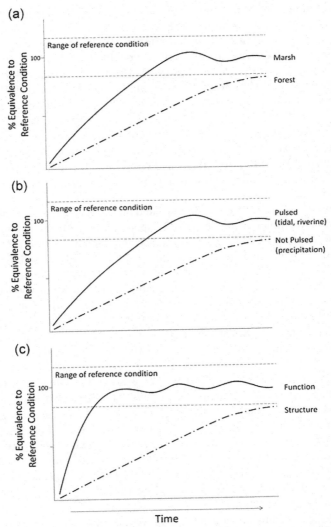

Figure 10.1 Trajectories describing changes during ecosystem development of (a) marsh versus forest; (b) pulsed versus not pulsed; and (c) wetland function and structure of constructed saline tidal marshes.

and Endress, 2008; Matthews and Spyreas, 2010). In these situations, management is needed to place the restoration site on its preferred trajectory and guide it toward the desired endpoint. Sometimes no trajectories are evident with time (Zedler and Callaway, 1999), perhaps because the temporal period of monitoring is insufficient (Dawe et al., 2000) or because there is an irreversible transition to an alternative stable state that is different from what existed originally (Lindig-Cisneros et al., 2003). Hobbs and Norton (1996) observed that, for many restorations, once a system crosses such a threshold to another stable state, substantial management and effort is needed to restore it to its original condition.

Transition of a restoration site to an alternative stable state suggests that the environmental template was not properly established or that the stressors were not removed. Stochastic events such as floods (Matthews et al., 2009), outbreaks of disease (Zedler and Callaway, 1999; Klotzli and Grootjans, 2001), or other unplanned disturbances (Zedler and Callaway, 2000) may hinder or stall trajectories toward the desired endpoint. Monitoring of restored wetlands in Germany revealed that, after 40 years, successional patterns varied among restored wetlands, marshes, and grasslands, and sometimes along unforeseen pathways (Klotzli and Grootjans, 2001). According to Klotzli and Grootjans, the unpredictability of trajectories observed in restored wetlands and in natural wetlands is not unusual and should be expected.

Trajectories are identified by making repeated measurements at the restoration and reference sites over time. Examples of this approach are shown in Simenstad and Thom (1996), Craft et al. (1999, 2002), Zedler and Callaway (1999), Mitsch et al. (2012), and in Chapter 11, Case Studies, Figure 11.2. A widely used method is the space-for-time or chronosequence approach (Pickett, 1989) where restoration sites with similar hydroperiod, water chemistry/salinity, and vegetation type, but different ages, are sampled. The advantage of the chronosequence approach is that one can compress time by sampling the different age wetlands in a relatively short period of time. Examples of the approach are illustrated in Craft (2000), Craft and Sacco (2003), Craft et al. (2003), Jinbo et al., 2007, Meyer et al. (2008), Matthews et al. (2009), Osland et al. (2012), and Berkowitz (2013).

Mitsch et al. (2012) periodically followed ecosystem development of two created freshwater marshes at the Olentangy River Wetland Research Park, Ohio State University, over a 15-year period. One marsh was planted with 13 species whereas the second marsh was unplanted. Plant species richness increased in both marshes. After 15 years, the planted marsh contained 101 species compared to 97 in the unplanted marsh. Soil bulk density in the surface 8–10 cm decreased over time, from $1.30\,g/cm^3$ at time zero to $0.61\,g/cm^3$ (planted marsh) and $0.71\,g/cm^3$ (unplanted marsh). Soil carbon increased during the 15-year period, from 1.6% to 3.6% in the planted marsh and to 4.6% in the unplanted marsh.

Some of the best-documented trajectories of wetland ecosystem development come from saline tidal marshes. Craft et al. (2003) employed the chronosequence approach to investigate ecosystem development of eight constructed saline tidal marshes along the North Carolina coast. The eight sites were inundated twice daily by the astronomical tides, had floodwater salinities of 20–30 psu, and were planted with smooth cordgrass, *Spartina alterniflora*. At the time of sampling, sites ranged in age from 1 to 28 years old. Each site was paired with a nearby natural, mature reference marsh for comparison. A number of performance standards were measured, mostly those related to function.

Many performance standards, above- and belowground biomass, benthic invertebrate density, taxa richness (Figure 10.2), and trophic groups (see Craft and Sacco, 2003) exhibited asymptotic trajectories with time. Sediment deposition exhibited a

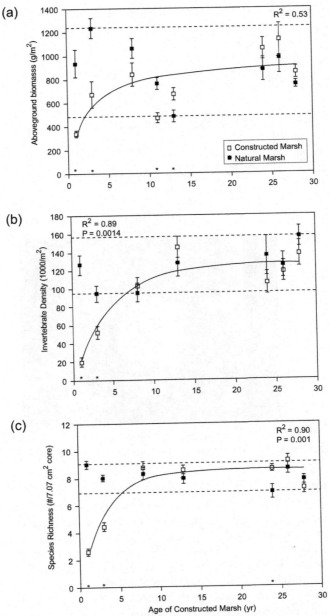

Figure 10.2 Asymptotic trajectory describing the increase in (a) aboveground biomass, benthic invertebrate (b) density and (c) species richness with time along a chronosequence constructed saline tidal marsh.
Adapted with permission from Craft et al. (2003). Reproduced with permission of the Ecological Society of America.

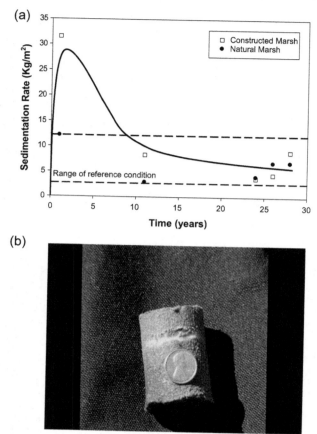

Figure 10.3 (a) Peaked trajectory of sediment deposition with time in a constructed saline tidal marsh. (b) Sediment deposition was measured atop feldspar marker layers placed in the marshes.

peaked trajectory (Figure 10.3(a)) as it initially was high as vegetation colonized the site and then declined to a level that was controlled by the long-term rate of sea level rise. Soil properties such as bulk density, percent organic C and N, and C and N pools exhibited linear trajectories. Soil development proceeded faster in surface than in sub-surface layers (Figure 10.4). Enrichment of the soil with fine particles such as silt and clay also was linear (Craft et al., 2003). Standards describing *Spartina* stem height, algal biomass (chlorophyll a), algae and diatom taxa richness, and some invertebrate taxa also exhibited a linear increase with time (Craft and Sacco, 2003; Zheng et al., 2004) as did microbial processes, decomposition, CH_4 production, and denitrification (Craft et al., 2003; Broome and Craft, 2009; Cornell et al., 2007).

Some standards exhibited no clear trajectory. For example, stem density of *Spartina* did not vary with age though young constructed marshes had more stems than reference marshes (Craft et al., 2003). Soil P also did not vary with marsh age (Craft et al., 2003), indicating the relative abundance of P in these N-limited wetlands.

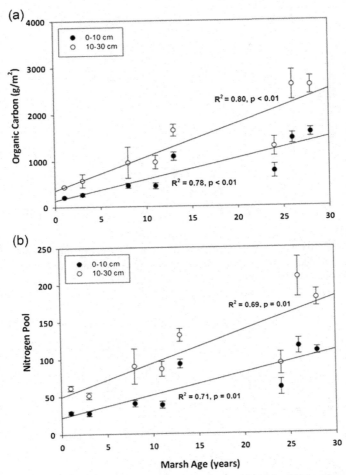

Figure 10.4 Increase in soil (a) organic C and (b) N pools (0–30 cm) with time in a constructed saline tidal marsh. Note that the rate of C and N enrichment is greater in surface (0–10 cm) than in subsurface (10–30 cm) soil.

Many performance standards describing heterotrophic activity were linked to soil organic matter. Microbial activity measured as CO_2 production increased linearly with soil organic C (Craft et al., 2003; Cornell et al., 2007) as did CH_4 production (Figure 10.5) and denitrification (Broome and Craft, 2009). Benthic invertebrates, density and diversity, also were positively related to soil organic C (Craft, 2000) but less so than for microorganisms (Craft and Sacco, 2003). Studies of freshwater marshes also suggest that soil organic matter is useful for tracking development of other standards of C and N cycling following restoration (Meyer et al., 2008).

A conceptual model describing ecosystem development of constructed saline tidal marshes was developed by Craft et al. (2003) (Figure 10.6). Physical processes driven by hydrology developed quickly and exhibited a peaked trajectory.

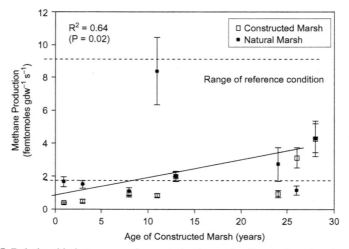

Figure 10.5 Relationship between methane production and soil organic carbon in constructed and natural saline tidal marshes. Other microbial processes, decomposition, CO_2 production, and denitrification, also are positively related to soil organic C.

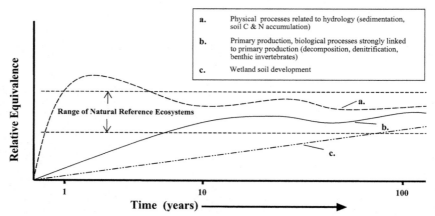

Figure 10.6 Trajectories describing changes in structural and functional performance standards during ecosystem development of constructed saline tidal marshes. See text for detailed explanation of the curves.
Adapted with permission from Craft et al. (2003). Reproduced with permission of the Ecological Society of America.

Sediment deposition and rates of organic matter and N accumulation increased to super-optimum levels as soon as hydrology was reintroduced and vegetation was reestablished, then declined. Standards describing biological processes developed more slowly and exhibited asymptotic or linear trajectories. Soil properties developed very slowly along linear trajectories, perhaps requiring decades to a century for subsoils to develop characteristics found in reference marshes.

All reference marsh soils (0–30 cm) contained at least 1000 g organic C/m^2 and $100 \, g \, N/m^2$ whereas the younger, less than 10–15 years old, constructed marshes contained less C and N. Bradshaw (1983) suggested that restoration of terrestrial ecosystems such as surface-mined lands required at least $100 \, g \, N/m^2$ to develop a self-sustaining system. Since N often limits plant growth, the $100 \, g \, N/m^2$ in the soil represents the N capital needed to restore plant-based functions (i.e., productivity) on the site. For detritus-based ecosystems like wetlands, the $1000 \, g \, C/m^2$ in the soil represents the organic matter (C) capital needed to support a self-sustaining heterotrophic community.

In summary, a properly restored wetland is capable of becoming a fully functioning system given time. Different performance standards require different amounts of time to develop. Some performance standards serve as proxies for development of other standards. A good example is soil organic matter that predicts development of the heterotrophic community, microorganisms, and benthic infauna. Inability of organisms to disperse to the site may slow ecosystem development and stochastic events such as floods and drought may slow, halt, or even reverse their development. Finally, invasive species may push the site toward an undesirable and perhaps irreversible trajectory necessitating the need for active management to guide the system toward its desired endpoint.

References

Anisfeld, S.C., 2012. Biogeochemical responses to tidal restoration. In: Roman, C.T., Burdick, D.M. (Eds.), Tidal Marsh Restoration: A Synthesis of Science and Management. Island Press, Washington, D.C., pp. 39–58.

APHA (American Public Health Association), 1998. Standard Methods for the Examination of Water and Wastewater. American Public Health Association, Washington, D.C.

Berkowitz, J.F., 2013. Development of restoration trajectory metrics in reforested bottomland hardwood forests applying a rapid assessment approach. Ecological Indicators 34, 600–606.

Blake, G.R., Hartge, K.H., 1986. Bulk density. In: Klute, A. (Ed.), Methods of Soil Analysis, Part I. Physical and Mineralogical Methods: Agronomy Monograph No. 9, second ed. American Society of Agronomy, Madison, Wisconsin, pp. 363–375.

Bonar, S.A., Hubert, W.A., Willis, D.W. (Eds.), 2009. Standard Methods for Sampling North American Freshwater Fishes. American Fisheries Society, Bethesda, Maryland.

Bradley, P.M., Morris, J.T., 1990. Influence of oxygen and sulfide concentration on nitrogen uptake kinetics in Spartina alterniflora. Ecology 71, 282–287.

Bradshaw, A.D., 1983. The reconstruction of ecosystems. Journal of Applied Ecology 20, 1–17.

Bremner, J.M., 1996. Nitrogen – total. In: Sparks, D.L. (Ed.), Methods of Soil Analysis. Part 3– Chemical Methods. Book Series No. 5. Soil Science Society of America. American Society of Agronomy, Madison, Wisconsin, pp. 1083–1121.

Bridgham, S.D., Ye, R., 2013. Organic matter mineralization and decomposition. In: DeLaune, R.D., Reddy, K.R., Richardson, C.J., Megonigal, J.P. (Eds.), Methods in Biogeochemistry of Wetlands. SSSA Book Series 10. Soil Science Society of America, Madison, Wisconsin, pp. 385–406.

Brinson, M.M., Rheinhardt, R., 1996. The role of reference wetlands in functional assessment and mitigation. Ecological Applications 6, 69–76.

Brittain, R.A., Meretsky, V., Craft, C.B., 2010. Avian communities of the Altamaha River estuary in Georgia USA. The Wilson Journal of Field Ornithology 122, 532–544.

Broome, S.W., Craft, C.B., 2009. Tidal marsh creation. In: Perillo, G.M.E., Wolanski, E., Cahoon, D., Brinson, M.M. (Eds.), Coastal Wetlands. Elsevier B.V., Amsterdam, The Netherlands, pp. 715–736.

Burgin, A.J., Hamilton, S.K., Gardner, W.S., McCarthy, M.J., 2013. Nitrate reduction, denitrification, and dissimilatory nitrate reduction to ammonium in wetland sediments. In: DeLaune, R.D., Reddy, K.R., Richardson, C.J., Megonigal, J.P. (Eds.), Methods in Biogeochemistry of Wetlands. SSSA Book Series 10. Soil Science Society of America, Madison, Wisconsin, pp. 519–537.

Callaway, J.C., Cahoon, D.R., Lynch, J.C., 2013. The surface elevation table-marker method for measuring wetland accretion and elevation dynamics. In: DeLaune, R.D., Reddy, K.R., Richardson, C.J., Megonigal, J.P. (Eds.), Methods in Biogeochemistry of Wetlands. SSSA Book Series 10. Soil Science Society of America, Madison, Wisconsin, pp. 901–917.

Chambers, R.M., 1997. Porewater chemistry associated with *Phragmites* and *Spartina* in a Connecticut tidal marsh. Wetlands 17, 360–367.

Chambers, R.M., Mozdzer, T.J., Ambrose, J.C., 1998. Effects of salinity and sulfide on the distribution of *Phragmites australis* and *Spartina alterniflora* in a tidal salt marsh. Aquatic Botany 62, 161–169.

Conner, W.H., Cherry, J.A., 2013. Plant productivity–bottomland hardwood forests. In: DeLaune, R.D., Reddy, K.R., Richardson, C., Megonigal, J.P. (Eds.), Methods in Biogeochemistry of Wetlands. SSSA Book Series. Soil Science Society of America, Madison, Wisconsin, pp. 225–242.

Cornell, J.A., Craft, C., Megonigal, J.P., 2007. Ecosystem gas exchange across a created salt marsh chronosequence. Wetlands 27, 240–250.

Cornwell, J.C., 2013. Measurement of sulfate reduction in wetland soils. In: DeLaune, R.D., Reddy, K.R., Richardson, C.J., Megonigal, J.P. (Eds.), Methods in Biogeochemistry of Wetlands. SSSA Book Series 10. Soil Science Society of America, Madison, Wisconsin, pp. 765–773.

Craft, C.B., 2000. Co-development of wetlands soils and benthic invertebrate communities following salt marsh creation. Wetlands Ecology and Management 8, 197–207.

Craft, C.B., 2001. Biology of wetland soils. In: Richardson, J.L., Vepraskas, M.J. (Eds.), Wetland Soils: Their Genesis, Hydrology, Landscape and Separation into Hydric and Nonhydric Soils. CRC Press, Boca Raton, Florida, pp. 107–135.

Craft, C.B., 2013. Emergent macrophyte biomass production. In: DeLaune, R.D., Reddy, K.R., Richardson, C., Megonigal, J.P. (Eds.), Methods in Biogeochemistry of Wetlands. SSSA Book Series. Soil Science Society of America, Madison, Wisconsin, pp. 137–154.

Craft, C.B., Sacco, J.N., 2003. Long-term succession of benthic infauna communities on constructed *Spartina alterniflora* marshes. Marine Ecology–Progress Series 257, 45–58.

Craft, C., Broome, S., Campbell, C., 2002. Fifteen years of vegetation and soil development after brackish-water marsh creation. Restoration Ecology 10, 248–258.

Craft, C., Reader, J., Sacco, J.N., Broome, S.W., 1999. Twenty five years of ecosystem development of constructed *Spartina alterniflora* (Loisel) marshes. Ecological Applications 9, 1405–1419.

Craft, C.B., Megonigal, J.P., Broome, S.W., Cornell, J., Freese, R., Stevenson, R.J., Zheng, L., Sacco, J., 2003. The pace of ecosystem development of constructed *Spartina alterniflora* marshes. Ecological Applications 13, 1417–1432.

Dawe, N.K., Bradfield, G.E., Boyd, W.S., Trethewey, D.E.C., Zolbrod, A.N., 2000. Marsh creation in a northern Pacific estuary: is thirteen years of monitoring vegetation dynamics enough? Conservation Ecology 4, 12.

DeLaune, R.D., Reddy, K.R., Richardson, C.J., Megonigal, J.P., 2013. Methods in Biogeochemistry of Wetlands. SSSA Book Series 10. Soil Science Society of America, Madison, Wisconsin.

Donath, T.W., Hozel, N., Otte, A., 2003. The impact of site conditions and seed dispersal on restoration success in alluvial meadows. Applied Vegetation Science 6, 13–22.

Fahey, T.J., Knapp, A.K. (Eds.), 2007. Principles and Standards for Measuring Primary Production. Oxford University Press, New York.

Findley, S.E.G., Kiviat, E., Nieder, W.C., Blair, E.A., 2002. Functional assessment of a reference wetland set as a tool for science, management and restoration. Aquatic Science 64, 107–117.

Freeman, B.J., Greening, H.S., Oliver, J.D., 1984. Comparison of three methods for sampling fishes and macroinvertebrates in a vegetated freshwater marsh. Journal of Freshwater Ecology 2, 603–609.

Galatowitsch, S.M., 1996. Restoring prairie pothole wetlands: does the species pool concept offer decision-making guidance for re-vegetation? Applied Vegetation Science 9, 261–270.

Groffman, P.M., Holland, E.A., Myrold, D.D., Robertson, G.P., Zou, X., 1999. Denitrification. In: Robertson, G.P., Coleman, D.C., Bledsoe, C.S., Sollins, P. (Eds.), Standard Soil Methods for Long-term Ecological Research. Oxford University Press, New York. pp. 272–288.

Hedges, J., Stern, J., 1984. Carbon and nitrogen determinations of carbonate-containing solids. Limnology and Oceanography 29, 657–663.

Hobbs, R.J., 2007. Setting effective and realistic restoration goals: key directions for research. Restoration Ecology 15, 354–357.

Hobbs, R.J., Norton, D.A., 1996. Towards a conceptual framework for restoration ecology. Restoration Ecology 4, 93–110.

Holland, E.A., Robertson, G.P., Greenberg, J., Groffman, P.M., Boone, R.D., Gosz, J.R., 2007. Soil CO_2, N_2O and CH_4 exchange. In: Robertson, G.P., Coleman, D.C., Bledsoe, C.S., Sollins, P. (Eds.), Standard Soil Methods for Long-term Ecological Research. Oxford University Press, New York, pp. 185–201.

Hopple, A., Craft, C., 2013. Managed disturbance enhances biodiversity of restored wetlands in the agricultural Midwest. Ecological Engineering 61, 505–510.

Inglett, K.S., Chanton, J.P., Inglett, P.W., 2013. Methanogenesis and methane oxidation in wetland soils. In: DeLaune, R.D., Reddy, K.R., Richardson, C.J., Megonigal, J.P. (Eds.), Methods in Biogeochemistry of Wetlands. SSSA Book Series 10. Soil Science Society of America, Madison, Wisconsin, pp. 407–425.

Jenkinson, D.S., Powlson, D.S., 1976. The effects of biocidal treatments on metabolism in soil–V: a method for measuring the soil biomass. Soil Biology and Biochemistry 8, 209–213.

Jinbo, Z., Changchun, S., Shenmin, W., 2007. Dynamics of soil organic carbon and its fractions after abandonment of cultivated wetlands in northeast China. Soil and Tillage Research 96, 350–360.

Johnson, Y.B., Shear, T.H., James, A.L., 2014. Novel ways to assess forest wetland restoration in North Carolina using ecohydrological patterns from reference sites. Ecohydrology 7, 692–702.

Kentula, M.E., 2000. Perspectives on setting success criteria for wetland restoration. Ecological Engineering 15, 199–209.

Klotzli, F., Grootjans, A.P., 2001. Restoration of natural and semi-natural wetland systems in central Europe: progress and predictability of developments. Restoration Ecology 9, 209–219.

Konisky, R.A., Burdick, D.M., Dionne, M., Neckles, H.A., 2006. A regional assessment of salt marsh restoration and monitoring in the Gulf of Maine. Restoration Ecology 14, 516–525.

Kuo, S., 1996. Phosphorus. In: Sparks, D.L. (Ed.), Methods of Soil Analysis: Part 3. Chemical Methods. Soil Science Society of America, Madison, Wisconsin, pp. 869–919.

Lindig-Cisneros, R., Desmond, J., Boyer, K.E., Zedler, J.B., 2003. Wetland restoration thresholds: can a degradation transition be reserved with increased effort? Ecological Applications 13, 193–205.

Matthews, J.W., Endress, A.G., 2008. Performance criteria, compliance success, and vegetation development in compensatory mitigation wetlands. Environmental Management 41, 130–141.

Matthews, J.W., Spyreas, G., 2010. Convergence and divergence in plant community trajectories as a framework for monitoring wetland restoration progress. Journal of Applied Ecology 47, 1128–1136.

Matthews, J.W., Spyreas, G., Endress, A.G., 2009. Trajectories of vegetation-based indices to assess wetland restoration progress. Ecological Applications 19, 2093–2107.

McCall, B.D., Pennings, S.C., 2012. Disturbance and recovery of salt marsh arthropod communities following BP *Deepwater Horizon* oil spill. PLoS One 7 (3), E 32735.

Meyer, C.K., Bauer, S.G., Whiles, M.R., 2008. Ecosystem recovery across a chronosequence of restored wetlands in the Platte River valley. Ecosystems 11, 193–208.

Minello, T.J., Webb Jr., J.W., 1997. Use of natural and created *Spartina alterniflora* salt marshes by fishery species and other aquatic fauna in Galveston Bay, Texas. USA. Marine Ecology Progress Series 151, 165–179.

Mitsch, W.J., Zhang, L., Stefanik, K.C., Nahlik, A.M., Anderson, C.J., Bernal, B., Hernandez, M., Song, K., 2012. Creating wetlands: primary succession, water quality changes and self-design over fifteen years. BioScience 62, 237–250.

Morgan, P.A., Short, F.C., 2002. Using functional trajectories to track constructed salt marsh development in the Great Bay Estuary, Maine/New Hampshire. U.S.A. Restoration Ecology 10, 461–473.

Morris, J.T., 2007. Estimating net primary production of salt marsh macrophytes. In: Fahey, T.J., Knapp, A.K. (Eds.), Principles and Standards for Measuring Primary Production. Oxford University Press, New York, pp. 106–119.

Mulvaney, R.L., 1996. Nitrogen – inorganic forms. In: Sparks, D.L. (Ed.), Methods of Soil Analysis: Part 3. Chemical Methods. Soil Science Society of America, Madison, Wisconsin, pp. 1123–1200.

Neckles, H.A., Dionne, M., Burdick, D.M., Roman, C.T., Buchsbaum, R., Hutchins, E., 2002. A monitoring protocol to assess tidal restoration of salt marshes on local and regional scales. Restoration Ecology 10, 556–563.

Nelson, D.W., Sommers, L.E., 1996. Total carbon, organic carbon, and organic matter. In: Sparks, D.L. (Ed.), Methods of Soil Analysis: Part 3. Chemical Methods. Soil Science Society of America, Madison, Wisconsin, pp. 961–1010.

Noe, G.B., Hupp, C.R., 2005. Carbon, nitrogen and phosphorus accumulation in floodplains of Atlantic Coastal Plain rivers. Ecological Applications 15, 1178–1190.

Odum, E.P., Barrett, G.W., 2005. Fundamentals of Ecology, fifth ed. Thomson, Brooks/Cole, Belmont, California.

Osland, M.J., Spivak, A.C., Nestlerode, J.A., Lessman, J.M., Almario, A.E., Heitmuller, P.T., Russell, M.J., Krauss, K.W., Alvarez, F., Dantin, D.D., Harvey, J.E., From, A.S., Cormier, N., Stagg, C.L., 2012. Ecosystem development after mangrove wetland creation: plant-soil change across a 20-year chronosequence. Ecosystems 15, 848–868.

Paul, E.A., Harris, D., Klug, M.J., Reuss, R.W., 1999. The determination of microbial biomass. In: Robertson, G.P., Coleman, D.C., Bledsoe, C.S., Sollins, P. (Eds.), Standard Soil Methods for Long-term Ecological Research. Oxford University Press, New York, pp. 291–317.

Pentrak, M., Pentrakova, L., Stucki, J.W., 2013. Iron and manganese reduction-oxidation. In: DeLaune, R.D., Reddy, K.R., Richardson, C.J., Megonigal, J.P. (Eds.), Methods in Biogeochemistry of Wetlands. SSSA Book Series 10. Soil Science Society of America, Madison, Wisconsin, pp. 701–721.

Pickett, S.T.A., 1989. Space-for-time substitution as an alternative to long-term studies. In: Likens, G.E. (Ed.), Long-term Studies in Ecology: Approaches and Alternatives. Springer-Verlag, New York. pp. 110–135.

Rinklebe, J., Langer, U., 2013. Soil microbial biomass and phospholipid fatty acids. In: DeLaune, R.D., Reddy, K.R., Richardson, C.J., Megonigal, J.P. (Eds.), Methods in Biogeochemistry of Wetlands. SSSA Book Series 10. Soil Science Society of America, Madison, Wisconsin, pp. 331–348.

Robertson, G.P., Coleman, D.C., Bledsoe, C.S., Sollins, P., 1999. Standard Soil Methods for Long-term Ecological Research. Oxford University Press, New York.

Rozas, L.P., Minello, T.J., 2009. Using nekton growth as a metric for assessing habitat restoration by marsh terracing. Marine Ecology Progress Series 394, 179–193.

Ruiz-Jaen, M.C., Aide, T.M., 2005. Restoration success: how is it being measured? Restoration Ecology 13, 569–577.

Ryan, M.G., Melillo, J.M., Ricca, A.C., 1990. A comparison of methods for determining proximate carbon fractions of forest litter. Canadian Journal of Forest Research 20, 166–171.

Sacco, J.N., Seneca, E.D., Wentworth, T.R., 1994. Infaunal community development of artificially established salt marshes in North Carolina. Estuaries 17, 489–500.

Silliman, B.R., van de Koppel, J., Bertness, M.D., Stantion, L.E., Mendelssohn, I.A., 2005. Drought, snails, and large-scale die-off of southern U.S. salt marshes. Science 310, 1803–1806.

Simenstad, C.A., Thom, R.M., 1996. Functional equivalency trajectories of the restored Gog-Le-Hi-Te estuarine wetland. Ecological Applications 6, 38–56.

Simenstad, C.A., Reed, D., Ford, M., 2006. When restoration is not? Incorporating landscape-scale processes to restore self-sustaining ecosystems in coastal wetland restoration. Ecological Engineering 26, 27–39.

Sinsabaugh, R.L., Klug, M.J., Collins, H.P., Yeager, P.E., Petersen, S.O., 1999. Characterizing soil microbial communities. In: Robertson, G.P., Coleman, D.C., Bledsoe, C.S., Sollins, P. (Eds.), Standard Soil Methods for Long-Term Ecological Research. Oxford University Press, New York, pp. 318–348.

Sparks, D.L., 1996. In: Methods of Soil Analysis. Part 3–Chemical Methods. Book Series No. 5. Soil Science Society of America. American Society of Agronomy, Madison, Wisconsin.

Suding, K.N., 2011. Toward an era of restoration in ecology: successes, failures and opportunities ahead. Annual Review of Ecology, Evolution and Systematics 42, 463–487.

Taylor, D.I., Allanson, B.R., 1993. Impacts of dense crab populations on carbon exchanges across the surface of a salt marsh. Marine Ecology Progress Series 101, 119–129.

Teal, J.M., 1958. Distribution of fiddler crabs in Georgia salt marshes. Ecology 39, 185–193.

Tierney, G.L., Fahey, T.J., 2007. Estimating belowground primary production. In: Fahey, T.J., Knapp, A.K. (Eds.), Principles and Standards for Measuring Primary Production. Oxford University Press, New York, pp. 120–141.

Thomas, G.W., 1996. Soil pH and soil acidity. In: Sparks, D.L. (Ed.), Methods of Soil Analysis. Part 3 – Chemical Methods. Book Series No. 5. Soil Science Society of America. American Society of Agronomy, Madison, Wisconsin, pp. 475–490.

Ulrich, G.A., Krumholz, L.R., Suflita, J.M., 1997. A rapid and simple method for estimating sulfate reduction activity and quantifying inorganic sulfides. Applied and Environmental Microbiology 63, 1627–1630.

White, P.S., Walker, J.L., 1997. Approximating nature's variation: selecting and using reference information in restoration ecology. Restoration Ecology 5, 338–349.

Zedler, J.B., Callaway, J.C., 1999. Tracking wetland restoration: do mitigation sites follow desired trajectories? Restoration Ecology 7, 69–73.

Zedler, J.B., Callaway, J.C., 2000. Evaluating the progress of engineered tidal wetlands. Ecological Engineering 15, 211–225.

Zheng, L., Stevenson, R.J., Craft, C., 2004. Changes in benthic algal attributes during salt marsh restoration. Wetlands 24, 309–323.

Case Studies

11

Chapter Outline

Introduction

Goals of various wetland restoration projects differ depending on the historical and contemporary extent of the resource, ecosystem services and benefits they provide, laws and regulations to support their restoration, public interest, and financial resources. Historical and contemporary loss of wetlands and availability of potential restoration sites often determines amounts and types of wetlands to be restored. Ecosystem services they deliver to the local community and the larger public determines the goals and implementation of restoration projects. In the US, environmental laws that protect wetlands require restoration when wetland resources are lost to human activities. Financial resources to implement restoration projects is essential to ensure that projects are adequately designed, properly implemented, and monitored so that the restoration goals are met. In this chapter, five case studies are presented to illustrate restoration of wetlands including saline tidal marsh, freshwater marsh, peatland, forested wetland, and brackish marsh, in different landscape settings.

Saline Tidal Marshes

The first experimental saline tidal marsh creation and restoration projects were established in the 1960s along the North Carolina (USA) coast. Projects were funded by the U.S. Army Corps of Engineers to stabilize dredge material (Figure 11.1(a)) and eroding shorelines (Figure 11.1(b); Seneca et al., 1976; Broome et al., 1986). At these sites, attributes describing wetland community structure and ecosystem processes were measured and, today, are widely used as performance standards for restoration projects (see Chapter 10, Measuring Success: Performance Standards and Trajectories

Creating and Restoring Wetlands. http://dx.doi.org/10.1016/B978-0-12-407232-9.00011-7

Figure 11.1 Saline tidal marsh establishment on (a) dredged material and (b) eroding shoreline.
Photo credits: S.W. Broome.

of Ecosystem Development). With implementation of the Clean Water Act in the 1970s, saline tidal marshes increasingly were created and restored to mitigate for wetland loss from coastal development. Some projects involved creating wetlands where no wetland existed before, by grading terrestrial soils to intertidal elevation, then connecting them to the estuary (Broome et al., 1982, 1988). The effort was led by Dr. W.W. Woodhouse, Jr. and coworkers E.D. Seneca and S.W. Broome at North Carolina State University who pioneered the development of techniques to create and restore saline tidal marshes (Woodhouse, 1979; Woodhouse et al., 1974) and coastal dune habitat (Savage and Woodhouse, 1968; Woodhouse et al., 1976; Woodhouse, 1982).

Techniques (reestablishing hydrology, fertilization, planting techniques, and appropriate spacing of plants) used today were developed from these projects. Appropriate hydrology was determined by visiting natural saline tidal marshes and observing elevations at which species to be planted grow relative to the tidal frame (Broome et al., 1988). Fertilization trials were conducted on a number of sites and the outcome of these experimental plantings was that N, and sometimes P, was needed to quickly establish plant cover (Broome et al., 1975, 1982). This was particularly true for excavated terrestrial soils where organic matter and nutrient stocks were removed during excavation. Transplanting seedlings, especially greenhouse grown seedlings, was more reliable as compared to seeding (Broome et al., 1988). Greenhouse grown transplants were planted in pots containing peat and fertilizer and produced a well-developed root mat that survived planting shock better than field dug transplants. Furthermore, potting medium held water which was advantageous when planting during periods of low rainfall (Broome et al., 1988). Spacing experiments revealed that 45- and 60-cm spacing were successful for establishment on low elevation sites whereas 90-cm spacing was adequate on medium elevation sites (Broome et al., 1986). Closer spacing performed better on sites exposed to waves and boat traffic. Creating or restoring sites on high energy shorelines, those with long fetch and much boat traffic, should not be undertaken unless protective measures like nearshore breakwaters are incorporated into the project.

Plant species diversity is low in saline tidal marshes (see Chapter 8, Tidal Marshes), so most plantings consist of only a few species that grow as monocultures in distinct zones that correspond to duration of inundation during high tides and saline tidal marsh creation. Restoration projects should strive to reproduce these spatial patterns. Once vegetation is established, marsh vegetation will grow and spread as long as disturbances such as wrack and trash deposited by storm tides are removed. Herbivory by geese and mammals such as nutria also may be a problem, especially when transplants are young and tender.

Once established, the plant community quickly matured and aboveground biomass developed to levels comparable to or greater than that of natural reference marshes within 3 to 5 years (Figure 11.2(a,b)). Development of belowground biomass took longer, gradually increasing and achieving equivalence after 5–15 years. Different species accumulated biomass stocks at different rates with fast-growing C_4 species such as *Spartina* accruing biomass faster than C_3 species such as *Juncus*. Benthic invertebrates, an important component of saline tidal marsh food webs, also developed to levels found in natural marshes within 3 to 5 years (Table 11.1). Trophic groups such as surface and subsurface deposit feeding benthic infauna developed at rates comparable to the overall community (Table 11.1) though taxa that do not disperse readily (i.e., oligochaete worms), required longer (Craft and Sacco, 2003).

Soil properties developed more slowly than plant and invertebrate communities. For marshes established on eroding shorelines and dredged material, bulk density, organic C, and total N and P achieved levels comparable to natural marshes 10–20 years following restoration (Table 11.2). In contrast, soil formation was much slower on marshes created on terrestrial soils. Since their creation involves excavation and topsoil removal, the planting substrate was exposed subsoil with high bulk density, low hydraulic conductivity, and little organic C and N (Table 11.2). Following restoration, porosity, pH, and reduced forms of Fe and Mn increased (Craft et al., 2002). Porewaters also contained high dissolved Fe and Mn and low dissolved organic C, dissolved organic N and NH_4-N relative to reference marshes (Craft et al., 1991).

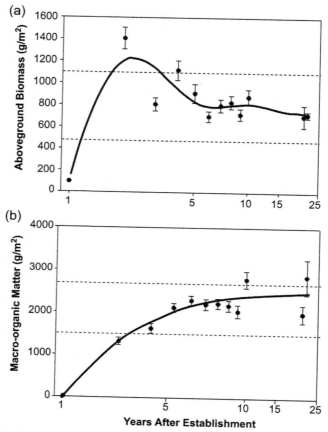

Figure 11.2 Change in (a) above- and (b) belowground biomass over time of saline tidal marsh vegetation planted on an eroding shoreline.
Modified from Craft et al. (1999).

Table 11.1 **Benthic Invertebrate Density and Feeding Groups Over Time on Saline Tidal Marshes Created on Dredged Material and on Eroding Shorelines Compared with Natural Reference Marshes**

Marsh	Age (year)	Density (no./m²)	Surface Feeders (%)	Subsurface Feeders (%)
Dredged material	16	49,000	41	59
	25	102,000	54	44
Reference marsh	–	31,000–69,000	34–66	34–37
Eroding shoreline	12	20,000	45	55
	21	32,000	28	62
Reference marsh	–	18,000–19,000	5–62	38–71

Modified from Craft et al. (1999).

Table 11.2 Change in Soil Properties (Bulk Density, Organic C and N, Total P, 0–30 cm) Over Time on Saline Tidal Marshes Created on Dredged Material, Eroding Shoreline, and Excavated Upland

Marsh	Age (year)	Bulk density (g/cm³)	Carbon (%)	Nitrogen (%)	Phosphorus (µg/g)
Dredged	14	1.06	0.95	0.04	138
material[a]	25	0.90	1.67	0.11	213
Eroding	14	1.40	0.31	0.02	434
shoreline[a]	25	1.28	0.75	0.04	437
Excavated	5	1.35	0.5	0.03	200
upland[b]	15	1.21	0.9	0.09	140

[a]Craft et al. (1999).
[b]Craft et al. (1991, 2002).

Freshwater Marshes

Freshwater marshes are restored in the US through federal programs, especially the Conservation and Wetland Reserve Programs (Fennessy and Craft, 2011; Marton et al., 2014b). The programs are administered by the US Department of Agriculture and pay farmers to voluntarily idle some of their cultivated acreage. Such lands often are marginally productive, being too wet in spite of drainage to produce an economical harvest in some years. In the Corn Belt of the Midwest US, large acreages have been restored to freshwater marshes by plugging drainage ditches and breaking tile drains (see Chapter 5, Freshwater Marshes).

In northwestern Indiana, the Nature Conservancy, a global nongovernmental conservation organization, acquired marginal agricultural lands beginning in the late 1990s with the goal of restoring the prairie–oak savanna–freshwater wetland landscape, historically known as the *Great Kankakee* marsh. The nearly 3000-ha Kankakee Sands Preserve has restored more than 640 ha of wetlands since 1998 (Ted Anchor, TNC, personal communication). Restoration involved, first, taking the land out of cultivation then reestablishing hydrology by plugging drainage ditches (Figure 11.3(a)). Reestablishing vegetation involved natural colonization and active management by introducing seeds and planting native wetland vegetation (Hopple and Craft, 2013). Removal of woody vegetation, prescribed fire, and mowing also were needed to reduce abundance of woody saplings and invasive herbaceous species that colonized in the interim between cessation of tillage and initiation of restoration activities. The culmination of these efforts was the establishment of a diverse assemblage of wetland vegetation within 10 years (Figure 11.3(b)). A comparison with reference marshes revealed no difference in species richness among restored and natural marshes (Table 11.3). There also was no difference in the Floristic Quality Assessment Index (Andreas and Lichvar, 1995), a measure of the quality of the plant community.

With active management, it was not surprising that overall plant species richness developed quickly. However, after 10 years, some differences in species composition of the restored marshes and natural marshes were noted. Restored marshes contained

Figure 11.3 (a) Freshwater marsh restoration at Kankakee Sands (Indiana, USA) by plugging and filling drainage ditches. (b) A marsh restored 10 years earlier. Note the diverse assemblage of vegetation made possible by active management (see text for explanation). Photo credits: Christopher Craft.

more low quality (weedy) species than reference marshes (Hopple and Craft, 2013). Restored marshes also contained fewer hydrophytic plant species (see Chapter 2, Definitions, Table 2.2), species that are more commonly found in wetlands with extended hydroperiod. Seeding clearly enhanced species diversity. On restored sites, seeding introduced eight new species as compared to 25 species that colonized unassisted (Hopple and Craft, 2013).

Compared to vegetation, there were greater differences in soil properties between the 10-year-old restored marshes and natural marshes. Restored marsh soils contained less organic matter, total N, total P, and moisture and greater bulk density, pH, and available P than reference marshes (Table 11.4). Water quality improvement functions such as P sorption and denitrification potential also were lower in restored marshes (Table 11.4). Compared to cultivated soils in the area, restored marshes contained less soil pH and available P (Marton et al., 2014a), suggesting that land-use legacies,

Table 11.3 Species Richness and Quality (Coefficient of Conservatism, Hydrophytic Indicator Status) of 10-Year-Old Restored Marshes and Natural Marshes

	Restored	Natural
Species richness (m²)	6.2	5.5
(Site)	34	27
Coefficient of Conservatism (% of total species)		
Opportunistic	14[a]	5
Tolerant	50[a]	36
Moderately tolerant	30[a]	51
Intolerant	6	7
Sensitive	0	1
		104
Hydrophytic Indicator Status (% of total species)		
Obligate	51[a]	65
Facultative wet	36	32
Facultative	5	2
Facultative upland	7[a]	<1
Upland	<1	<1

[a]Significantly different from the reference marshes.
From Hopple and Craft (2013).

Table 11.4 Soil Properties and Water Quality Improvement Functions of 10-Year-Old Restored Marshes and Natural Marshes in the Corn Belt of Northwestern Indiana, USA

	Restored	Natural
Soil Properties		
Bulk density (g/cm³)	1.05[a]	0.49
Moisture (%)	36[a]	60
pH	6.8[a]	5.2
Organic matter (%)	4.4[a]	15
Organic C (%)	2.2[a]	9.2
Total N (%)	0.18[a]	0.78
NH_4-N (µg/g)	2.4	2.7
Total P (µg/g)	230[a]	620
Available P (µg/g)	0.42[a]	0.15
Water Quality Improvement		
PSI[b] (mg/kg)	114[a]	297
Denitrification (ng N/g soil/h)	107[a]	329

[a]Significantly different from the reference marshes.
[b]Phosphorus sorption potential.
From Marton et al. (2014a).

liming and P fertilization, were declining following restoration. Changes in soil C, N, and P, though, were not evident after 10 years, perhaps due to the short hydroperiod characterized by summer drawdown and use of prescribed fire to maintain prairie vegetation. Fluxes of greenhouse gases (CO_2, CH_4, and N_2O) were low and were attributed to the relatively short hydroperiod (Richards and Craft, 2014).

In agricultural landscapes, restoring wetland functions such as denitrification and P sorption can help offset nutrient loadings from intensively cultivated agricultural lands that encompass the region. Even with the slow rate of (functional) development, after 10 years, the restored wetlands performed at a lower level than natural wetlands. Restoring wetlands on agricultural land is essential if we hope to offset the large nutrient loads that originate from surrounding landscape.

Peat Bogs

Active restoration of peatlands is a recent phenomenon even though they have been harvested for fuel and abandoned for centuries. In the Czech Republic and elsewhere in Europe, peatlands were used to produce fuel for glassblowing activities beginning in the 1700s. Peat typically was hand cut. As the harvested sites were small in size, on the order of several to tens of hectares, the size of the disturbance was small but intense as several meters thickness of peat were removed. Cervene Blato, the Red Mud or Red Bog, was one such site. Peat was harvested from areas of the bog beginning in 1774. The site was abandoned in the mid-twentieth century and left to revegetate on its own. In 1953, the 400-ha site was declared a protected area and the hydrology was restored by plugging ditches and streams that drained it. Since 2000, vegetation and peat-(forming) properties of Cervene Blato were investigated, comparing the plant community and peat properties of the harvested area of the bog with areas where peat was not harvested. On the restored site, a sequence of successional communities exists. On the most recently harvested sites, open water is present though herbaceous species like cotton grass (*Eriophorum vaginatum*) encroach (Figure 11.4(a)). On older sites, cotton grass dominates while woody vegetation colonizes the edges (Figure 11.4(b)). On the oldest sites, vegetation consists of open forest of bog pine (*Pinus mugo* Turra) with an understory of Labrador tea (*Ledum palustre*, today known as *Rhododendron tomentosum*) (Figure 11.4(c)).

Peat cores were collected from a portion of the bog where peat harvesting ceased in the late 1800s and from an area that never was harvested (the reference site). Succession on the restored site proceeded to the point where an open forest now dominated the site (Figure 11.4(c)). Visual observation of peat from restored and reference sites revealed a clear difference in quality of surface peat. The upper 10–20 cm of the restored bog peat contained more live *Sphagnum* than the unharvested peat bog due to lower elevation that created wetter conditions. Below this depth and down to 1 m, the two peat cores were similar in appearance.

Properties of the surface (0–30 cm) peat of the restored and the reference bog, however, were similar and were typical of bog soils (Table 11.5). Bulk density was low, 0.10–0.13 g/cm^2, and percent organic C was high, 52%. Total P was low, less than

Figure 11.4 Chronosequence of sites where peat was harvested at Cervene Blato, the Red Bog, in southern Bohemia of the Czech Republic from the seventeenth to twentieth century. (a) The youngest sites that were abandoned last contain open water with *Eriophorum* (cotton grass) encroaching. (b) Older sites are dominated by cotton grass. (c) The oldest sites consist of open forest. Photo credits: Christopher Craft.

Table 11.5 **Peat Properties of Restored and Natural Areas of Cerve Blato, the Red Bog, in South Bohemia, Czech Republic. Bulk Density, Percent Organic C and N, and Total P Represent the Surface (0–30 cm) Peat**

Peat Property	Harvested/Restored	Unharvested/Intact
Bulk density (g/cm^3)	0.10	0.13
Organic C (%)	52	52
Total N (%)	1.1	1.3
Total P (µg/g)	310	325
137Cs		
Peat accretion rate (mm/year)	–	1.1
C sequestration (g/m^2/year)	–	60
N accumulation (g/m^2/year)	–	1.6
P accumulation (g/m^2/year)	–	0.02
210Pb		
Peat accretion rate (mm/year)	2.3	2.9
C sequestration (g/m^2/year)	114	148
N accumulation (g/m^2/year)	2.5	3.8
P accumulation (g/m^2/year)	0.07	0.05

C, Craft unpublished data.

350 µg/g, reflecting the lack of mineral input to these precipitation-fed wetlands. At both sites, peat was actively accreting. The rate of peat accretion was 1–3 mm/year in the reference portion of the bog, while in the restored bog, the rate of peat accretion was comparable, 2.3 mm/year (Table 11.5). Carbon sequestration in restored and reference areas also was comparable, ranging from 60 to nearly 150 g/m^2/year. Accumulation of N in peat also was comparable. Phosphorus accumulation was very low, 0.02–0.07 g/m^2/year, typical of ombrotrophic peat-accumulating wetlands.

Similarities in peat accretion, C sequestration, and nutrient (N, P) accumulation in restored and reference portions of the bog indicate that peat-forming processes are functioning at comparable levels. With cessation of peat harvesting, hydrologic restoration, and adequate time, even slow processes like peat accretion and C sequestration can be restored.

Forested Wetlands

On the campus of Duke University (Durham, NC, USA), a complex of riparian forest, stormwater reservoir, and treatment wetland was constructed to reduce nutrient loading and improve water quality in Jordan Lake, a municipal drinking water supply downstream (Richardson and Pahl, 2005). The Stream and Wetland Assessment Management Park (SWAMP) included restoration of 600 m of degraded stream and riparian forest, construction of an earthen dam and 1.6-ha stormwater reservoir, and building a 0.5-ha treatment wetland on upper Sandy Creek that runs through the campus (Flanagan et al., 2008).

Figure 11.5 Photographs of the Stream Wetland Assessment Management Park (SWAMP) on the campus of Duke University. (a) Highly eroded Sandy Creek channel (2005) prior to restoration. (b) Newly created channel and riparian vegetation 2 years after restoration. Photo credit: Richardson et al. (2011). Reproduced with permission of Elsevier.

The watershed is urbanized with about 21% of its area consisting of impervious surface (Unghire et al., 2011). Prior to restoration, the stream was deeply incised (Figure 11.5(a)) with no connectivity to the riparian zone. The site also was dominated by a number of exotic species (Table 11.6). The riparian area was restored by grading the site, restoring meanders of the stream channel, reconnecting the floodplain with the stream, and replanting vegetation (Figure 11.5(b); Richardson and Pahl, 2005).

Ho and Richardson (2013) followed floristic succession for 5 years following restoration. A number of native species reestablished including species such as *Juncus* and *Carex* characteristic of freshwater wetlands. Many invasive species disappeared though Japanese stilt grass (*Microstegium*) was still present at higher elevations of the wetland. Unghire et al. (2011) compared soil properties of the site before restoration (2003) and 5 years later (2008). Soil organic matter decreased whereas extractable P and moisture increased following restoration (Figure 11.6). Soil organic matter and

Table 11.6 Dominant Ground Cover of the Stream and Wetland Assessment Management Park (SWAMP) Riparian Wetland before and after Restoration

Species	Invasive	Prerestoration	Postrestoration
Microstegium vimineum	X	X	X
Lonicera japonica	X	X	
Hedera helix	X	X	
Euonymus fortune	X	X	
Toxicodendron radicans		X	
Parthenocissus quinquefolia		X	
Smilax rotundifolia		X	
Amphicarpaea bracteata		X	
Juncus effusus			X
Juncus acuminatus			X
Boehmeria cylindrica			X
Carex lurida			X
Mimulus ringens			X
Cardamine hirsuta	X		X

From Ho and Richardson (2013).

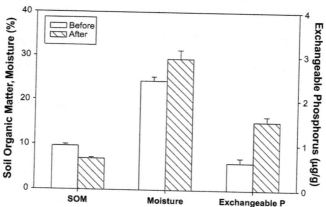

Figure 11.6 Soil organic matter (SOM), moisture, and extractable P before and after restoration of a riparian forest.
Adapted from Unghire et al. (2011).

P also were more spatially homogeneous across the site following restoration. These changes were attributed to earth moving activities to remove exotic vegetation during restoration. Denitrification potential was measured in experimental plots planted with one, four, or eight species in the restoration site (Sutton-Grier et al., 2011). Denitrification was more strongly linked to soil properties than to species richness and plots with higher organic matter and moisture content exhibited the highest rate of denitrification.

Restoration projects such as SWAMP provide a number of benefits, including improving water quality in urban catchments, research and teaching opportunities for students, and outreach to local and state officials and the general public.

Urban Wetlands

Restoring wetlands in urban landscapes is challenging. Urban environments contain stressors, both on site and off, not found in agricultural or natural landscapes (see Chapter 4, Considerations of the Landscape). Alteration of hydrology, especially the magnitude and timing of inundation driven by runoff from impervious surfaces, is a uniquely urban problem. Pollutants such as hydrocarbons, heavy metals, and pharmaceutical products also are distinctive of urban environments. Last, but not least, urban wetlands exist in highly fragmented watersheds. Riparian corridors are nonexistent and vegetated patches for habitat are few and far between (see Table 4.1). The cost of land acquisition for restoration in urban environments also is much greater than in agricultural and natural landscapes. Finally, concerns of the general public such as mosquitoes and pests must be placated. In spite of the constraints, the benefits of restoring wetlands in urban landscapes are great. The creation of green space for recreation, bird watching, and wildlife habitat provides incalculable benefits to people living in cities.

In greater metropolitan New York City (NYC) and elsewhere, wetlands are restored to provide these benefits. The NYC region is home to more than 16 million people. During the past 300 years nearly all wetlands in the area were drained or filled. In addition, many wetlands were degraded by accidental release of pollutants, including heavy metals from industrial operations or oil from pipeline spills and refinery operations. Restoration of wetlands in the region frequently is undertaken to compensate or mitigate for loss of the functions and values associated with such activities.

At Woodbridge River, New Jersey, a 10-ha tidal brackish marsh was restored to mitigate for damage to a natural *Spartina alterniflora* marsh caused by oil from a pipeline spill at a nearby refinery. Prior to restoration, the site experienced a number of alterations to its hydrology. It was partially enclosed by a levee that isolated it from the Woodbridge River, restricting tidal inundation (Figure 11.7), ditches were constructed to drain it for mosquito control (Figure 11.7), and earthen fill was placed on portions of it. In addition, the site was overrun by *Phragmites*, an aggressive plant that displaced native marsh vegetation. In 2007, hydrology was restored by removing the levee and filling the mosquito ditches. Portions of the site were excavated to remove fill that was placed on it (Figure 11.8(a)). By lowering the elevation of the marsh surface, tidal exchange was enhanced, increasing duration and depth of inundation.

Prior to restoration, the site was brackish water "high" marsh inundated on spring and storms tides and dominated by the shrub, *Iva*, and graminoids *Spartina patens* and *Distichlis spicata*. Because *Phragmites* is a strong competitor in this habitat, having overtaken a number of high marshes in the region (Marks et al., 1994; Galatowitsch et al., 1999), the decision was made to restore the site to a regularly flooded *S. alterniflora* marsh. By creating a *S. alterniflora* marsh with its longer hydroperiod, anaerobic

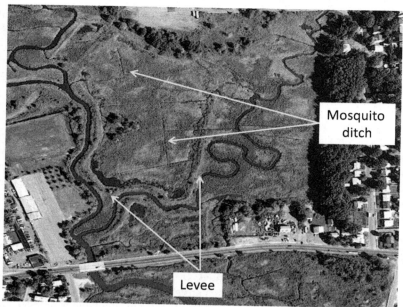

Figure 11.7 Aerial view of the Woodbridge River brackish marsh restoration site prior to restoration. Note the levee and mosquito ditches.
Photo credit: Carl Alderson.

processes like sulfate reduction would be enhanced and more hydrogen sulfide produced. Hydrogen sulfide inhibits N uptake by *Phragmites* relative to *S. alterniflora* (Chambers, 1997) thereby limiting its distribution in tidal salt marshes (Chambers et al., 1998). In the restored marsh, it was expected that enhanced tidal inundation with brackish water would favor establishment of *S. alterniflora* over *Phragmites*. Furthermore, greater depth and duration of inundation would support more sediment and P deposition and increased access by nekton, finfish, and shellfish to the marsh (Figure 11.9).

To reduce *Phragmites* coverage initially, herbicide was applied to the site following excavation. Then the site was planted with *S. alterniflora*. After planting, wire mesh was placed over the young transplants to keep geese from eating the tender vegetation (Figure 11.8(b)). Within 2 years, native vegetation covered the site (Figure 11.8(c)).

Comparison of the newly restored marsh with a nearby natural reference marsh revealed that the restored marsh was inundated longer and to a greater depth (Table 11.7). Porewater sulfide also was greater in the restored marsh as compared to the site before restoration and to the reference marsh. Thus, the goal of creating a marsh with extended hydrology and anaerobic conditions including sulfate reduction and hydrogen sulfide was achieved. Restored marsh vegetation also was more productive than the reference marsh (Table 11.7). Furthermore, species richness was comparable to the reference marsh and it was attributed to preservation of remnant areas of high marsh with their high species richness during site preparation. Greater depth and duration of

Figure 11.8 (a) Removal of fill to lower the elevation of the marsh surface during restoration; (b) Netting to discourage herbivory of young *Spartina alterniflora* transplants by geese; and (c) The site 2 years following restoration.
Photo credits: Carl Alderson (a,b), Chris Craft (c).

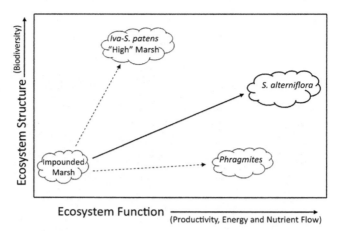

Figure 11.9 Possible trajectories of ecosystem development of the Woodbridge River restoration project. See text for explanation of the goals and constraints on restoring the site.

Table 11.7 Comparison of Wetland Hydrology, Porewater Chemistry, and Vegetation/Nekton Community Structure of a Restored Brackish Water Marsh with the Site Prior to Restoration and with a Nearby Natural Reference Marsh in the Urban Landscape of Greater New York City

	Restored Marsh		Reference Marsh
	Prerestoration	Postrestoration	
Hydrology			
Inundation depth (m)	–	0.22	0.18
Inundation duration (%)	–	23	10
Porewater			
Salinity (ppt)	18–22	14	26
Sulfide (µg/l)	5–6	12	8
Vegetation			
Aboveground biomass (g/m²)	430–590	1450	1070
Species richness (no./marsh)	7	9	9
Nekton[a]			
Density (no./marsh)	–	4893	440
Diversity (no./marsh)	–	7	5

[a]Includes fish, shrimp, and crab.

Prerestoration data are from Sturdevant et al. (2002). Postrestoration and Reference Marsh Data are from HDR Engineering Inc. (2011).

inundation also enhanced nekton utilization of the restored marsh, and the density of finfish, crabs, and shrimp was 10 times that recorded in the reference marsh (Table 11.7).

The Woodbridge River restoration marsh is a successful example of creating *S. alterniflora* marsh for improved fisheries habitat in a highly urbanized environment. To maintain the site, annual spraying with herbicide is needed to keep *Phragmites* from colonizing the remnant high marsh areas. Urban wetland restoration projects such as this illustrates the ongoing struggle to control invasive species before, during, and following restoration.

References

Andreas, B.K., Lichvar, R.W., 1995. Floristic Index for Establishing Assessment Standards: A Case Study for Northern Ohio. Technical Report WRP-DE-8. U.S. Army Corps of Engineers. Waterways Experiment Station, Vicksburg, MS.

Broome, S.W., Woodhouse Jr., W.W., Seneca, E.D., 1975. The relationship of mineral nutrients to growth of *Spartina alterniflora* in North Carolina: II. The effects of N, P and Fe fertilizers. Soil Science Society of America Proceedings 39, 301–307.

Broome, S.W., Craft, C.B., Seneca, E.D., 1988. Creation and development of brackish-, water marsh habitat. In: Zelazny, J., Feierabend, J.S. (Eds.), Proceedings of the Conference on Increasing Our Wetlands Resources. National Wildlife Federation, Washington, D.C., pp. 197–205.

Broome, S.W., Seneca, E.D., Woodhouse Jr., W.W., 1982. Establishing brackish marshes on graded upland sites in North Carolina. Wetlands 2, 152–178.

Broome, S.W., Seneca, E.D., Woodhouse Jr., W.W., 1986. Long-term growth and development of transplants of the salt-marsh grass *Spartina alterniflora*. Estuaries 9, 63–74.

Chambers, R.M., 1997. Porewater chemistry associated with *Phragmites* and *Spartina* in a Connecticut tidal marsh. Wetlands 17, 360–367.

Chambers, R.M., Mozdzer, T.J., Ambrose, J.C., 1998. Effects of salinity and sulfide on the distribution of *Phragmites australis* and *Spartina alterniflora* in a tidal saltmarsh. Aquatic Botany 62, 161–169.

Craft, C.B., Sacco, J.N., 2003. Long-term succession of benthic infauna communities on constructed *Spartina alterniflora* marshes. Marine Ecology–Progress Series 257, 45–58.

Craft, C., Broome, S., Campbell, C., 2002. Fifteen years of vegetation and soil development after brackish-water marsh creation. Restoration Ecology 10, 248–258.

Craft, C.B., Seneca, E.D., Broome, S.W., 1991. Porewater chemistry of natural and created marsh soils. Journal of Experimental Marine Biology and Ecology 152, 187–200.

Craft, C., Reader, J., Sacco, J.N., Broome, S.W., 1999. Twenty five years of ecosystem development of constructed *Spartina alterniflora* (Loisel) marshes. Ecological Applications 9, 1405–1419.

Fennessy, S., Craft, C., 2011. Agricultural conservation practices increase wetland ecosystem services in the Glaciated Interior Plains. Ecological Applications 21, S49–S64.

Flanagan, N.E., Richardson, C.J., Ho, M., Pahl, J.W., Roberts, B.T., Medley, L., 2008. Quantification of Water Quality Improvement in Sandy Creek, a Tributary Watershed of Jordan Lake in the Cape Fear River Basin, after Stream and Riparian Restoration and Wetland Treatment Cell Creation. Final Report of the Scientific Findings to the Nonpoint Source Management Program, Division of Water Quality, NC Department of Environment and Natural Resources. EPA 319 Program Grant No. EW05040. Duke University Wetland Center. Nicholas School of the Environment and Earth Sciences. Duke University, Durham, NC.

Galatowitsch, S.M., Anderson, N.O., Ascher, P.D., 1999. Invasiveness in wetland plants in temperate North America. Wetlands 19, 733–755.

RN: NFFKHC30-10-18275. Year One Progress Report. Prepared for: National Oceanic and Atmospheric Administration (NOAA). NOAA-NMFS, 74 Magruder Rd., Highlands, NJ. 07732. 77 pp. HDR Engineering Inc, 2011. Woodbridge Creek Salt Marsh Restoration, Post-restoration Monitoring.

Ho, M., Richardson, C.J., 2013. A five year study of the floristic succession in a restored urban wetland. Ecological Engineering 61B, 511–518.

Hopple, A., Craft, C., 2013. Managed disturbance enhances biodiversity of restored wetlands in the agricultural Midwest. Ecological Engineering 61, 505–510.

Marks, M., Lapin, B., Randall, J., 1994. *Phragmites australis (P. communis)*: threats, management and monitoring. Natural Areas Journal 14, 285–294.

Marton, J.M., Fennessy, M.S., Craft, C.B., 2014a. Functional differences between natural and restored wetlands in the Glaciated Interior Plains. Journal of Environmental Quality 43, 409–417.

Marton, J.M., Fennessy, M.S., Craft, C.B., 2014b. USDA conservation practices increase carbon storage and water quality improvement functions: an example from Ohio. Restoration Ecology 22, 117–124.

Richards, B., Craft, C., 2014. Greenhouse gas fluxes from restored agricultural wetlands and natural wetlands, northwestern Indiana. In: Vymazal, J. (Ed.), The Role of Natural and Constructed Wetlands in Nutrient Cycling and Retention on the Landscape. Springer, New York, pp. 17–32.

Richardson, C.J., Flanagan, N.E., Ho, M., Pahl, J.W., 2011. Integrated stream and wetland restoration: a watershed approach to improved water quality on the landscape. Ecological Engineering 37, 25–39.

Richardson, C.J., Pahl, J.W., 2005. The Duke Forest Stormwater Improvement and Wetlands Restoration Project. Final Report to the North Carolina Clean Water Management Trust Fund and the North Carolina Ecosystem Enhancement Program. Duke University Wetland Center. Nicholas School of the Environment and Earth Sciences. Duke University, Durham, NC.

Savage, R.P., Woodhouse Jr., W.W., 1968. Creation and stabilization of coastal barrier dunes. In: Proceedings of the Eleventh Conference on Coastal Engineering. American Society of Civil Engineers, New York. pp. 671–700.

Seneca, E.D., Broome, S.W., Woodhouse Jr., W.W., Cammen, L.M., 1976. Establishing *Spartina alterniflora* marshes in North Carolina. Environmental Conservation 3, 185–188.

Sturdevant, A., Craft, C.B., Sacco, J.N., 2002. Effects of impoundment on ecological functions of estuarine marshes along Woodbridge River, NY/NJ Harbor. Urban Ecosystems 6, 163–181.

Sutton-Grier, A.E., Wright, J.P., McGill, B.M., Richardson, C., 2011. Environmental conditions influence the plant functional diversity effects on potential denitrification. PLoS One 6, e16584.

Unghire, J.M., Sutton-Grier, A.E., Flanagan, N.E., Richardson, C.J., 2011. Spatial impacts of stream and wetland restoration on riparian soil properties in the North Carolina Piedmont. Restoration Ecology 19, 738–746.

Woodhouse Jr., W.W., 1979. Building Salt Marshes along the Coasts of the Continental United States. Special Report No. 4. U.S. Army Corps of Engineers. Coastal Engineering Research Center. Fort Belvoir, Virginia, 96 pp.

Woodhouse Jr., W.W., 1982. Coastal sand dunes of the U.S. In: Lewis III, R.R. (Ed.), Creation and Restoration of Coastal Plant Communities. CRC Press, Boca Raton, FL, pp. 1–44.

Woodhouse Jr., W.W., Seneca, E.D., Broome, S.W., 1974. Propagation of *Spartina Alterniflora* for Substrate Stabilization and Salt Marsh Development. Technical Memorandum No. 46. U.S. Army Corps of Engineers. Coastal Engineering Research Center. Fort Belvoir, Virginia, 155 pp.

Woodhouse Jr., W.W., Seneca, E.D., Broome, S.W., 1976. Ten years of development of man-initiated coastal barrier dunes in North Carolina. Agricultural Experiment Station Bulletin, 453. North Carolina State University, Raleigh, NC, 52 pp.

Restoration on a Grand Scale

12

Chapter Outline

Saline Tidal Marshes (Delaware Bay, USA)

An intensively studied large-scale wetland restoration project is the Delaware Bay (USA) marshes where large tracts of diked salt hay farms were restored in the 1990s. More than 5000 ha of degraded wetlands were restored to offset losses of eggs, larvae, and fish entrained in cooling waters of the Salem Generating Station, two nuclear reactor units owned by Public Service Enterprise Group (PSEG) of New Jersey (Weinstein et al., 2001). Studies indicated that, with withdrawal of cooling waters from the Bay, five commercially and recreationally important species of fish (striped bass, weakfish, white perch, spot and bay anchovy) were at risk with long-term declines in their populations (Balletto et al., 2005).

Perimeter dikes placed around salt marshes in the Bay nearly a century ago led to subsidence, elimination of tidal creeks, and reduced acreage of salt marsh cordgrass, *Spartina alterniflora*, along with dramatic expansion of *Phragmites* (Weishar et al., 2005a). The Estuary Enhancement Program of PSEG relied on self-design (Mitsch et al., 2000) to restore the marshes with the goal of providing fisheries habitat, including sinuous tidal creeks with shallow and deep areas, vegetated marsh with many embedded tidal creeks to provide marsh edge and access for small fishes, and larger tidal channels to connect restoration sites to Delaware Bay (Weishar et al., 2005a). Two types of marsh restoration were implemented, restoration of diked salt hay (*Spartina patens*) farms and marshes dominated by the aggressive common reed, *Phragmites australis* (Teal and Peterson, 2005). Beginning in 1996, hay farms were reconnected to the estuary by breaching dikes, creating and enhancing major channels, and excavation to lower soil surface elevation (Weishar et al., 2005a). Smaller channels within the restoration sites were allowed to reestablish naturally (Balletto et al., 2005). *Phragmites*-infested marshes were treated with aerial application of herbicides followed by prescribed fire. Following restoration, mowing, removal of relic dikes, modification of microtopography, and intensive herbicide applications were performed over selected marsh areas

Creating and Restoring Wetlands. http://dx.doi.org/10.1016/B978-0-12-407232-9.00012-9

from 1999 to 2003 (Phillipp and Field, 2005). Restorations were implemented on a series of parcels on both sides of the Bay. Three tracts of diked salt hay farms, totaling about 1800 ha, were restored on the New Jersey side. Seven *Phragmites*-infested tracts, two in New Jersey and five in Delaware, totaling about 2700 ha, also were restored.

Previously, several diked salt hay farms were restored in New Jersey, providing a guide for expected restoration outcomes (Shisler, 1989). The best case scenario assumed the sites would achieve 80–100% cover of desired vegetation within about 10 years, the mid-range scenario assumed complete coverage in about 20 years, while the worst-case scenario assumed incomplete, about 40%, coverage of desired vegetation (Weinstein et al., 1997). A major component of the restoration plan was biological monitoring of restoration and reference sites beginning in 1996 and continuing at least through 2005 (Kimball and Able, 2007).

Restoration of the salt hay farms was completed in 1996 and small stream channels quickly developed. By 1999, the number of small channels at one site (a 149-ha salt hay farm) increased from 65 to more than 200 (Weishar et al., 2005b). Restored salt hay farms also quickly revegetated with *S. alterniflora* (Table 12.1). By 2001, the 149- and 459-ha sites, had 70–80% coverage of *S. alterniflora* (Hinkle and Mitsch, 2005). A larger site, 1171 ha, was slower to colonize, with about 30–40% cover of *S. alterniflora* in 2003. Rapid establishment of *S. alterniflora* on the smaller sites was attributed to their (small) size, proximity to seed sources, and restoration of tidally driven hydrology (Hinkle and Mitsch, 2005). Restoration sites where *Phragmites* was treated with herbicides were less successful (Table 12.1). Without continued use of herbicides, *Phragmites* grew back, becoming reestablished over extensive areas within 1 or 2 years (Phillipp and Field, 2005).

Restoration of diked salt hay farms quickly led to use by resident, transient, and migratory fish (Table 12.1) (Able et al., 2000). Reintroducing tidal inundation resulted in a dramatic increase in species abundance and richness at two restored sites as

Table 12.1 **Timeline of Restoration of Desired Vegetation (*Spartina alterniflora*), Nekton (Blue Crab), Resident (mummichog), Transient Marsh Fish and Migratory Fish (Striped Bass) on Restored Salt Marsh Farms in Delaware Bay**

	Time (Years) to Equivalence to Reference Marshes
Spartina alterniflora colonization	>5 years
Phragmites decline	Problematic without repeated use of herbicides
Blue crab (*Callinectes sapidus*)	1–2 years
Mummichog (*Fundulus heteroclitus*)	1–2 years
Weakfish (*Cynoscion regalis*)	–
Spot (*Leiostomus xanthurus*)	3 years
Atlantic croaker (*Micropogonias undulatus*)	–
Striped bass (*Morone saxatilis*)	1–2 years

See text for details.

compared to prerestoration (Able et al., 2004). Abundance, prey type consumed and stomach fullness of three important species, weakfish (*Cynoscion regalis*), spot (*Leiostomus xanthurus*), and Atlantic croaker (*Micropogonias undulatus*), were measured at the 149-ha restoration site 1–3 years following restoration (Nemerson and Able, 2005). After 3 years, use of the restored site by the three species was comparable to or exceeded levels measured in reference marshes. Growth rates of young-of-the-year Atlantic croaker also were comparable in restored and reference marshes (Miller and Able, 2002). Use by blue crab (*Callinectes sapidus*) also compared favorably in restored and reference marshes (Table 12.1; Kimball and Able, 2007). Jivac and Able (2003) compared blue crab abundance, size, sex ratio, and molt stage at two 1- to 2-year-old restored salt hay farms with reference marshes in the region. For the four measured parameters, the restored marshes performed comparably to or exceeded levels measured in reference marshes.

The marsh killifish or mummichog, *Fundulus heteroclitus*, perhaps the most conspicuous macrofauna of tidal marshes in the region, was extensively monitored in restored and reference marshes. Teo and Able (2003b) compared *Fundulus* use in a restored salt hay farm and an adjacent created tidal creek 2 years after restoration in 1998. Large numbers of mummichogs were found in the tidal creek during low tide then moving onto the restored marsh during high tide. Habitat use, movement, and production of mummichogs in the restored marsh were similar to published findings from natural marshes in the region (Teo and Able et al., 2003a,b). Removing *Phragmites* and creating small water-filled pools on the restored marsh surface led to greater abundance of mummichogs relative to unrestored *Phragmites*-dominated marshes (Able et al., 2003).

Large predatory fish also used restored marshes in sizable numbers. Tupper and Able (2000) compared striped bass (*Morone saxatilis*) abundance and feeding habits in a restored salt hay farm with a reference marsh. Abundance of striped bass was greater in the restored marsh while there was little difference in their feed habits among restored and reference marshes (Table 12.1). Overall, fauna assemblages, including fish, crabs, benthic infauna, and diamondback terrapins, of the restored marshes developed quickly and faster than marsh vegetation (Able et al., 2008), suggesting that fauna response depends mostly on access to the marsh surface.

Large-scale restoration of Delaware Bay tidal marshes was deemed successful. It was attributed to a combination of ecological engineering, construction of primary drainage systems, self-design, and adaptive management using postrestoration monitoring to make mid-course corrections when needed (Teal and Weishar, 2005). Nearly 20 years following restoration, management in the form of herbicidal control of *Phragmites* still is needed to keep it from overtaking restored sites (R.T. Kneib, personal communication).

Mesopotamian Marshes (Iraq)

The freshwater marshes at the confluence of the Euphrates and Tigris Rivers, known as the Mesopotamian Marshes, often are described as the cradle of western civilization and even the Garden of Eden (Lawler, 2005). In the 1970s, before efforts to drain them, the area consisted of shallow water lakes, freshwater and brackish marshes, and shrublands (Al-Hilli et al., 2009). *Phragmites australis* was the dominant plant species with luxuriant growth and producing as much as $5000\,g/m^2$ aboveground biomass.

Degradation of the marshes is attributed to a number of factors dating back to the 1950s when river flow was diverted to create lakes in Iraq and, later, with the construction of dams upstream (UNEP, 2001; Lawler, 2005). With these diversions, the flood pulse of freshwater needed to sustain the marshes was severely muted. Degradation of the marshes was compounded by three wars, the Iran–Iraq War (1980–1988), the Gulf War (1991), and the Second Gulf War (2003) when Saddam Hussein, the former President of Iraq, drained them to drive the local people, known as Marsh Arabs, out (Stevens and Ahmed, 2011). Dessication of the marshes resulted in disappearance of endemic animal species including the smooth-coated otter (*Lutra perspicillata*) and the barbel (*Barbus sharpeyi*) (UNEP, 2001). Migratory bird use of the marshes also declined following drainage (UNEP, 2001).

When the Hussein regime collapsed in 2003, the remaining Marsh Arabs and water ministers broke open dikes and reflooded large areas of marsh (Lawler, 2005; Richardson and Hussain, 2006). The war itself also damaged or destroyed dikes, releasing water into formerly drained areas. However, adding water alone did not necessarily lead to restoration of the wetlands. In the desert landscape, high salinity and sulfides following reflooding made it difficult to reestablish marsh vegetation (Richardson et al., 2005). Reflooding also posed other problems, including release of toxins from soils contaminated with chemicals, mines, and military ordnances (Richardson and Hussain, 2006).

By March 2004, nearly 20% of the original 15,000-km^2 marsh area was reflooded and common reed, *P. australis*, quickly reestablished (Richardson et al., 2005). Other species including *Schoenoplectus littoralis*, *Typha domingensis*, and *Ceratophyllum demersum* also reestablished (Table 12.2) (Richardson and Hussain, 2006).

Table 12.2 Wetland Plant Species Found in Reflooded and Natural Mesopotamian Marshes in Surveys Conducted from 2003 to 2005

	Reflooded Marsh	Natural Marsh
Phragmites australis	X	X
Typha domingensis	X	X
Ceratophyllum demersum	X	X
Shoenoplectus littoralis	X	
Panicum repens	X	
Potamogeton pectinatus	X	
Salvinia natans		X
Lemna minor		X
Myriophyllum verticillatum	X	

From Richardson and Hussain (2006). Reproduced with permission of Oxford University Press.

By September 2005, nearly 39% of the original marsh land was inundated mostly as a result of 2 years of record snowpack melt in the headwaters of Turkey and Iran (Richardson and Hussain, 2006) and a number of species of birds, fish, and macroinvertebrates recolonized the marshes. Hussain et al. (2009) collected 31 fish species from the restored Al-Hammar marsh, including 14 freshwater, 11 marine, and 6 invasive species. The 11 marine diadromous fish species consisted of mostly juveniles that relied on the wetlands for nursery and forage grounds (Mohamed et al., 2009). Compared to records from the 1970s, the restored marshes contained fewer species of macrophytes and birds but comparable species of fish (Richardson and Hussain, 2006). Long-term recovery of wetland vegetation has been slow and hindered by high levels of salinity as compared to pre-drainage measurements made in the 1970s (Hamdan et al., 2010). Aboveground biomass and plant species diversity also were low in 2006 relative to pre-drainage conditions.

The experience in Iraq suggests that, with adequate and reliable freshwater, Mesopotamian Marshes can be restored to some semblance of their pre-drainage condition (Figure 12.1). However, the long-term success of this restoration project is unknown because human needs for water in this arid landscape will take precedence. Remote sensing revealed that, in the period 2009 to 2012, the area of vegetated marsh declined to levels reported prior to the Second Gulf War in 2003 with the reduction attributed to construction of additional dams and water diversions upstream (Al-Yamani et al., 2007; Al-Handal and Hu, 2014). To date, there is no clear plan or policy by the Iraqi government to guide marsh restoration in the reflooded areas (Douabul et al., 2012) including a guaranteed annual allocation of water to sustain them.

Freshwater Marshes (Yellow River Delta, China)

The Yellow River Delta contains abundant seasonal wetlands, including highly saline–alkaline wetlands, *P. australis* marshes, wet meadows, and tree–shrub wetlands that cover 565 km² (Cui et al., 2009). In the last two decades, reduced river flow, saltwater intrusion, and anthropogenic activities such as land reclamation for agriculture led to loss of marshes from drying and hypersalinity (Huang et al., 2012; Guan et al., 2013). Since 1986, construction of roads to support oil production further isolated the wetlands from their water supply (Cui et al., 2009).

More than 35% of tidal wetlands and wet meadows and 17% of reed marshes have been lost (Li et al., 2009) with 67 km²/year disappearing from 1989 to 2000 (Coleman et al., 2008). The remaining wetlands dried out as water diversions upstream reduced discharge to the point where the river flows seasonally to Bohai Sea (Xu, 2004).

Beginning in 2002, restoration of 50 km² of degraded wetlands was undertaken by constructing dikes and four reservoirs to retain river water and one channel to divert freshwater from the Yellow River into the wetlands (Cui et al., 2009). Approximately 3 million m³ of water was pumped into the wetlands annually. The benefits of reflooding were dramatic. In 2001, prior to reflooding, soil salinities were 21 g/kg, but, by 2007, it decreased to 6 g/kg (Cui et al., 2009). Between 2001 and 2007,

Figure 12.1 (a) Undrained Al-Hawizeh reference marsh; (b) drained central marsh; and (c) reflooded Abu-Zaraeg marsh in Iraq.
Photo credits: (a) and (b) Richardson and Hussain (2006); (c) Curtis J. Richardson. From Richardson and Hussain (2006). Reproduced with permission of Oxford University Press. (c) Reproduced with permission of Curtis J. Richardson, Director, Duke University Wetland Center, Duke University, Durham, NC.

the ecosystem shifted from a hypersaline mudflat to a plant community dominated by oligohaline and brackish water species including *P. australis* and *Typha orientalis* (Figure 12.2). Plant species richness increased from 14 species in 2001 to 18 species in 2007 (Cui et al., 2009). Reintroduction of freshwater and reestablishment of vegetation led to increased use by shorebirds and waterbirds. Prior to reflooding, 15 species were identified. By 2007, the number of bird species increased to more than 35 (Cui et al., 2009). Wading birds increased in numbers following reflooding though shorebirds that use mudflats decreased (Hua et al., 2012). Swimming birds increased the first year after reflooding but declined thereafter as wetland vegetation covered the site.

Figure 12.2 (a) Unrestored hypersaline mudflat and (b) reflooded area dominated by open water and *Phragmites australis* in the Yellow River delta.
Photo credits: Chris Craft.

Soil properties changed markedly with reflooding (Wang et al., 2011). Seven years following hydrologic restoration, soil moisture, organic matter, and total N were greater and pH and electrical conductivity were less than in unflooded wetlands (Figure 12.3). At the restored sites, soil organic carbon was greater under herbaceous (0.4–0.7%) than woody vegetation (0.1%). Amendment (reed debris) and planting with *Suaeda salsa* also improved soil conditions (Guan et al., 2011). Organic amendments and planting reduced soil sodium levels, increasing *Suaeda* biomass compared to other treatments, plowing and fertilization.

Li et al. (2011) compared waterbird community composition of two natural, two degraded, and two 1- to 5-year-old restored marshes in the Yellow River Delta.

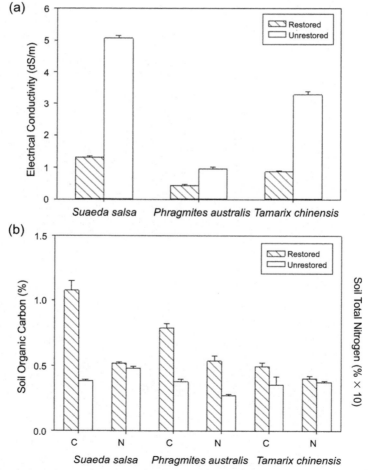

Figure 12.3 (a) Soil electrical conductivity and (b) organic C and total N (0–20-cm deep) in three plant communities of restored (reflooded) and unrestored marshes of the Yellow River delta, China.

Adapted from Wang et al. (2011).

Community composition varied between restored and natural marshes and was linked to differences in vegetation, with *S. salsa* dominant in the saline natural marshes and *P. australis* dominant in reflooded marshes.

Li et al. (2015) compared macrobenthic fauna in two marshes reflooded 6 and 10 years earlier with an unflooded mudflat. The 10-year-old reflooded marsh was dominated by *P. australis* and contained greater species richness and density of macrobenthos than the 6-year-old reflooded marsh and mudflat. The 6-year-old reflooded marsh did not appear to follow the same trajectory of development as the 10-year-old marsh. Higher salinity in the 6-year-old marsh impeded vegetation growth, leading to more open water.

Wang et al. (2012) compared plant community characteristics along a chronosequence of four reflooded *Phragmites* marshes in the Delta. Marshes were reflooded in 2002, 2005, 2007, and 2009, respectively, and were sampled in 2010. Plant cover increased with site age, reaching 99% in the 8-year-old marsh. Stem height and diameter also increased with time. In spite of these trends, overall plant community similarity was less than 35% in all restored sites relative to reference sites. The restoration is considered successful because, for nearly all measured parameters, the restored marshes had better developed plant communities relative to degraded (unrestored) *Phragmites* marshes (Wang et al., 2012).

Coastal Marshes (Mississippi River Delta, Louisiana)

The Mississippi River delta is a vast area of tidal saline, brackish, and freshwater marshes, and forests. Since the beginning of the twentieth century, more than $4500 \, km^2$ of wetlands have disappeared as a result of natural causes and human activities (Salinas et al., 1986; Boesch et al., 1994; Day et al., 2000). Natural causes of wetland loss include subsidence and the life cycle of subdeltas that build for several hundred years then deteriorate as new subdeltas form elsewhere (Wells and Coleman, 1987). Human activities include construction of flood control levees, canal dredging, and spoil banks, mostly associated with the oil and gas industry, that dramatically reduced river flooding and sediment deposition needed to support plant growth, organic matter accumulation, and wetland surface elevation increase in the face of subsidence and rising sea level (Deegan et al., 1984; Swenson and Turner, 1987). Construction of dams on main stem tributaries of the river and smaller engineered structures that retain sediment resulted in a 50–70% decline in suspended sediment transport to the delta (Kesel, 1988; Meade and Moody, 2010; Blum and Roberts, 2012). Limited hydrologic connectivity to the river promoted waterlogging and saltwater intrusion, converting freshwater wetlands to brackish and saline wetlands and open water (Boesch et al., 1994; DeLaune et al., 1994). Between 1985 and 1997, the annual loss of wetlands in the delta was more than $30 \, km^2$/year (Coleman et al., 2008).

A primary restoration strategy to protect the remaining wetlands is to reintroduce river flooding to provide much needed sediment and freshwater to counteract saltwater intrusion (DeLaune et al., 2003; Lane et al., 2006; Day et al., 2009). This is accomplished using crevasse splays, where the levee is deliberately breached, allowing river water and sediment to flow into a degraded or subsiding wetland (Boyer et al., 1997; Cahoon et al., 2011), and river diversions that mimic crevasse splays but rely on engineered structures to control water and sediment inputs (Lane et al., 2006). Using a modeling approach, Kim et al. (2009) suggested that such diversions, properly sized and placed, could create 700–1200 km^2 of new (wet)land over the next century.

Three river diversions were established prior to 2000—Violet (1979), Caernarvon (1991), and West Pointe a la Hache (1993)—with discharges ranging from 8 to 21 m^3/s (Lane et al., 2006). Measurements of wetland surface elevation, vertical accretion, and subsidence revealed that wetlands downstream of the Caernarvon and West Pointe a la Hache were keeping pace with sea level rise with greatest accretion and elevation gain proximal to culverts where river water is diverted. Wetlands downstream of the Violet diversion, however, were not keeping pace with sea level rise because of high rates of subsidence, low river discharge (8 m^3/second), and a fire that burned off about 4 cm of surface material.

DeLaune et al. (2003) measured vertical accretion and accumulation of mineral sediment, organic matter, and nutrients along transects along the Caernarvon Diversion downstream. Vertical accretion, mineral sediment deposition, and organic matter accumulation were greater near the diversion, decreasing with distance downstream. Soil iron (Fe) and extractable phosphorus (P) also decreased with distance downstream. Extractable sodium was lower at sites located nearest the diversion. Both Fe and P are essential elements needed to support healthy plant growth. Iron also counteracts the toxic effects of sulfide by precipitating with it to produce pyrite (FeS, FeS_2). DeLaune et al. (2003) suggested that river diversions, by reducing salinity and adding sediment and nutrients, may slow or even reverse the rate of wetland loss in the delta.

There is concern that river diversions may hasten the decline and disappearance of freshwater marshes in the delta. Freshwater marshes in the Mississippi River Delta are underlain by organic soils or exist as floating marshes (Swarzenski et al., 2008). Changes in water quality of Mississippi River water since the 1950s include a two- to threefold increase in nitrate from agricultural runoff from the Corn Belt of the Midwest US (Turner and Rabalais, 1991). Sulfate concentrations also doubled during that time (Swarzenski et al., 2008). In anaerobic soils, nitrate and sulfate serve as terminal electron acceptors to support decomposition of organic matter by heterotrophic bacteria. Swarzenski et al. (2008) compared the biogeochemical response of porewater and soils of two organic-rich marshes, one receiving river water for the past 30 years and one isolated from river water during that time. Soils of the marsh receiving river water were more decomposed and contained more sulfur. Porewater ammonium and orthophosphorus were at an order of magnitude greater in marsh soils receiving river water, and sulfide and alkalinity were two times greater.

Kearney et al. (2011) reported dramatic loss of wetland vegetation and coverage below three river diversions (Caernarvon, West Pointe a la Hache, Naomi) following Hurricanes Katrina and Rita since 2000. Loss of wetland vegetation in reference marshes was less and recovery was faster than in marshes below the diversions. Kearney and others (see Darby and Turner, 2008; Turner, 2011) suggested nutrients in river water reduce rhizome and root growth that is needed to maintain marsh elevation and tensile strength to anchor them against wind, waves, and storms. An unintentional river diversion was established in the 1970s when the Wax Lake outlet was constructed to reduce flooding along the Atchafalaya River. The Atchafalaya carries approximately 30% of the Mississippi River water and the Wax Lake outlet was designed to capture 30% of the Atchafalaya flow. By 1984, signs of delta formation were evident. By 1994, a well-developed delta lobe formed, and, by 2014, the lobe contained a number of distributary extensions into the Gulf of Mexico with more than 39 km^2 of new marsh (Figure 12.4).

Nyman (2014) suggests that river diversions serve two purposes: (1) building new marsh by flooding with river water high in sediment, and (2) slowing the rate of marsh deterioration by flooding with freshwater to counteract saltwater intrusion. To achieve (1), river diversions can be operated during periods of high flow and suspended sediment concentrations. Nyman (2014) further suggests that the natural cycle of delta building and abandonment be used to guide the use of river diversions. Diversions

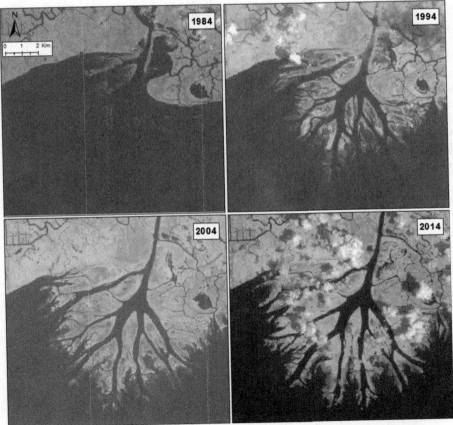

Figure 12.4 Formation of the Wax Lake Delta, Louisiana, over a 30-year period. The delta formed as the result of an unplanned river diversion to reduce flooding along the Atchafalaya River in the 1970s.
Photos were taken by Landsat 5 (1994–2004) and Landsat 8 (2014) satellites.

with high suspended sediment loads mimic the active phase of delta building and creation of new (wet)land. Such diversions are applicable to the creation of saline tidal marshes at the mouths of large deltas where salinity is high and the Fe-rich sediment reduces sulfide toxicity (DeLaune and Pezeshki, 2003). Diversions of freshwater flow in suspended sediment are appropriate for inactive deltas where freshwater reduces saltwater intrusion and enhance plant growth, organic matter accumulation, and vertical accretion to slow the rate of wetland loss.

Mangrove Reforestation (Mekong River Delta, Vietnam)

The Nature Reserve of Can Gio Mangroves encompasses $700 \, km^2$ of land and water and contains about $400 \, km^2$ of planted and natural mangroves (Nam et al., 2014). Located only a few kilometers from Ho Chi Minh City, the reserve is home to a number of diverse wetland habitats, including mangrove, salt marsh, seagrass meadows, and mudflats. The reserve contains 56 true and associated mangrove species.

During the second Indochina war in 1965–1969, the mangrove forests of Can Gio were almost completely destroyed by the spraying of herbicides and other chemical agents conducted by the U.S. Air Force (Nam et al., 2014). Nearly 57% of the Can Gio mangroves were destroyed (Nam and Sinh, 2014). Defoliation and deforestation were exacerbated by tree cutting by the local people for fuel and house construction. During the next 10 years, there was little natural regeneration except for some colonization by *Avicennia* spp. and *Nypa fruticans* (Nam and Sinh, 2014). At the same time, the fern, *Acrostichum aureum*, and the palm, *Phoenix paludosa*, colonized about $45 \, km^2$ of barren land while another $100 \, km^2$ of land remained barren (Nam et al., 2014).

Beginning in 1978, a reforestation effort was undertaken by the Ho Chi Minh City Forestry Department (Nam and Sinh, 2014). Nearly $20 \, km^2$ of mangroves were planted with *Rhizophora apiculata*. *Ceriops*, *Nypa*, *Aviennia*, *Kandelia*, and *Bruguiera* also were planted but over much smaller areas (Table 12.3). Sites infested with *A. aureum* were cut and burned and then planted with *R. apiculata*. At higher elevations, *P. paludosa* was cut, burned, and planted with *Ceriops tagal* and *Lumnitzera*

Table 12.3 **Mangrove Afforestation Efforts at the Can Gio Nature Reserve between 1978 and 2000**

	Area Planted (km²)
Rhizophora apiculata	211
Rhizophora mucronata	0.7
Ceriops spp.	6
Nypa fruticans	3
Avicennia alba	<1
Kandelia obovata	<1
Bruguiera sexangula	<1

From Nam and Sinh (2014).

racemosa. In abandoned salt pans, levees were breached to reintroduce tidal inundation, then planted with *C. tagal.* As of 2014, the oldest plantings were about 35 years in age (see Chapter 9, Mangroves, Figure 9.5(c)).

The Reserve supports an activity reforestation research, development, and management program. Mangroves are grown from seedlings in a nursery established within the forest (Figure 12.5(a)). A botanical garden containing more than 30 species of native and exotic mangroves also was established. Trees in the garden are labeled in Vietnamese and English (Figure 12.5(b)).

In 1991, the Can Gio mangroves were designated as an Environmental Protection Forest that emphasizes their role in shoreline stabilization, breeding, and nursery

Figure 12.5 (a) Nursery established to produce saplings for planting in the Can Gio mangrove forest. (b) Mangrove botanical garden containing more than 30 species.
Photo credits: Chris Craft.

grounds for marine fauna, and habitats for wildlife (Nam and Sinh, 2014). Newly accreted forests along the coast are strictly protected areas. Along riverbanks, a 10- to 20-m wide buffer of trees is maintained for erosion control. Inland of this buffer, the forest is managed for aquaculture and forestry. The rotation age for fuelwood is about 20 years and longer for timber.

References

Al-Handal, A., Hu, C., 2014. MODIS observations of human-induced changes in the Mesopotamian marshes in Iraq. Wetlands 35, 31–40.

Al-Hilli, M.R.A., Warner, B.G., Asada, T., Douabul, A., 2009. An assessment of vegetation and environmental controls in the 1970s of the Mesopotamian wetlands of southern Iraq. Wetlands Ecology and Management 17, 207–223.

Al-Yamani, F.Y., Bishop, J.M., Al-Rifaie, K., Ismail, W., 2007. The effects of river diversion, Mesopotamian marsh drainage and restoration, and river damming on the marine environment of the northwestern Arabian Gulf. Aquatic Ecosystem Health and Management 10, 277–289.

Able, K.W., Hagan, S.M., Brown, S.A., 2003. Mechanisms of marsh habitat restoration due to *Phragmites*: response of young-of-the-year mummichog (*Fundulus heteroclitus*) to treatment for *Phragmites* removal. Estuaries 26, 484–494.

Able, K.W., Nemerson, D.M., Gothues, T.M., 2004. Evaluating salt marsh restoration in Delaware Bay: analysis of fish response at former salt hay farms. Estuaries 27, 58–69.

Able, K.W., Nemerson, D.M., Light, P.R., Bush, R.O., 2000. Initial response of fishes to marsh restoration at a former salt hay farm bordering Delaware Bay. In: Weinstein, M.P., Kreeger, D.A. (Eds.), Concepts and Controversies in Tidal Marsh Ecology. Kluwer Academic Publishers, Dordrecht, The Netherlands, pp. 749–773.

Able, K.W., Grothues, T.M., Hagan, S.M., Kimball, M.E., Nemerson, D.M., Taghon, G.L., 2008. Long-term response of fishes and other fauna to restoration of former salt hay farms: multiple measures of restoration success. Reviews in Fish Biology and Fisheries 18, 65–97.

Balletto, J.H., Heimbuch, M.V., Mahoney, H.J., 2005. Delaware Bay salt marsh restoration: mitigation for a power plant cooling water system in New Jersey, USA. Ecological Engineering 25, 204–213.

Blum, M.D., Roberts, H.H., 2012. The Mississippi Delta region: past, present and future. Annual Review of Earth and Planetary Sciences 40, 655–683.

Boesch, D.F., Josselyn, M.N., Mehta, A.J., Morris, J.T., Nuttle, W.K., Simenstad, C.A., Swift, D.J.P., 1994. Scientific assessment of coastal wetland loss, restoration, and management. Journal of Coastal Research 20, 1–103 (Special issue).

Boyer, M.E., Harris, J.O., Turner, R.E., 1997. Constructed crevasses and land gain in the Mississippi River delta. Restoration Ecology 7, 85–92.

Cahoon, D.R., White, D.A., Lynch, J.C., 2011. Sediment infilling and wetland formation dynamics in an active crevasse splay of the Mississippi River delta. Geomorphology 131, 57–68.

Coleman, J.M., Huh, O.K., Braud Jr., D., 2008. Wetland loss in world deltas. Journal of Coastal Research 24, 1–14.

Cui, B., Yang, Q., Yang, Z., Zhang, K., 2009. Evaluating ecological performance of wetland restoration in the Yellow River delta, China. Ecological Engineering 35, 1090–1103.

Darby, F.A., Turner, R.E., 2008. Effects of eutrophication on salt marsh root and rhizome biomass accumulation. Marine Ecology Progress Series 363, 63–70.

Day, J.W., Britsch, L.D., Hawes, S., Shaffer, G., Reed, D.J., Cahoon, D., 2000. Pattern and process of land loss in the Mississippi River delta: a spatial and temporal analysis of wetland habitat change. Estuaries 23, 425–438.

Day, J.W., Cable, J.E., Cowan Jr., J.W., DeLaune, R., de Mutsert, K., Fry, B., Mashriqui, H., Justic, D., Kemp, P., Lane, R.R., Rick, J., Rick, S., Rozas, L.P., Snedden, G., Swenson, E., Twilley, R.R., Wissel, B., 2009. The impacts of pulsed reintroduction of river water on a Mississippi delta coastal basin. Journal of Coastal Research 54, 225–243.

Deegan, L.A., Kennedy, H.M., Neill, C., 1984. Natural factors and human modifications contributing to marsh loss in Louisiana's Mississippi River deltaic plain. Environmental Management 8, 519–528.

DeLaune, R.D., Pezeshki, S.R., 2003. The role of soil organic carbon in maintaining surface elevation in rapidly subsiding U.S. Gulf of Mexico coastal marshes. Water, Air and Soil Pollution 3, 167–179.

DeLaune, R.D., Nyman, J.A., Patrick Jr., W.H., 1994. Peat collapse, ponding and wetland loss in a rapidly submerging coastal marsh. Journal of Coastal Research 10, 1021–1030.

DeLaune, R.D., Jugsujinda, A., Peterson, G.W., Patrick Jr., W.H., 2003. Impact of Mississippi River freshwater reintroduction on enhancing marsh accretionary processes in a Louisiana estuary. Estuarine, Coastal and Shelf Science 58, 653–662.

Douabul, A.A.Z., Al-Mudhafer, N.A., Alhello, A.A., Al-Saad, H.T., Al-Maarofi, S.S., 2012. Restoration versus reflooding: Mesopotamian marshlands. Hydrology Current Research 3 (5), 140.

Guan, B.J., Yu, J., Lu, Z., Xie, W., Chen, X., Wang, X., 2011. The ecological effects of *Suaeda salsa* on repairing heavily degraded coastal saline-alkaline wetlands in the Yellow River delta. Acta Ecological Sinica 31, 4835–4840.

Guan, B., Yu, J., Cao, D., Li, Y., Han, G., Mao, P., 2013. The ecological restoration of heavily degraded saline wetland in the Yellow River delta. CLEAN–Soil, Air, Water 41, 690–696.

Hamdan, M.A., Asada, T., Hassan, F.M., Warner, B.G., Douabul, A., Al-Hilli, M.R.A., Alwan, A.A., 2010. Vegetation response to re-flooding in the Mesopotamian wetlands, Southern Iraq. Wetlands 30, 177–188.

Hinkle, R.L., Mitsch, W.J., 2005. Salt marsh vegetation recovery at salt hay farm wetland restoration sites on Delaware Bay. Ecological Engineering 25, 240–251.

Hua, Y., Cui, B., He, W., 2012. Changes in waterbirds habitat suitability following wetland restoration in the Yellow Sea delta, China. CLEAN–Soil, Air, Water 40, 1076–1084.

Huang, L., Bai, J., Chen, B., Zhang, K., Huang, C., Liu, P., 2012. Two-decade wetland cultivation and its effects on soil properties in salt marshes of the Yellow Riverdelta, China. Ecological Informatics 10, 49–55.

Hussain, N.A., Mohamed, A.R.M., Al Noo, S.S., Mutlak, F.M., Aben, I.M., Coad, B.W., 2009. Structure and ecological indices of fish assemblages in the recently restored Al-Hammar marsh, southern Iraq. BioRisk 3, 173–186.

Jivac, P.R., Able, K.W., 2003. Evaluating salt marsh restoration in Delaware Bay: the response of blue crabs, *Callinectes sapidus*, at former salt hay farms. Estuaries 26, 709–719.

Kearney, M.S., Riter, J.C.A., Turner, R.E., 2011. Freshwater river diversions for marsh restoration in Louisiana: twenty-six years of changing vegetative cover and marsh area. Geophysical Research Letters 38, L16405.

Kesel, R.H., 1988. The decline in the suspended sediment load of the lower Mississippi River and its influence on adjacent wetlands. Environmental Geology and Water Science 11, 271–281.

Kim, W., Mohrig, D., Twilley, R., Paola, C., Parker, G., 2009. Is it feasible to build new land in the Mississippi River delta? Eos 90, 373–375.

Kimball, M.E., Able, K.W., 2007. Tidal utilization of nekton in Delaware Bay restored and reference intertidal salt marsh creeks. Estuaries and Coasts 30, 1075–1087.

Lane, R.R., Day Jr., J.W., Day, J.N., 2006. Wetland surface elevation, vertical accretion, and subsidence at three Louisiana estuaries receiving diverted Mississippi River water. Wetlands 26, 1130–1142.

Lawler, A., 2005. Reviving Iraq's wetlands. Science 307, 1186–1188.

Li, D., Chen, S., Guan, L., Lloyd, H., Liu, Y., Lv, J., Zhang, Z., 2011. Patterns of waterbird community composition across a natural and restored wasteland landscape mosaic, Yellow River delta, China. Estuarine, Coastal, and Shelf Science 91, 325–332.

Li, S., Wang, G., Deng, W., Hu, Y., Hu, W., 2009. Influence of hydrology process on wetland landscape patterns: a case study in the Yellow River delta. Ecological Engineering 35, 1719–1726.

Li, S., Cui, B., Xie, T., Zhang, K., 2015. Diversity pattern of macrobenthos associated with different stages of wetland restoration in the Yellow River delta. Published online February 2015. Wetlands. http://dx.doi.org/10.1007/s13157-015-0641-7.

Meade, R.H., Moody, J.A., 2010. Causes for decline of suspended-sediment discharge in the Mississippi River system. 1940-2007. Hydrological Processes 24, 35–49.

Miller, M.J., Able, K.W., 2002. Movements and growth of tagged young-of-the-year Atlantic croaker (*Micropogonias undulatus* L.) in restored and reference marsh creeks in Delaware Bay, USA. Journal of Experimental Marine Biology and Ecology 267, 15–33.

Mitsch, W.J., Wu, X.B., Nairn, R.W., Wang, N., 2000. To plant or not to plant: a response by Mitsch et al. BioScience 50, 189–190.

Mohamed, A.R.M., Hussain, N.A., Al-Noor, S.S., Coad, B.W., Mutlak, F.M., 2009. Status of diadromous fish species in the restored East Hammar marsh in southern Iraq. American Fisheries Society Symposium 69, 577–588.

Nam, V.N., Sinh, L.V., 2014. Destruction, Restoration and Management of Can Gio Mangroves. In Studies in Can Gio Mangrove Biosphere Reserve, Ho Chi Minh City, Viet Nam. Mangrove Ecosystems. Technical Report No. 6. International Society for Mangrove Ecosystems, Okinawa, Japan, pp. 9–13.

Nam, V.N., Sinh, L.V., Miyagi, T., Baba, S., Chan, H.T., 2014. An overview of Can Gio District and Mangrove Biosphere Reserve. In: Studies in Can Gio Mangrove Biosphere Reserve, Ho Chi Minh City, Viet Nam. Mangrove Ecosystems Technical Report No. 6. International Society for Mangrove Ecosystems, Okinawa, Japan, pp. 1–7.

Nemerson, D.M., Able, K.W., 2005. Juvenile sciaenid fishes respond favorably to Delaware Bay marsh restoration. Ecological Engineering 25, 260–274.

Nyman, J.A., 2014. Integrating successional ecology and the delta lobe cycle in wetland research and restoration. Estuaries and Coasts 37, 1490–1505.

Phillipp, K.R., Field, R.T., 2005. *Phragmites australis* expansion in Delaware Bay salt marshes. Ecological Engineering 25, 275–291.

Richardson, C.J., Hussain, N.A., 2006. Restoring the Garden of Eden: an ecological assessment of the marshes of Iraq. BioScience 56, 477–489.

Richardson, C.J., Reiss, P., Hussain, N.A., Alwash, A.J., Pool, D.J., 2005. The restoration potential of the Mesopotamian marshes of Iraq. Science 307, 1307–1311.

Salinas, L.M., DeLaune, R.D., Patrick, W.H., 1986. Changes occurring along a rapidly submerging coastal area: Louisiana. Journal of Coastal Research 2, 269–284.

Shisler, J., 1989. Creation and restoration of coastal wetlands of the northeastern United States. In: Kusler, J.A., Kentula, M.E. (Eds.), Wetland Creation and Restoration: The Status of the Science. EPA/600/3089/038. U.S. Environmental Protection Agency, Corvallis, Oregon, pp. 145–165.

Stevens, M., Ahmed, H.K., 2011. Eco-cultural restoration of the Mesopotamian marshes, southern Iraq. In: Egan, D. (Ed.), Human Dimensions of Ecological Restoration: Integrating Science, Nature and Culture. Island Press/Center for Resource Economics, Washington, D.C., pp. 289–298.

Swarzenski, C.M., Doyle, T.W., Fry, B., Hargis, T.G., 2008. Biogeochemical response of organic-rich freshwater marshes in the Louisiana delta plain to chronic river water influx. Biogeochemistry 90, 49–63.

Swenson, E.M., Turner, R.E., 1987. Spoil banks: effects on a coastal marsh water level regime. Estuarine, Coastal and Shelf Science 24, 599–609.

Teal, J.M., Peterson, S.B., 2005. Introduction to the Delaware Bay salt marsh restoration. Ecological Engineering 25, 199–203.

Teal, J.M., Weishar, L., 2005. Ecological engineering, adaptive management, and restoration management in Delaware Bay salt marsh restoration. Ecological Engineering 25, 304–314.

Teo, S.L.H., Able, K.W., 2003a. Growth and production of the mummichog (*Fundulus heteroclitus*) in a restored salt marsh. Estuaries 26, 51–63.

Teo, S.L.H., Able, K.W., 2003b. Habitat use and movement of mummichog (*Fundulus heteroclitus*) in a restored salt marsh. Estuaries 26, 720–730.

Tupper, M., Able, K.W., 2000. Movements and food habits of striped bass (*Morone saxatilis*) in Delaware Bay (USA) salt marshes: comparison of a restored and a reference marsh. Marine Biology 137, 1049–1058.

Turner, R.E., 2011. Beneath the salt marsh canopy: loss of soil strength with increasing nutrient loads. Estuaries and Coasts 34, 1084–1093.

Turner, R.E., Rabalais, N.N., 1991. Changes in Mississippi River water quality this century. BioScience 41, 140–147.

UNEP (United Nations Environmental Programme), 2001. The Mesopotamian Marshes: Demise of an Ecosystem. UNEP, Nairobi, Kenya.

Wang, H., Wang, R., Yu, Y., Mitchell, M.J., Zhang, L., 2011. Soil organic carbon of degraded wetlands treated with freshwater in the Yellow River delta, China. Journal of Environmental Management 92, 2628–2633.

Wang, X., Yu, J., Zhou, D., Dong, H., Li, Y., Lin, Q., Guan, B., Wang, Y., 2012. Vegetative ecological characteristics of restored reed (*Phragmites australis*) wetlands in the Yellow River delta, China. Wetlands Ecology and Management 49, 325–333.

Weinstein, M.P., Balletto, J.H., Teal, J.M., Ludwig, D.F., 1997. Success criteria and adaptive management for a large-scale wetland restoration project. Wetlands Ecology and Management 4, 111–127.

Weinstein, M.P., Teal, J.M., Balletto, J.H., Strait, K.A., 2001. Restoration principles emerging from one of the world's largest tidal marsh restoration projects. Wetlands Ecology and Management 9, 387–402.

Weishar, L.L., Teal, J.M., Hinkle, R., 2005a. Designing large-scale wetland restoration for Delaware Bay. Ecological Engineering 25, 231–239.

Weishar, L.L., Teal, J.M., Hinkle, R., 2005b. Stream order analysis in marsh restoration on Delaware Bay. Ecological Engineering 25, 252–259.

Wells, J.T., Coleman, J.M., 1987. Wetland loss and the subdelta cycle. Estuarine, Coastal and Shelf Science 25, 111–125.

Xu, J., 2004. A study of anthropogenic seasonal rivers in China. Catena 55, 17–32.

Future of Wetland Restoration

13

Chapter Outline

Introduction

In the twenty-first century, wetland restoration will depend on the rate of growth of the human population and its demands for water. Global population is expected to reach more than 9 billion people by 2050 and more than 10 billion by 2100. With increasing population comes increasing demand for food and water to grow it. In the last century, wetlands and aquatic ecosystems around the world, including the Colorado River Delta, Aral Sea, Yellow River Delta, Murray–Darling Basin, and elsewhere, suffered dramatic losses as freshwater was diverted for agriculture (Aladin and Potts, 1995; Grafton et al., 2013; Guan et al., 2013). Intensification of agriculture on existing lands, including greater use of fertilizers and pesticides, will only exacerbate eutrophication of aquatic ecosystems and wetlands. Increasing urbanization, especially in coastal regions, will accelerate loss of tidal marshes and mangroves. Their loss will be compounded by rising sea levels as wetland migration inland will be thwarted by hard structures of the "built" environment. Climate change will have a number of effects on freshwater and estuarine wetlands. Increased warming and interannual variability of temperature and precipitation may lead to drying of wetlands at regional scales, especially in regions where water scarcity already exists. Species ranges and composition may shift as warming may increase species ranges and increasing CO_2 may favor some plant species over others. A more densely populated, warmer, and weedier world will lead to wetland degradation and loss, but, at the same time, provide opportunities for restoration.

Population Growth and Global Food Demand

The world's population will grow from more than 7 billion people today to 10–11 billion by 2100 (Figure 13.1(a)) (Lee, 2011; United Nations, 2013) and will be accompanied by changing demographics including an aging population, lower fertility rates, and more people living in urban areas (Cohen, 2003). The population will be wealthier, consuming more resources, food, water, and energy (Pimentel and Pimentel, 2006). Studies suggest that 70–100% more food will be needed by 2050 (Godfrey et al., 2010). With increasing population, more food will be needed, requiring more agricultural and pasture land as well as intensification of crop production on existing farmland. Agriculture, cropland and pasture, is the largest use of land, covering about 38% of the earth's terrestrial surface (Foley et al., 2011). Worldwide, more than 70% of grassland, 50% of savannah, 45% of temperate forest, and 27% of tropical forests have been converted to agriculture. It is estimated that an additional 10 million km^2 of natural ecosystems will be converted to agriculture by 2050 (Figure 13.1(b)) (Tilman et al., 2001). Conversion of wetlands and other natural ecosystems will lead to loss of the ecosystem services they provide, including freshwater and forest resources, climate and air quality, and infectious diseases (Foley et al., 2005).

Agriculture is the single largest user of water, accounting for about 70–90% of human water use (Wallace, 2000; Oki and Kanae, 2006; Foley et al., 2011). Intensification of agriculture on existing farmland will require greater use of water, fertilizer, and pesticides (Figure 13.1(b)). In the past 50 years, the area of irrigated land has doubled while rates of fertilizer use increased 500–800% (Foley et al., 2011). It is estimated that irrigated land produces more than 1/3 of the world's crops (Gilland, 2002). Between 2015 and 2030, irrigated land is expected to increase 10% (Bruinsma, 2003), and, between 2010 and 2050, fertilizer (N, P, K) use is predicted to increase more than 50% (Alexandratos and Bruinsma, 2012). Greater water and fertilizer use will deprive wetlands of much needed water, exacerbate eutrophication, and contribute to declining biodiversity.

Eutrophication

One consequence of increasing food production is greater fertilizer use, expanding eutrophication of aquatic ecosystems and wetlands. In addition to well-documented reductions in plant biodiversity (Moore et al., 1989; Bedford et al., 1999; Davis, 1991; Craft and Richardson, 1997), nutrient enrichment destabilizes the root and rhizome mat that, in coastal environments, provides stability against wind and wave action. A number of studies show that belowground biomass declines in response to fertilizer (N) additions (Darby and Turner, 2008a,b; Ket et al., 2011) and nutrient-enriched (NO_3) river diversions (Swarzenski et al., 2008; Kearney et al., 2011). Soil strength also is reduced and decomposition is increased with enrichment (Turner et al., 2009; Turner, 2011; Wigand et al., 2014). In a whole ecosystem enrichment experiment in Massachusetts, additions of nutrients (NO_3) commonly observed in eutrophic coastal water reduced belowground biomass of bank stabilizing roots,

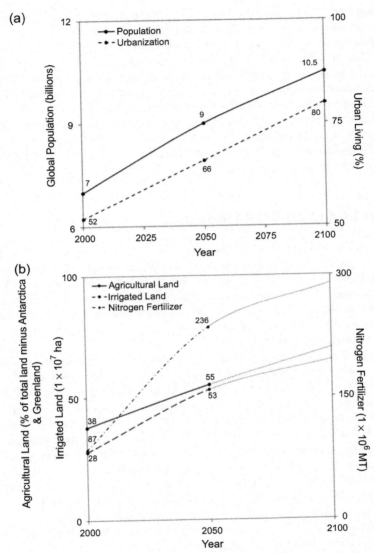

Figure 13.1 Projected change in (a) global population and urbanization and (b) agricultural land, irrigated land, and nitrogen fertilizer use between 2000 and 2100.
Compiled from United Nations (2013) (population), United Nations (2014) (urbanization), Tilman et al. (2001), Foley et al. (2011) (agricultural land), and Tilman et al. (2001) (irrigated land and N fertilizer use).

increased organic matter decomposition, and led to creek bank collapse (Deegan et al., 2012). Root biomass is important for building elevation in many coastal marshes so any reduction may increase marsh susceptibility to sea level rise (see Section Sea Level Rise).

Nutrient enrichment also may have a positive effect on marsh stability by changing species composition, potentially buffering the effects of sea level rise. Fox et al. (2012) monitored changes in plant cover and elevation in plots fertilized annually from 1970 to 2005 in a New England tidal marsh. After 35 years of fertilization, *Distichlis spicata* was the dominant species, supplanting *Spartina alterniflora*. The shift in species composition was accompanied by increased marsh surface elevation. Graham and Mendelssohn (2014) fertilized oligohaline marsh plots in Louisiana for 13 years. After 13 years, live root biomass decreased in response to fertilization yet there was no difference in total (live + dead) belowground biomass. Overall, fertilization led to greater soil organic matter (SOM) accumulation, shear strength, and surface elevation relative to unfertilized plots.

Restoration in an Urban World

In the coming century, people will continue to move to urban areas. In 2014, 54% of the world's population lived in cities, and, by 2050, more than 66% are expected to live there (Figure 13.1(a)) (United Nations, 2014). Today, 11 of the world's 15 largest cities are located in coastal environments (Chapman and Underwood, 2009) with great effects on tidal marshes, mangroves, and floodplain wetlands. Effects of urbanization include land reclamation, aquaculture, increased fertilizer use, and sewage effluent and other consequences (He et al., 2014).

Wetlands in urban environments are among the severely degraded (see Chapter 4, Consideration of the Landscape). Hydrology of existing wetlands is dramatically altered by the amount of impervious surface, resulting in flashy flows. The duration of flooding and soil saturation is short, leading to greater aerobic conditions and organic and N mineralization (Faulkner, 2004). Denitrification is less leading to less retention and greater export of N (Pouyat et al., 2009). Sediment and nutrient loads are increased and propagule availability is reduced (Baldwin, 2004). Habitats are highly fragmented, resulting in low native plant, macroinvertebrate, amphibian abundance, and increased numbers of invasive species (Faulkner, 2004). Many shorelines in urban coastal environments are armored, inhibiting a wetland's ability to migrate (Figure 13.2) with rising water levels. To successfully restore wetlands in urban environments, off-site activities such as reducing stormwater flows, stream restoration, riparian buffer reforestation, and land purchases for conservation are needed (Guntenspergen et al., 2009).

Climate Change

Increasing greenhouse gas emissions, including CO_2, CH_4, and N_2O, from human activities will lead to a warmer planet, increasing atmospheric water with changes in precipitation and evapotranspiration patterns. Globally, changes include: (1) greater warming at high latitudes, (2) warming of surface ocean waters, (3) shrinking glaciers worldwide and sea ice and ice sheets in the northern hemisphere, (4) greater precipitation, (5) ocean acidification, and (6) interannual variability in temperature

Figure 13.2 *Spartina alterniflora* "flowerbox" on the Hudson River, New York City.
Photo credit: Willis Elkins.

and precipitation with increased frequency of drought, flood, and heat waves (Stocker et al., 2013). Climate models predict increased aridity in the twenty-first century over many parts of the world, including most of Africa, southern Europe, the Middle East, most of the Americas, Australia, and Southeast Asia (Dai, 2010). Carbon dioxide itself may affect plant species composition by acting as a fertilizer. The effects of climate change on wetlands may be manifested in four ways with both potential positive or negative impacts: (1) changes in plant production and C cycling; (2) mineralization rates of organic matter; (3) distribution, diversity, and succession of species; and (4) emissions of greenhouse gases, especially CH_4 (Bridgham et al., 1995).

Warming

Warming across most of the earth is given with potential effects on species distributions and ranges. Using experimental wetland plant communities with seedbanks collected from low-, middle-, and high-latitude regions, Baldwin et al. (2014) demonstrated that warming of 2.8 °C increased biomass and decreased species richness. The response was greater with middle and north latitude seedbanks. Some invasive species (e.g., *Lytrum salicaria*) responded more to warming than other species. Introducing seedbanks to soils from different latitudes to simulate migration led to large (60–100%) increases in species richness, suggesting that if species can disperse to different climates, they will colonize it. Warming experiments in New England salt marshes show that warming of 4 °C or less reduces species diversity as the C_4 plant, *Spartina patens*, replaced C_3 forbs (Gedan and Bertness, 2009).

Climate change models predict that warming will be more pronounced at higher latitudes (IPCC, 2013) where extensive peatlands exist. Bridgham et al. (1999) conducted a field mesocosm experiment in Minnesota where intact soil monoliths (2.1 m² surface area, 0.5–0.7 m deep) removed from bog and fen were exposed to three levels of warming using infrared lamps and three water table treatments. Warming increased monthly growing season soil temperature (15 cm deep) by 1.6–4.1 °C and increased production of shrubs in bogs and graminoids in fens (Weltzin et al., 2000). Water level manipulation had no effect on soil temperature but drawdown favored growth of shrubs over mosses. In the bog, the combination of increasing temperature and lowered water table increased cover of shrubs by 50% (Weltzin et al., 2003). Overall, peatland plant production was unaffected by water table elevation as aboveground production was greatest in the wet treatment while belowground production was greatest in the dry treatment. Plant response to warming and drying was species- and life-form specific (Weltzin et al., 2003), highlighting the complexities associated with predicting the effects of climate change on these wetlands.

Effects of warming and water table manipulation on C and N cycling were mixed. Updegraff et al. (2001) evaluated the response of ecosystem respiration (ER) and CH_4 emissions in the same warming/water table manipulation experiment. ER was positively related to soil warming. CH_4 emissions were not strongly related to either treatment in the first year but were strongly related to plant community composition with bog monoliths emitting three times more CH_4 than fen monoliths. At the end of the study, 5 and 6 years of the experiment, CH_4 fluxes were positively related to water level but not to warming (White et al., 2008). Keller et al. (2004) measured C mineralization in incubated cores collected at the end of the 6-year experiment. There was no effect of warming on C (CO_2, CH_4) mineralization whereas increased water level increased CO_2 production in bog soil and CH_4 production in fen soil. Dissolved organic carbon concentration increased with warming in the bog but not the fen (Pastor et al., 2003). After 8 years of warming and water table manipulation, fen mesocosms had no change or lost soil carbon with greatest losses occurring in warmer and drier mesocosms (Bridgham et al., 2008). In contrast, bog mesocosms gained C during the first 3 years but with no further change in C after that time. These studies demonstrate that peatlands respond quickly to climate change but in different ways.

Climate Variability

Greater interannual variation of temperate and precipitation is predicted to have substantial effects on wetlands. In the Prairie Pothole region of North America, with its millions of wetlands and the largest population of breeding waterfowl in the world, modeling has been used to predict the effects of climate change on them (Johnson et al., 2005). Wetlands in the region are strongly linked to cycles of drought and deluge with vegetation exhibiting changes from dry to shallow to deep marsh then to open water that span a period of decades. Large environmental gradients exist across the region as mean annual temperature ranges from 1 °C in the north and 10 °C in the south and precipitation varies from 300 mm/year in the west to 900 mm/year in the east (Millett et al., 2009). Johnson et al. (2005) modeled changes in wetland vegetation in response to a 3-°C increase in temperature along with a ±20% change in precipitation. Increased temperature alone resulted in a longer dry down phase for most marshes. The combination of increased temperature and precipitation had a counterbalancing effect, producing only small changes in wetland cover. Not surprisingly, increased temperature and decreased precipitation had the greatest effect, leading to the dry marsh condition over much of the region.

Climate variability, especially storms, affects coastal wetlands. In California, Zedler (2010) documented a dramatic decline in species richness of a saline tidal marsh following a sequence of extreme events, winter storms that closed the mouth of Tijuana estuary, filling it with sand, followed by a drought later in the year. Other extreme events such as intense hurricanes may interact with sea level rise (SLR) to accelerate the loss of coastal wetlands. Studies show that the storm surge of hurricanes leads to collapse of low salinity tidal wetlands (Howes et al., 2010). In the Mississippi River Delta, the storm surge of Hurricanes Katrina and Rita in 2005 led to widespread erosion of these wetlands whose vegetation is shallow rooted and whose losses were thought to be exacerbated nutrient enrichment from agricultural runoff in the Midwest (Kearney et al., 2011).

CO_2 Enrichment

A number of CO_2 enrichment experiments have been conducted in terrestrial and wetland ecosystems. One of the longest continuous running experiments, beginning in 1987, is in a brackish marsh of the Chesapeake Bay. Initial studies (the first 1–3 years) demonstrated that C_3 plants (*Scirpus olneyi*) grew more in response to CO_2 than C_4 plants (*S. patens*) (Arp et al., 1993), a response typically seen in CO_2 enrichment experiments. The response(s) of decomposition and biogeochemistry mirrored that of vegetation. CO_2 enrichment of plots dominated by C_3 vegetation exhibited greater decomposition as reflected in higher porewater dissolved inorganic and organic carbons, and dissolved CH_4 (Marsh et al., 2005; Keller et al., 2009).

Elevated CO_2 not only stimulates plant growth, it also increases microbial activity as measured by mineralization of SOM and emissions of CH_4. In studies using growth chambers, enhanced plant growth under elevated CO_2 released oxygen into the soil, increasing redox potential and mineralization of SOM (Wolf et al., 2007). Methane emissions by herbaceous and woody wetland species also increased indicating the tight coupling between plant and microbial C cycling (Megonigal and Schlesinger, 1997; Vann and Megonigal, 2003).

The effects of CO_2 enrichment will not operate in isolation. Rather, it will interact with other anthropogenic stressors like excess nutrients. In the same long-term experimental plots of the brackish marsh, the combined effects of CO_2 and N fertilization led to a shift from C_3 to C_4 vegetation, diminishing the fertilizer effect of CO_2 enrichment (Langley and Megonigal, 2010). The combined effects of CO_2 and N yielded unexpected benefits. For example, fine root production increased, increasing soil surface elevation that potentially could counterbalance the effects of sea level rise (Langley et al., 2009). CO_2 and N enrichment also increased the C_3 species (*Schoenoplectus americanus*) tolerance to sea level rise by increasing its productivity (Langley et al., 2013).

Sea Level Rise

Sea level is predicted to increase from 45 to 125 cm in the coming century (Stocker et al., 2013) with greater increases in areas of subsidence such as the Mississippi River Delta (Blum and Roberts, 2009) and other deltaic regions. Climate change-driven SLR is expected to threaten tidal wetlands unless they can maintain their elevation as sea level rises or migrate inland. Tidal wetlands are somewhat resilient to SLR as they trap mineral sediment and add organic matter to the soil (Morris et al., 2002). In this way, they grow vertically or accrete at rates that enable them to maintain their elevation as sea level rises. Factors such as elevation within the tidal frame, tide range, suspended sediment, and plant productivity determine the rate of accretion. Modeling the future effects of SLR on tidal wetlands shows varying degrees of wetland loss by 2100 (Nicholls, 2004; Craft et al., 2009; Kirwan et al., 2010; Fagherazzi et al., 2012). Increased saltwater intrusion will cause changes in habitats as fresh forest and marshes convert to brackish marsh or open water (DeLaune et al., 1987; Craft et al., 2009; Conner et al., 2007; Hackney and Avery, 2015) leading to reduced plant diversity and productivity (Callaway et al., 2007). Mesocosm studies suggest that such shifts will occur more in response to chronic saltwater intrusion than seasonal pulses associated with low river flow or drought (Sharpe and Baldwin, 2012).

Furthermore, development will limit migration of tidal wetlands inland (Linhoss et al., 2015), especially on developed shorelines. In urban environments, shoreline armoring, docks and piers, roads and other infrastructure will allow little to no migration (Figure 13.2). On undeveloped shorelines, restoration activities may slow the rate of wetland loss in areas susceptible to SLR. In Sri Lanka, by planting mangroves at higher densities ($7/m^2$), sediment deposition increased relative to low density plantings ($1–3/m^2$) (Kumara et al., 2010).

Wetland Restoration in the Twenty-First Century

In the coming century, human activities, especially agricultural land use and intensification, water abstraction, and climate change, will lead to drying and salinization of freshwater wetlands and loss of coastal wetlands (Finlayson et al., 2013; Gopal, 2013; Junk, 2013; Mitchell, 2013). In northern latitudes, the effects of warming may lead to greater evapotranspiration and lower organic matter accumulation in boreal soils (Cizkova et al., 2013). Anthropogenic nutrient enrichment will exacerbate eutrophication, increasing dominance by invasive species and reduced biodiversity.

Restoring wetlands in the twenty-first century will be increasingly challenging, so it will be critical to set priorities. Erwin (2009) offers suggestions for successful restoration of wetlands in the face of climate change and population growth (Table 13.1). As global warming is uncontrollable for now, it is imperative to reduce other stressors such as water availability and nutrient enrichment. This means guaranteeing adequate allocation of water, including buying water rights, to sustain restoration projects. Protect coastal wetlands by purchasing setbacks and lands upstream, also known as dynamic buffers, to allow migration as sea level rises (Rogers et al., 2014). Monitoring is essential to understand variability including baseline conditions, the range of temporal variability and whether it is shifting, and encroachment by invasive species. Incorporate known climate oscillations such as ENSO (El Nino Southern Oscillation) and NAO (North Atlantic Oscillation) to understand temporal variability and "fine-tune" management strategies. Select restoration sites in anticipation of changes in future temperature and precipitation patterns to accommodate species ranges in the future. This can be done by placing sites at cooler (or wetter) edges of their existing range.

Table 13.1 Key Concepts for Successful Restoration of Wetlands in the Twenty-First Century

- Reduce nonclimate stressors
- Purchase/guarantee water allocations for wetlands and aquatic ecosystems
- Establish setbacks and dynamic buffers to enable coastal wetlands migration to accommodate rising sea level
- Support monitoring to determine baseline conditions, temporal variability, and trends over time
- Incorporate known climate oscillations into restoration expectations
- Develop strategies for selecting and managing restoration areas in anticipation of climate change

Adapted from Erwin (2009).

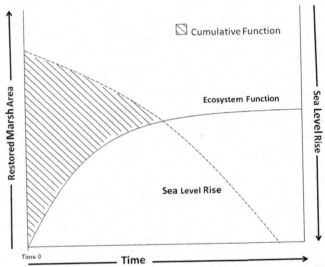

Figure 13.3 Change in cumulative functions of restored coastal wetlands as sea level rises. Restoration of large acreages of wetlands initially provides high levels of function despite the fact that young sites provide lower levels than mature wetlands. Over time, functions increase as restored marshes mature. As sea level rise accelerates, functions decline as wetlands are converted to less productive habitats such as higher salinity wetlands, mudflats, and open water. Eventually, accelerated SLR will lead to loss of coastal wetlands and the ecosystem function and services they provide.

Perhaps the biggest threat to the sustainability of coastal wetlands is shoreline development that will prevent landward marsh expansion as sea level rises (Orr et al., 2003). Unfortunately, this is not taken into consideration in recent restoration projects along the US east, Gulf, and west coasts (Ravit and Weis, 2014). Recognize that full restoration of wetland structure, function, and ecosystem services takes time, and the life expectancy of many projects will be reduced as sea level rises (Figure 13.3). The future of wetland restoration is full of uncertainty and the only thing of which we can be certain is change. With wetland restoration, we can help the world's wetlands roll with it.

References

Aladin, N.V., Potts, W.T.W., 1995. The Aral Sea dessication and possible ways of rehabilitating and conserving its northern part. Environmetrics 6, 17–29.

Alexandratos, N., Bruinsma, J., 2012. In: World Agriculture towards 2030/2060: The 2012 Revision. ESA Working Paper No. 12-03. United Nations Food and Agricultural Organization, Rome.

Arp, W.J., Drake, B.G., Pockman, W.T., Curtis, P.S., Whigham, D.F., 1993. Interactions between C_3 and C_4 salt marsh plant species during four years of elevated atmospheric CO_2. Vegetatio 104/105, 133–143.

Baldwin, A.H., 2004. Restoring complex vegetation in urban settings: the case of tidal freshwater marshes. Urban Ecosystems 7, 125–137.

Baldwin, A.H., Jensen, K., Schonfeldt, M., 2014. Warming increases plant biomass and reduces diversity across continents, latitudes, and species migration scenarios in experimental wetland communities. Global Change Biology 20, 835–850.

Bedford, B.L., Walbridge, M.R., Aldous, A., 1999. Patterns in nutrient availability and plant diversity in temperate North American wetlands. Ecology 80, 2151–2169.

Blum, M.D., Roberts, H.H., 2009. Drowning of the Mississippi River Delta due to insufficient sediment supply and global sea level rise. Nature Geoscience 2, 488–491.

Bridgham, S.D., Johnston, C.A., Pastor, J., Updegraff, K., 1995. Potential feedbacks of northern wetlands to climate change. BioScience 45, 262–274.

Bridgham, S.D., Pastor, J., Dewey, B., Weltzin, J.F., Updegraff, K., 2008. Rapid carbon response of peatlands to climate change. Ecology 89, 3041–3048.

Bridgham, S.D., Pastor, J., Updegraff, K., Malterer, T., Johnson, K., Harth, C., Chen, J., 1999. Ecosystem control over temperature and energy flux in northern peatlands. Ecological Applications 9, 1345–1358.

Bruinsma, J., 2003. World Agriculture: Towards 2015/2030. Earthscan Publications Ltd, London.

Callaway, J.C., Parker, V.T., Vasey, M.C., Schile, L.M., 2007. Emerging issues for the restoration of tidal marsh ecosystems in the context of predicted climate change. Madroño 54, 234–248.

Chapman, M.G., Underwood, A.J., 2009. Comparative effects of urbanization in marine and terrestrial habitats. In: McDonnell, M.J., Hahs, A.K., Breuste, J.A. (Eds.), Ecology of Cities and Towns: A Comparative Approach. Cambridge University Press, Cambridge, UK, pp. 51–70.

Cizkova, H., Kvet, J., Comin, F.A., Laiho, R., Pokorny, J., Pithart, D., 2013. Actual state of European wetlands and their possible future in the context of climate change. Aquatic Sciences 75, 3–26.

Cohen, J.E., 2003. Human population: the next half century. Science 302, 1172–1175.

Conner, W.H., Doyle, T.W., Krauss, K.R., 2007. Ecology of Tidal Forested Wetlands of the Southeastern United States. Springer, New York.

Craft, C.B., Richardson, C.J., 1997. Relationships between soil nutrients and plant species composition in Everglades peatlands. Journal of Environmental Quality 26, 224–232.

Craft, C.B., Clough, J., Ehman, J., Joye, S., Park, D., Pennings, S., Guo, H., Machmuller, M., 2009. Forecasting the effects of accelerated sea level rise on tidal marsh ecosystem services. Frontiers in Ecology and the Environment 7, 73–78.

Dai, A., 2010. Drought under global warming: a review. Wiley Interdisciplinary Reviews: Climate Change 2, 45–65.

Darby, F.A., Turner, R.E., 2008a. Below- and aboveground biomass of *Spartina alterniflora*: response to nutrient addition in a Louisiana salt marsh. Estuaries and Coasts 31, 326–334.

Darby, F.A., Turner, R.E., 2008b. Effects of eutrophication on salt marsh root and rhizome biomass accumulation. Marine Ecology Progress Series 363, 63–70.

Davis, S.M., 1991. Growth, decomposition and nutrient retention of *Cladium jamaicense* Crantz and *Typha domingensis* Pers. in the Florida Everglades. Aquatic Botany 40, 203–224.

Deegan, L.A., Johnson, D.S., Warren, R.S., Peterson, B.J., Fleeger, J.W., Fagherazzi, S., Wolheim, W.M., 2012. Coastal eutrophication as a driver of salt marsh loss. Nature 490, 388–392.

DeLaune, R.D., Patrick Jr., W.H., Pezeshki, S.R., 1987. Foreseeable flooding and death of coastal wetland forests. Environmental Conservation 14, 129–133.

Erwin, K.L., 2009. Wetlands and global climate change: the role of wetland restoration in a changing world. Wetlands Ecology and Management 17, 71–84.

Fagherazzi, S., Kirwan, M., Mudd, S., Temmerman, S., D'Alpaos, A., van de Koppel, J., Rybczyk, J., Reyes, E., Craft, C., Clough, J., Guntenspergen, G., 2012. Numerical models of salt marsh evolution. Reviews of Geophysics 50, RG1002.

Faulkner, S., 2004. Urbanization impacts on the structure and function of forested wetlands. Urban Ecosystems 7, 89–106.

Finlayson, C.M., Davis, J.A., Gell, P.A., Kingsford, R.T., Parton, K.A., 2013. The status of wetlands and the predicted effects of climate change: the situation in Australia. Aquatic Sciences 75, 73–93.

Foley, J.A., DeFries, R., Asner, G.P., Barford, C., Bonan, G., Carpenter, S.R., Chapin, F.S., Coe, M.T., Daily, G.C., Gibbs, H.K., Helkowski, J.H., Holloway, T., Howard, E.A., Kucharik, C.J., Monfreda, C., Patz, J.A., Prentice, I.C., Ramankutty, N., Snyder, P.K., 2005. Global consequences of land use. Science 309, 570–574.

Foley, J.A., Ramankutty, N., Brauman, K.A., Cassidy, E.S., Gerber, J.S., Johnston, M., Mueller, N.D., O'Connell, C., Ray, D.K., West, P.C., Balzer, C., Bennett, E.M., Carpenter, S.R., Hill, J., Monfreda, C., Polasky, S., Rockstrom, J., Sheehan, J., Siebert, S., Tilman, D., Zaks, D.P.M., 2011. Solutions for a cultivated planet. Nature 478, 337–342.

Fox, L., Valielea, I., Kinney, E.L., 2012. Vegetation cover and elevation in long-term experimental nutrient-enrichment plots in Great Sippewissett salt marsh, Cape Cod, Massachusetts: implications for eutrophication and sea level rise. Estuaries and Coasts 35, 445–458.

Gedan, K.B., Bertness, M.D., 2009. Experimental warming causes rapid loss of plant diversity in New England salt marshes. Ecology Letters 12, 842–848.

Gilland, B., 2002. World population and food supply: can food production keep pace with population growth in the next half-century. Food Policy 27, 47–63.

Godfrey, H.C.J., Beddington, J.R., Crute, I.R., Haddad, L., Lawrence, D., Muir, J.F., Pretty, J., Robinson, S., Thomas, S.M., Toulmin, C., 2010. Food security: the challenge of feeding 9 billion people. Science 327, 812–818.

Gopal, B., 2013. Future of wetlands in tropical and subtropical Asia, especially in the face of climate change. Actual state of European wetlands and their possible future in the context of climate change. Aquatic Sciences 75, 39–61.

Grafton, R.Q., Pittock, J., Davis, R., Williams, J., Fu, G., Warburton, M., Udall, B., McKenzie, R., Yu, X., Che, N., Connell, D., Jiang, Q., Kompas, T., Lynch, A., Norris, R., Possingham, H., Quiggin, J., 2013. Global insights into water resources, climate change and governance. Nature Climate Change 3, 315–321.

Graham, S.A., Mendelssohn, I.A., 2014. Coastal wetland stability maintained though counterbalancing accretionary responses to chronic nutrient enrichment. Ecology 95, 3271–3283.

Guan, B., Yu, J., Cao, D., Li, Y., Han, G., Mao, P., 2013. The ecological restoration of heavily degraded saline wetland in the Yellow River Delta. Clean–Soil, Air, Water 41, 690–696.

Guntenspergen, G.R., Baldwin, A.H., Hogan, D.M., Neckles, H.A., Nielsen, M.G., 2009. Valuing urban wetlands: modification, restoration and preservation. In: McDonnell, M.J., Hahs, A.K., Breuste, J.A. (Eds.), Ecology of Cities and Towns: A Comparative Approach. Cambridge University Press, Cambridge, UK, pp. 503–520.

Hackney, C.T., Avery, G.B., 2015. Tidal wetland community response to varying levels of flooding by saline water. Wetlands 35, 227–236.

He, Q., Bertness, M.D., Bruno, J.F., Li, B., Chen, G., Coverdale, T.C., Altieri, A.H., Bai, J., Sun, T., Pennings, S.C., Liu, J., Ehrlich, P.R., Cui, B., 2014. Economic Development and Coastal Ecosystem Change in China. Scientific Reports 4. Article 5995. http://dx.doi.org/10.1038/srep05995.

Howes, N.C., Fitzgerald, D.M., Hughes, Z.J., Georgiou, I.Y., Kulp, M.A., Minor, M.D., Sith, J.M., Barras, J.A., 2010. Hurricane-induced failure of low salinity wetlands. Proceedings of the National Academy of Sciences 107, 14014–14019.

Johnson, W.C., Millett, B.V., Gilmanov, T., Voldseth, R.A., Guntenspergen, G.R., Naugle, D.E., 2005. Vulnerability of northern prairie wetlands to climate change. BioScience 55, 863–872.

Junk, W.J., 2013. Current state of knowledge regarding South America wetlands and their future under global climate change. Aquatic Sciences 75, 113–131.

Kearney, M.S., Riter, J.C.A., Turner, R.E., 2011. Freshwater river diversions for marsh restoration in Louisiana: twenty-six years of changing vegetative cover and marsh area. Geophysical Research Letters 38, LI6405.

Keller, J.K., White, J.R., Bridgham, S.D., Pastor, J., 2004. Climate change effects on carbon and nitrogen mineralization in peatlands through changes in soil quality. Global Change Biology 10, 1053–1064.

Keller, J.K., Wolf, A.A., Weisenhorn, P.B., Drake, B.G., Megonigal, J.P., 2009. Elevated CO_2 affects porewater chemistry in a brackish marsh. Biogeochemistry 96, 101–117.

Ket, W.A., Schubauer, J.P., Craft, C.B., 2011. Effects of five years of nitrogen and phosphorus additions on a *Zizaniopsis miliacea* tidal freshwater marsh. Aquatic Botany 95, 17–23.

Kirwan, M.L., Guntenspergen, G.R., D'Alpaos, A., Morris, J.T., Mudd, S.M., Temmerman, S., 2010. Limits on the adaptability of coastal marshes to rising sea level. Geophysical Research Letters 37, L23401. http://dx.doi.org/10.1029/2010GL045489.

Kumara, M.P., Jayatissa, L.P., Krauss, K.W., Phillips, D.H., Huxham, M., 2010. High mangrove density enhances surface accretion, surface elevation change, and tree survival in coastal areas susceptible to sea-level rise. Oecologia 164, 545–553.

Langley, J.A., Megonigal, J.P., 2010. Ecosystem response to elevated CO_2 levels limited by nitrogen-induced plant shift. Nature 466, 96–99.

Langley, J.A., McKee, K.L., Cahoon, D.R., Cherry, J.A., Megonigal, J.P., 2009. Elevated CO_2 stimulates marsh elevation gain, counterbalancing sea-level rise. Proceedings of the National Academy of Sciences 106, 6182–6186.

Langley, J.A., Mozdzer, T., Shephard, K.A., Hagerty, S.B., Megonigal, J.P., 2013. Tidal plant responses to elevated CO_2, nitrogen fertilization, and sea level rise. Global Change Biology 19, 1495–1503.

Lee, R., 2011. The outlook for population growth. Science 333, 569–573.

Linhoss, A.C., Kiker, G., Shirley, M., Frank, K., 2015. Sea-level rise, inundation, and marsh migration: simulating impacts on developed lands and environmental systems. Journal of Coastal Research 31, 36–46.

Marsh, A.S., Rasse, D.P., Drake, B.G., Megonigal, J.P., 2005. Effect of elevated CO_2 on carbon pools and fluxes in a brackish marsh. Estuaries 28, 694–704.

Millett, B., Johnson, W.C., Guntenspergen, G., 2009. Climate trends of the northern prairie pothole region 1906–2000. Climatic Change 93, 243–267.

Megonigal, J.P., Schlesinger, W.H., 1997. Enhanced CH_4 emissions from a wetland soil exposed to elevated CO_2. Biogeochemistry 37, 77–87.

Mitchell, S.A., 2013. The status of wetlands, threats and the predicted effect of global climate change: the situation in Sub-Saharan Africa. Aquatic Sciences 75, 95–112.

Moore, D.R.J., Keddy, P.A., Gaudet, C.L., Wisheu, I.C., 1989. Conservation of wetlands: do infertile wetlands deserve a higher priority? Conservation Biology 47, 203–217.

Morris, J.T., Sundareshwar, P.V., Nietch, C.T., Kjerfve, B., Cahoon, D.R., 2002. Responses of coastal wetlands to rising sea level. Ecology 83, 2869–2877.

Nicholls, R.J., 2004. Coastal flooding and wetland loss in the 21st century: changes under the SRES climate and socioeconomic scenarios. Global Environmental Change 14, 69–86.

Oki, T., Kanae, S., 2006. Global hydrological cycles and world water resources. Science 313, 1068–1072.

Orr, M., Crooks, S., Williams, P.B., 2003. Will restored tidal marshes be sustainable? San Francisco Estuary and Water Science 6, 1–33.

Pastor, J., Solin, J., Bridham, S.D., Updegraff, K., Harth, C., Weishampel, P., Dewey, B., 2003. Global warming and the export of dissolved organic carbon from boreal peatlands. Oikos 100, 380–386.

Pimentel, D., Pimentel, M., 2006. Global environmental resources versus world population growth. Ecological Economics 59, 195–198.

Pouyat, R.V., Carreiro, M.M., Groffman, P.M., Pavao-Zuckerman, M.A., 2009. Investigative approaches to urban biogeochemical cycles: New York metropolitan and Baltimore as case studies. In: McDonnell, M.J., Hahs, A.K., Breuste, J.A. (Eds.), Ecology of Cities and Towns: A Comparative Approach. Cambridge University Press, Cambridge, UK, pp. 329–351.

Ravit, B., Weis, J.S., 2014. Clean Water Act Section 404 and coastal marsh stability. Coastal Management 42, 464–477.

Rogers, K., Saintilan, N., Copeland, C., 2014. Managed retreat of saline coastal wetlands: challenges and opportunities identified from the Hunter River estuary, Australia. Estuaries and Coasts 37, 67–78.

Sharpe, P.J., Baldwin, A.H., 2012. Tidal marsh plant community response to sea-level rise: a mesocosm study. Aquatic Botany 101, 34–40.

Stocker, T.F., Qin, D., Plattner, G.-K., Alexander, L.V., Allen, S.K., Bindoff, N.L., Breon, F.-M., Church, J.A., Cubasch, U., Emori, S., Forster, P., Friedlingstein, P., Gillet, N., Gregory, J.M., Hartmann, D.L., Jansen, E., Kirtman, B., Knutti, R., Krishna Kumar, K., Lemke, P., Marotzke, J., Masson-Delmotte, V., Meehl, G.A., Mokhov, I.I., Piao, S., Ramaswamy, V., Randall, D., Rhein, M., Rojas, M., Sabine, C., Shindell, D., Talley, L.D., Vaughn, D.G., Xie, S.-P., 2013. Technical Summary. In: Stocker, T.F., Qin, D., Plattner, G.-K., Tignor, M., Allen, S.K., Boschung, J., Nauels, A., Xia, Y., Bex, V., Midgley, P.M. (Eds.), Climate Change 2013: The Physical Science Basis. Contribution of Working Group L to the Fifth Assessment Report of the Intergovernmental Panel on Climate Change. Cambridge University Press, New York.

Swarzenski, C.M., Doyle, T.W., Fry, B., Hargis, T.G., 2008. Biogeochemical response of organic-rich freshwater marshes in the Louisiana deltaic plain to chronic river water influx. Biogeochemistry 90, 49–63.

Tilman, D., Fargione, J., Wolff, B., D'Antonio, C., Dobson, A., Howarth, R., Schindler, D., Schlesinger, W.H., Simberloff, D., Swackhamer, D., 2001. Forecasting agricultural driven global environmental change. Science 292, 281–284.

Turner, R.E., 2011. Beneath the salt marsh canopy: loss of soil strength with increasing nutrient loads. Estuaries and Coasts 34, 1084–1093.

Turner, R.E., Howes, B.L., Teal, J.M., Milan, C.S., Swenson, E.M., Goeringer-Toner, D.D., 2009. Salt marshes and eutrophication: an unsustainable outcome. Limnology and Oceanography 54, 1634–1642.

United Nations, Department of Economic and Social Affairs, Population Division, 2013. In: World Population Prospects: The 2012 Revision. Key Findings and Advance Tables. Working Paper No. ESA/P/WP.227. United Nations, New York.

United Nations, Department of Economic and Social Affairs, Population Division, 2014. In: World Urbanization Prospects: The 2014 Revision, Highlights (ST/ESA/SER.A/352). United Nations, New York.

Updegraff, K., Bridgham, S.D., Pastor, J., Weishampel, P., Harth, C., 2001. Response of CO_2 and CH_4 emissions from peatlands to warming and water table manipulation. Ecological Applications 11, 311–326.

Vann, C.D., Megonigal, J.P., 2003. Elevated CO_2 and water depth regulation of methane emissions: comparison of woody and non-woody wetland plant species. Biogeochemistry 63, 117–134.

Wallace, J.S., 2000. Increasing agricultural water use efficiency to meet future crop production. Agriculture, Ecosystems and Environment 82, 105–119.

Weltzin, J.F., Pastor, J., Harth, C., Bridgham, S.D., Updegraff, K., Chapin, C.T., 2000. Response of bog and fen plant communities to warming and water-table manipulations. Ecology 81, 3464–3478.

Weltzin, J.F., Bridgham, S.D., Pastor, J., Chen, J., Harth, C., 2003. Potential effects of warming and drying on peatland plant community composition. Global Change Biology 9, 141–151.

White, J.R., Shannon, R.D., Weltzin, J.F., Pastor, J., Bridgham, S.D., 2008. Effects of soil warming and drying on methane cycling in a northern peatland mesocosm study. Journal of Geophysical Research: Biogeosciences 113, G00A06.

Wigand, C., Roman, C.T., Davey, E., Stolt, M., Johnson, R., Hanson, A., Watson, E.B., Moran, S.B., Cahoon, D.R., Lynch, J.C., Rafferty, P., 2014. Below the disappearing marshes of an urban estuary: historic nitrogen trends and soil structure. Ecological Applications 24, 633–649.

Wolf, A.A., Drake, B.G., Erickson, J.E., Megonigal, J.P., 2007. An oxygen-mediated positive feedback between elevated carbon dioxide and soil organic matter decomposition in a simulated anaerobic wetland. Global Change Biology 13, 2036–2044.

Zedler, J.B., 2010. How frequent storms affect wetland vegetation: a preview of climate change impacts. Frontiers in Ecology and the Environment 8, 540–547.

Index

Note: Page numbers followed by "f" or "t" indicates figures and tables respectively.

Printed in the United States
By Bookmasters